THIS LAND
A GUIDE TO WESTERN NATIONAL FORESTS

THIS LAND

A GUIDE TO WESTERN NATIONAL FORESTS

Robert H. Mohlenbrock

UNIVERSITY OF CALIFORNIA PRESS

Berkeley   Los Angeles   London

This book is dedicated to my wife, Beverly Ann, who has accompanied me to every one of the national forests in the United States, who has taken copious notes for me, and who has typed many drafts of my manuscripts.

University of California Press
Berkeley and Los Angeles, California

University of California Press, Ltd.
London, England

Library of Congress Cataloging-in-Publication Data

Mohlenbrock, Robert H., 1931–
    This land : a guide to western national forests / by Robert H. Mohlenbrock.
        p.   cm.
    Includes bibliographical references and index.
    ISBN 0-520-23967-9 (pbk. : alk. paper)
    1. Forest reserves — West (U.S.) — Guidebooks. 2. West (U.S.) — Guidebooks. I. Title.

    SD428.A2W45 2005
    333.75'0978--dc22

                                                                                2004014329

Manufactured in Canada
10   09   08   07   06   05
10   9   8   7   6   5   4   3   2   1

The paper used in this publication meets the minimum requirements of ANSI/NISO Z39.48–1992 (R 1997) (*Permanence of Paper*). ♾

Cover: Falls along Boulder Chain Lakes Creek above Hammock Lake, White Cloud Peaks, Sawtooth National Recreation Area, Idaho. Photograph by Scott T. Smith.

CONTENTS

Plates follow page 120

FOREWORD

As the Civil War came to an end, the United States found itself positioned to become a leader among nations. A country of immigrants with a rich endowment of natural resources, America was already a land of opportunity, but the young nation lacked the cultural marks of achievement that characterized its Old World counterparts. Europe had great temples, cathedrals, and museums filled with artifacts. Asia had great dynasties that embodied its long and glorious past. Though short on history, America did have a powerful national spirit that was expressed especially well through its abundant and bountiful land, much of which was public domain, west of the Mississippi—the great frontier. Historian Frederick Jackson Turner referred to this land as the "greatest free gift ever bestowed on mankind."

Yet around the time of the Civil War, a number of influential Americans were becoming increasingly concerned that some of the country's public lands were being plundered. Although homesteading had served the nation well in bringing territory under effective national control, it had become clear that some of the public domain lands had to be set aside as a legacy for all Americans.

In 1864, Henry David Thoreau called for the establishment of "national preserves" of virgin forests, "not for idle sport or food, but for inspiration and our own true re-creation." That same year, President Abraham Lincoln signed legislation granting Yosemite Valley and the Mariposa Big Tree Grove to the state of California to hold forever "for public use, resort, and recreation." Also in the 1860s, Frederick Edwin Church painted *Twilight in the Wilderness,* which inspired artists to capture on canvas the grandeur of the American landscape.

In 1891, President Benjamin Harrison created the nation's first forest reserve, the 1.2-million-acre Yellowstone Park Timber Land Reservation, just south of Yellowstone National Park. Today this area comprises Shoshone and Teton National Forests. Before his term ended, President Harrison proclaimed another 13 million acres of forest reserves in the West, laying the foundation for a National Forest system. However, the reserves were little more than lines drawn on a map, or "paper parks," without managers, regulations, or budgets.

The National Academy of Sciences established a National Forestry Commission in 1896, which issued a report that became the blueprint for forest policy emphasizing that the federal forest reserves belonged to all Americans and should be managed for them and not for any particular class. It went on to say that "steep-sloped lands should not be cleared, the grazing of sheep should be regulated, miners should not be allowed to burn land over willfully, lands better suited for agriculture or mining should be eliminated from the reserves, mature timber should be cut and sold, and settlers and miners should be allowed to cut only such timber as they need."

Just before the close of the nineteenth century, President Grover Cleveland established another 21 million acres of forest reserves, and in 1897, the Forest Management Act, or Organic Act, was passed, which specified that forest reserves were "to improve and protect the forest, or for the purpose of securing favorable conditions of water flows, and to furnish a continuous supply of timber for the use and necessities of citizens of the U.S."

Although Lincoln was perhaps the first president to see the "people's land" as a legacy to preserve for posterity, President Theodore Roosevelt thrust the nation into its first conservation movement. Roosevelt acted aggressively to expand federal forest reserves and to establish the first national wildlife refuges and national monuments. In 1905, with Gifford Pinchot at the helm, the USDA Forest Service was established. Secretary of Agriculture James Wilson, in 1905, directed Forest Service Chief Pinchot to manage the national forests "for the greatest good for the greatest number for the long run." Thanks to such visionary leaders in the nineteenth and early twentieth centuries, today we have 192 million acres of national forests that are owned by the people and are the birthright of all American citizens.

The pages that follow in this volume by Robert H. Mohlenbrock, a distinguished botanist, natural historian, and conservationist, are an account of the national forests of the West. Mohlenbrock has spent more than 40 years visiting and working in all of the 155 national forests. State by state, Professor Mohlenbrock describes each of the region's national forests in detail, including their size, location, access routes, basic geology, hydrology, and biota, as well as things to see and do.

He describes the trails that take visitors to wilderness areas and features of special interest and concern. He discusses rare and endangered species, notable historical landmarks and events, and sites of scenic beauty. *This Land: A Guide to Western National Forests* contains a wealth of information presented in clear and concise language. It adeptly conveys the sense of awe that characterizes our national forests. In the end, this volume will help us and future generations understand and appreciate the wealth of this land and remind us of the importance of being responsible stewards of the people's land today and for future generations.

Mike Dombeck
Chief Emeritus of the U.S. Forest Service

PREFACE

My family and I began visiting the national forests in 1960, and we have spent all of our vacations and an enormous amount of days in them, eventually visiting each of the 155 national forests. We soon discovered that they contain millions of acres of habitats and scenery that are nearly on a par with those found in national parks and national monuments. Many of these marvelous areas in national forests are little known, and we had to do considerable research to find the most exciting and beautiful areas. I have tried to provide information for these areas in this book.

Because I am a professional botanist, my family and I spent considerable time in areas known as Research Natural Areas, which are part of a national network of ecological areas designated in perpetuity for research and education, to maintain biological diversity on National Forest System lands, or both. Although many of the Research Natural Areas are in remote areas that are not accessible to the ordinary person, others are more accessible. Special permission may be required to visit a few of them. Some of my favorites are included in this book.

When you begin your exploration of a national forest, you will need to obtain an up-to-date forest map showing where the major roads and back roads are located. In most forests I have not listed the forest road and forest highway numbers since these change occasionally and are often vandalized so that visitors to the forest may have difficulty even finding the forest service road markers. The reader is advised to obtain the latest forest service map for the national forest from the district ranger stations or from the forest supervisor's office. While you are at the ranger stations, you can usually pick up several brochures describing trails and other points of interest. Forest fires which may have occurred while this book was in press may have altered the forests to some extent.

In 1984, I received a call from Mr. Alan Ternes, then the editor of *Natural History* magazine, published by the American Museum of Natural History in New York. Mr. Ternes asked if I would be interested in writing a monthly column for the magazine about some areas in the national forests that I particularly liked. An area did not necessarily need to be pretty, but it should have a biological or geological story to tell. My first article appeared in the November 1984 issue of *Natural History* magazine in my This Land column. The articles that I published in *Natural History* that pertained to national forests in the western United States accompany some of the national forest descriptions in this book. I am grateful to *Natural History* magazine and its editor, Ellen Goldensohn, for allowing me to republish these articles in this book.

I am indebted to several forest personnel, some of whom directed me to some little known areas, and others who read drafts of the manuscript of their particular national forest. Any errors which may have crept into the book are strictly my own. I am also grateful to Blake Edgar of the University of California Press who suggested this series of books, and to Mr. Scott Norton who has worked untiringly as my editor.

Whatever your interest in the out-of-doors, you will undoubtedly be able to enjoy our national forests, which are truly a unique American treasure.

INTRODUCTION

During the rapid development of the United States after the American Revolution, and during most of the twentieth century, many forests in the United States were logged, with the logging often followed by devastating fires; ranchers converted the prairies and the plains into vast pastures for livestock; sheep were allowed to venture onto heretofore undisturbed alpine areas; and great amounts of land were turned over in an attempt to find gold, silver, and other minerals.

In 1875, the American Forestry Association was born. This organization was asked by Secretary of the Interior Carl Schurz to try to change the concept that most people had about the wasting of our natural resources. One year later, the Division of Forestry was created within the Department of Agriculture. However, land fraud continued, with homesteaders asked by large lumber companies to buy land and then transfer the title of the land to the companies. In 1891, the American Forestry Association lobbied Congress to pass legislation that would allow forest reserves to be set aside and administered by the Department of the Interior, thus stopping wanton destruction of forest lands. President Benjamin Harrison established forest reserves totalling 13 million acres, the first being the Yellowstone Timberland Reserve, which later became the Shoshone and Teton National Forests.

Gifford Pinchot was the founder of scientific forestry in the United States, and President Theodore Roosevelt named him chief of the Forest Service in 1898 because of his wide-ranging policy on the conservation of natural resources.

Pinchot pursuaded Congress to transfer the Forest Service to the Department of Agriculture, an event that transpired on February 1, 1905. He realized that the forest reserves were areas where timber production would be beneficial to the nation and where clear water, diverse wildlife, and scenic beauty could be maintained.

In 1964, the Wilderness Act was passed by Congress, authorizing the setting aside of vast areas that were still in pristine condition. Although the establishment of wilderness areas has preserved some of our most beautiful areas, it has also made these areas off-limits to anyone who is aged or physically handicapped or who just cannot backpack for miles and miles into an area.

Slowly, as people from all walks of life began to use the national forests for recreational purposes, the U.S. Forest Service adopted the multiple-use concept, where timber production, wildlilfe management, conservation of plants and animals, preservation of clear water, maintenance of historic sites, and recreation could be accommodated in the national forests and enjoyed year round. Although camping, picnicking, and scenic driving are the major recreational activities on national forest land, other activities include boating, swimming, fishing, hunting, whitewater rafting, horseback riding, nature study, photography, wilderness trekking, hang-gliding, rockhounding, and winter sports. Special areas are set aside for off-highway vehicle activity.

Today, we have 155 national forests, although to save administrative costs, some have been combined. The U.S. Forest Service administers other areas as well, such as national grasslands, the Columbia River Gorge National Scenic Area, and the Lake Tahoe Basin Management Area.

NATIONAL FORESTS IN ALASKA

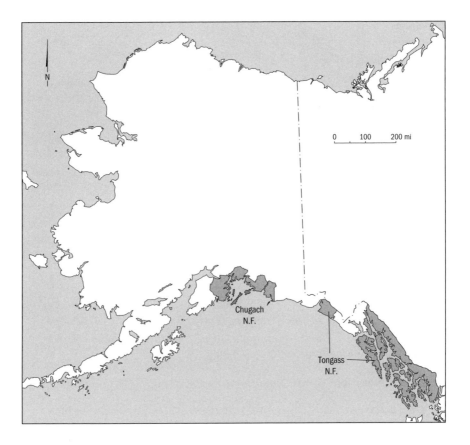

The national forests in Alaska are in Region 10 of the U.S. Forest Service. The address for the regional headquarters is 709 W. 9th Street, Juneau, AK 99802. The two national forests in Alaska occupy a total of 22.4 million acres and contain 19 wildernesses.

Chugach National Forest

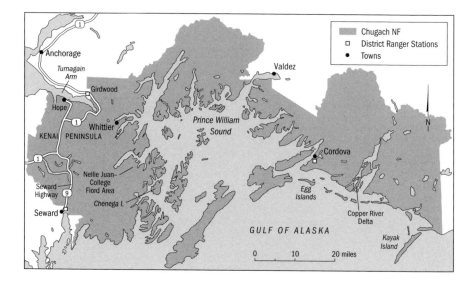

Map legend:
- Chugach NF
- □ District Ranger Stations
- • Towns

Map labels: Anchorage, Turnagain Arm, Girdwood, Hope, Whittier, KENAI PENINSULA, Nellie Juan–College Fiord Area, Seward Highway, Chenega I., Seward, Valdez, Prince William Sound, Cordova, Egg Islands, Copper River Delta, Kayak Island, GULF OF ALASKA, N

0   10   20 miles

SIZE AND LOCATION: 5.6 million acres in south-central Alaska between Anchorage and Cordova. Major highways are State Routes 1, 4, 9, 10, the Seward Highway, and the Hope Highway. District Ranger Stations: Girdwood, Cordova, Seward. Forest Supervisor's Office: 3301 C Street, Suite 300, Anchorage, AK 99503, www.fs.fed.us/r10/chugach.

SPECIAL FACILITIES: Visitor Center; winter sports areas.

SPECIAL ATTRACTIONS: Copper River Delta; Seward Scenic Byway; Nellie Juan–College Fiord Area.

Chugach is the most northern and most western national forest in the United States and is second in size only to Alaska's Tongass National Forest. Only 500 miles from the Arctic Circle, the Chugach is about the size of Massachusetts and Rhode Island combined. Well over half of the national forest is tundra and glaciers, although it has no shortage of forests, mountains, lakes, rivers, wildlife, and wetlands. The main attraction for visitors is the opportunity for world-class fishing and hunting. For fishing regulations, please contact the Alaska Department of Fish and Game at (907) 465-5999 or www.state.ak.us/adjg/adfghome.

People by the thousands come each year to the Chugach, some to fish for red salmon in the Russian River, others for chum salmon, king salmon, silver salmon, steelhead trout, Coho trout, lake trout, rainbow trout, and Dolly

Varden trout that are in any of the hundreds of streams that permeate the forest.

Major activity in the forest centers around the Kenai Peninsula, where Seward is located, Prince William Sound, the Copper River Delta, and the scenic Nellie Juan–College Fiord Area.

For a first glimpse of the national forest, we suggest driving the Seward Scenic Byway from Anchorage to Seward and through the heart of the Kenai Peninsula. This 127-mile route has incomparable scenery at every turn, with wildlife to observe and several opportunities to hike and camp. After leaving Anchorage and paralleling the north shore of Turnagain Arm, the highway enters the national forest at Milepost 87. (The mileposts are numbered from Seward to Anchorage.) Immediately on the national forest boundary is the popular Alyeska Ski Area. The scenic byway curves around the eastern end of Turnagain Arm where there is a must stop at the Begich, Boggs Visitor Center, located in picturesque Portage Valley at Milepost 78.8 (fig. 1). At the visitor center you may learn all about glaciers before taking a short boat tour that includes a close-up view of Portage Glacier (pl. 1) and some incredibly tall icebergs. The visitor center is built on remnants of the Portage Glacier terminal moraine that was deposited in 1893. From the visitor center you may hike a short and informative nature trail or take the five-mile-long Glacier View Trail for a glimpse of Byron Glacier.

Just south of Portage Valley, at Milepost 56.5, is the junction with Hope Highway. This side road travels southwest along Ingram and Granite Creeks

Figure 1. Portage Valley ice cave.

before abruptly heading north to the southern edge of Turnagain Arm. After a short distance along the arm, the highway comes to the village of Hope, where the Porcupine Campground is nearby. The easy five-mile Gull Rock Trail begins at the campground and parallels Turnagain Arm, first following an old wagon train road. Where the trail meets Johnson Creek are ruins of an old sawmill and the remains of a cabin and stable. On the way to Gull Rock, which protrudes into Turnagain Arm, you walk through a woods of Alaska paper birch, Kenai birch, and quaking aspen and then trek softly on carpets of moss in a forest of western hemlock. At one point the trail dips down to a gully lined by a dense growth of alder. The occasional patches of tundra permit you to see plants such as mountain cranberry, alpine bearberry, four species of cassiope, alpine azalea, Kamschatka rhododendron, three kinds of mountain heaths, and stunted black spruce.

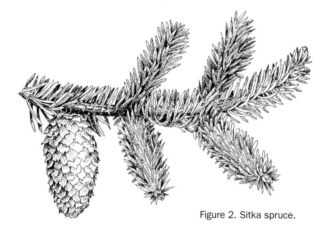

Figure 2. Sitka spruce.

The Hope Point Trail, half as long as the Gull Rock Trail, also begins at the Porcupine Campground but is steep after the first third of a mile along Porcupine Creek. Once you get to Hope Point, there are stunning views of Turnagain Arm and the Chugach Mountains. Even if you do not make it to the point, you are rewarded by impressive views of Resurrection Creek Valley.

At Milepost 19.7 on the Seward Scenic Byway is the roadside pullout for the Victor Creek Trail. The first half of this two-mile-long trail is steep, but it passes the remains of an old mining camp. Look for mountain goats (pl. 2) in the high country. Grayling Lake Trail is an easy two-mile family trail at Milepost 13.2. It goes through spruce forests and open meadows before reaching Grayling Lake.

On the western side of Prince William Sound is a vast wild area that has been under consideration for wilderness status for many years. Called the

Nellie Juan–College Fiord Area, the region extends from College Fiord at the northern edge of the Chugach to Cape Puget at the southern tip of the Kenai Peninsula and includes several adjacent islands. The Chugach Eskimos lived on Chenega Island until 1964 when a violent earthquake created a giant seismic sea wave and destroyed their village. Elevations range from sea level to tall mountains, the highest reaching 5,200 feet. Virgin forests of Sitka spruce (fig. 2), western hemlock, and mountain hemlock are interrupted by glaciers in the high country and by muskegs in the low.

On the east side of the national forest near the community of Cordova is the remarkable Copper River Delta (described in more detail below). This is the largest contiguous wetland in the United States. The Copper River has its origin in the rugged Chugach Mountains where many glaciers feed the river. Where the river flows into the Gulf of Alaska, it has deposited silt, gravel, and boulders, thus forming a 50-mile-wide delta. Here are tidal flats, brown mudflats, gray mudflats, marshes, freshwater ponds, and sloughs, many of them inundated by saltwater at high tide.

The dominant plants of the delta are sedges and grasses. A part of the Pacific Flyway migratory route, the Copper River Delta attracts millions of birds. Trumpeter swans, Canada geese, and many species of shorebirds and ducks stop to feed and rest on their migratory route, and thousands of pairs of waterfowl breed and raise their young in the marshes. When salmon begin their run up the Copper River, hundreds of bald eagles appear, along with hungry bears. The Alaganik Slough Boardwalk extends for 900 feet into a part of the wetland.

Copper River Delta

East of Cordova, Alaska, and continuing along the Gulf of Alaska for 50 miles is the vast wetland complex that makes up the Copper River Delta. This region extends up to 37 miles inland in some places and occupies low-lying terrain between sea level and 300 feet. This large plain is occasionally interrupted by spurs of the Chugach Mountains. Outside the eastern edge of the wetland is Bering Glacier, the largest in North America. A short distance offshore and in the Gulf of Alaska are small barrier islands and spits of sand, including the Egg Islands, Strawberry Reef, Kanak Island, and Softuck Spit.

The climate of the Copper River Delta consists of mild wet summers and cool wet winters. The Alaska Current helps maintain the stability of the climate, while the adjacent Chugach Mountains to the north tend to shelter the coastal area.

Geologists believe that recent tectonic movements, along with massive deposits of sediments and strong erosional forces, have shaped today's landscape. An earthquake in 1964 uplifted the area six to 12 feet. At that time, most of the delta was covered by a brackish marsh. The earthquake elevated the marshes above much of the influence of the tides.

The topography and wetland habitats today consist of glacial moraines, outwash plains, uplifted marshes, and linear dunes, with barrier islands, sand spits, and coastal dunes just off the mainland. The islands, spits, and coastal dunes are regulated by the currents, waves, and wind and shelter the delta from the ocean currents and waves. A major influence today is the great amount of sediment deposited as the Copper River heads to the ocean. One geologist estimates that 97 million tons of sediment are deposited by the river annually. As the more active river enters the calmer waters of the ocean, the velocity of the water lessens, resulting in the rapid dropping of the sediment load. Subsequent erosion caused by the tides, waves, and the Alaska Current further shapes the landscape.

OUTWASH PLAINS. Outwash plains are broad alluvial plains that consist of streams, abandoned stream channels, tree- or shrub-dominated terraces, and scattered ponds. At the inner region of the delta nearer the glaciers, the outwash plains have a rougher terrain with abrupt terraces and peatlands. Toward the ocean the outwash plains are more smooth. It is on these smoother surfaces that forests of Sitka spruce and western hemlock ultimately develop.

Botanist Keith Boggs has described the major and minor plant communities of the Copper River Delta. Four major communities occupy the outwash plains, the type depending on the topography. Sitka spruce is the dominant tree on alluvial soil, with a shrub layer that consists of devil's-club, salmonberry, and Alaska blueberry. Spinulose wood fern and oak fern are common, along with foamflower, twisted stalk, and five-leaved bramble. Most of the ground beneath the Sitka spruce is carpeted by a variety of mosses. A community in the alluvial soil of the outwash plain that is sometimes inundated when the river is up is dominated by black cottonwood above a shrub layer of Sitka alder. Another community is dominated by Sitka alder, with a fairly continuous ground cover of bluejoint grass and Sitka sedge. In areas that are a little drier, meadow horsetail is common beneath a dense growth of Sitka alder. In ponds of the outwash plain, Sitka sedge often forms dense stands, with bluejoint grass, meadow horsetail, buckbean, and five-finger potentilla usually present.

UPLIFTED MARSH. The uplifted marshes consist of ponds and levees maintained by freshwater streams, tidal creeks, and low sea cliffs. The ponds in the uplifted marshes usually have a depth of 30 to 45 inches; the sea cliffs may reach heights of six feet.

Where there is saturated peat in the uplifted marsh, a shrubby community of sweet gale occurs above a ground layer of Lyngby's sedge or Sitka sedge. In areas with fluctuating water levels, Barclay willow and sweet gale dominate above an understory of Sitka sedge, buckbean, and five-finger potentilla, although undergreen willow is often abundant. One common community that Boggs noted is found on mats of roots in the uplifted marshes. These mats support a growth of Lyngby's sedge, vetchling, bluejoint grass, Alaska bent grass, and a species of bedstraw, usually with a substantial amount of sphagnum. In areas with permanent or semipermanent water, the aquatic communities may be dominated by swamp horsetail, marestail, buckbean, or five-finger potentilla.

LINEAR DUNES. Linear dunes radiate from the mouth of the Copper River. These dunes, mixed with terraces, levees, and ponds, are long and narrow, becoming wider and steeper upwind. The sides of the linear dunes support vegetation, whereas the tops are usually bare of plants. The linear dunes range in height from 20 to 250 feet and may be as long as nine miles.

The linear dunes that radiate from the mouth of the Copper River usually have substantial communities of Sitka alder, salmonberry, and red elderberry above an herbaceous layer of common horsetail, cow parsnip, and lady fern.

TIDAL MARSH. The rhythmic movement of the tides creates a series of habitats that include mudflats, marshes, tidal creeks, and often a zone occupied by shrubs. At high tide, the tidal creeks are bank full.

The extent of the tidal marshes depends on the amount of sediment deposited. Because the barrier islands reduce the amount of wave action in the estuary of the Copper River Delta, the fine sediments are conducive to good tidal marsh development. The most common communities of the tidal marshes, according to Boggs, are sweet gale–Lyngby's sedge, Lyngby's sedge–seaside buttercup, and pure stands of dwarf alkali grass. Sweet gale and Lyngby's sedge occur on the drier parts of the tidal marsh, with Barclay willow, undergreen willow, and Hooker willow often present.

BARRIER ISLANDS AND SAND SPITS. Barrier islands, sand spits, and coastal dunes are all dune communities, with the elongated islands separated from the mainland by estuaries or bays. Sand spits are elongated strands of sand where a dune enters the ocean. Coastal dunes are beach ridges or dunes at the edge of the mainland.

Beach rye is the grass that first colonizes newly formed sand spits and new areas of the barrier islands. On the sand above the pure stands of beach rye, yarrow becomes a common associate with the grass. As one climbs above the beach rye–yarrow community, fireweed becomes a dominant species along with lady fern, bluejoint grass, and sea coast angelica. Sometimes a zone of wild strawberry occurs just below the fireweed community.

Primrose Trail

The Chugach National Forest is located northwest of Alaska's other national forest, the Tongass. The Chugach is adjacent to the coast from the Bering Glacier on the east, North America's largest glacier, to the Russian River on the west. Included in this region are the towns of Cordova and Seward. Valdez lies just to the north of the national forest and Anchorage to the northwest. The national forest is north of the Gulf of Alaska and Prince William Sound.

The Chugach National Forest land adjacent to the coast and along some of the rivers and streams supports an old-growth forest of Sitka spruce and associated species. As the altitude increases away from the coast, a vast alpine area of tundra and glaciers occurs.

The Seward Highway between Seward and Anchorage is a great way to get acquainted with the Chugach National Forest because the lower half of the highway penetrates the forest. At Milepost 17, a few miles north of Seward, is a 1.5-mile side road to the Primrose Campground where there are hiking trails. The Primrose Trail goes through a majestic forest and into the alpine area. Lost Lake, southwest of the campground, with Mt. Ascension forming its western boundary, is exceptionally scenic and is a popular fishing lake for rainbow trout. The first four miles of the Primrose Trail are moderate in ruggedness but in good condition, whereas the last four miles are through steep, eroded, often boggy terrain. The wildlife you may see as you hike the trail includes black bear, moose, Dall sheep, mountain goat, marmot, ptarmigan, and grouse. At Milepost 3 along the trail is a spur trail to a viewpoint of scenic Porcupine Creek Falls.

Sitka spruce forests along the Gulf of Alaska are dense, casting heavy shade, and are reminiscent of the rain forests of British Columbia and Washington. If you stray from the developed trail, you will find yourself in extremely rough topography. In many places in the Chugach National Forest, the Sitka spruce forest occupies a very narrow band between the ocean and the tundra-topped mountains above. In several places in the national forest, the scenery is enhanced by glaciers that reach the coast, picturesque mountain lakes, and occasional waterfalls. Some parts of the coast may receive as much as 222 inches of rain annually. Because of the high rainfall and the accompanying high humidity, mounds of mosses on the ground and on fallen trees are extensive. In many places, the mosses hang from the lower branches of living trees.

Where drainage is poor in the forest near the coast, wetlands known as muskegs are frequent. Although the muskegs are dominated by low-growing shrubs, sedges, and grasses, scattered scrubby lodgepole pines and Sitka spruces may occur.

Above timberline is alpine tundra, characterized by exposed ridges of bare rocks with mat-forming plants often crowded between the rocks.

SITKA SPRUCE FOREST. Although huge Sitka spruces usually occur in this low-elevation forest zone, western hemlock and mountain hemlock also may grow extremely large. Black cottonwood is the dominant tree along streams that penetrate the forest. A large number of shrubs often occupy a shrub zone in the Sitka spruce forest. Common plants include Barclay willow, Scouler willow, Sitka willow, salmonberry, trailing black currant, Sitka alder, Pacific red elder, and three members of the heath family—early blueberry, highbush cranberry, and rusty menziesia. The spiny devil's-club often forms impenetrable thickets.

MUSKEGS. Muskegs are open, boggy areas underlain by at least 16 inches of sphagnum, and they usually have a thick cover of sphagnum and various kinds of lichens. Several shrubs are characteristic of the coastal muskegs, including bog rosemary, bog cranberry, bog blueberry, rusty menziesia, crowberry, Barclay willow, and undergreen willow.

ALPINE TUNDRA. Although the environmental conditions in the alpine tundra are severe and the growing season very short, a surprisingly large number of mat-forming flowering plants occur. Several of these plants are actually dwarf shrubs with woody stems, such as the alpine bearberry, narrow-leaf Labrador tea, two kinds of mountain avens, dwarf Arctic birch, Beauverd spiraea, bog crowberry, bog blueberry, alpine azalea, and several species of willows.

Tongass National Forest

SIZE AND LOCATION: 16.8 million acres in southeastern Alaska, from Yakutat Bay in the north to Dall Island in the south, encompassing the towns of Juneau, Sitka, and Ketchikan. Few roads are in the area except for Glacier Highway, Hollis-Klawock Highway, and Salmon River Road. Visitors may reach the region by airplane or a variety of boats. District Ranger Stations: Craig, Hoonah, Juneau, Ketchikan, Petersburg, Sitka, Thorne Bay, Wrangell, Yakutat. Forest Supervisor's Office: Federal Building, Ketchikan, AK 99901, www.fs.fed.us/r10/tongass.

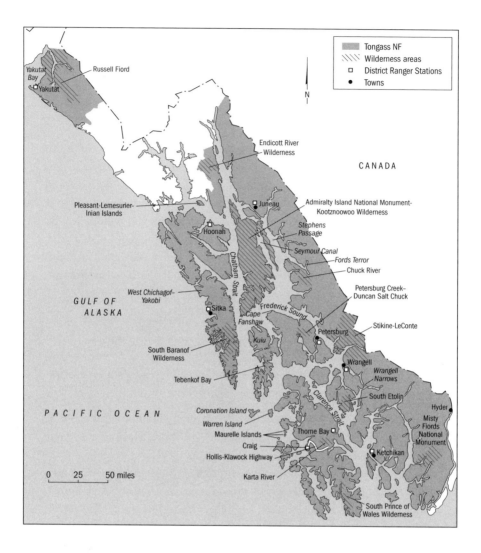

SPECIAL FACILITIES: Landing areas; boat docks; backcountry cabins; observation platforms.

SPECIAL ATTRACTIONS: Misty Fiords National Monument; Admiralty Island National Monument.

WILDERNESS AREAS: Coronation Island (19,232 acres); Endicott River (98,729 acres); Kootznoowoo (988,050 acres); Maurelle Islands (4,937 acres); Misty Fiords (2,142,907 acres); Petersburg Creek–Duncan Salt Chuck (46,849 acres); Russell Fiord (348,701 acres); South Baranof (319,568 acres); South Prince of Wales (91,018 acres); Stikine-LeConte (449,951 acres); Tebenkof Bay (66,839 acres); Tracy Arm–Ford Terror (653,179 acres); Warren Island

(11,181 acres); West Chichagof–Yakobi (265,529 acres); Chuck River (74,990 acres); Karta River (39,984 acres); Kuiu (60,581 acres); Pleasant-Lemesurier-Inian Islands (23,151 acres); South Etolin (83,371 acres).

Majestic is the only way to describe the scenery in southeastern Alaska. Whether you see it from a cruise ship, a kayak, a float plane, or while hiking, the fiords, steep sea cliffs, glaciers, temperate rain forests, and waterways will leave an indelible impression on you.

Most of the area of southeastern Alaska is in the Tongass National Forest, and much of it is inaccessible to the average sightseer and explorer of nature. The enormous Tongass covers 16.8 million acres, an area slightly larger than West Virginia. It stretches for hundreds of miles from Yakutat Bay in the north to Dall Island some distance south of Ketchikan and from the Gulf of Alaska to British Columbia, Canada. Hundreds of islands, some very tiny to the largest, Prince of Wales, are in the national forest. A few lowlands are found around Sitka and Ketchikan, but most of the Tongass is mountainous. On these mountains is a continuous cover of temperate rain forest. The more than 120 inches of annual precipitation in Ketchikan and the 90 inches annually in Juneau are conducive to these lush forests.

Most of the forests in the Tongass National Forest are dominated by western hemlock and Sitka spruce, some of them attaining tremendous sizes. A few of the spruces are more than 200 feet tall with diameters exceeding eight feet. The hemlocks usually top out at about 190 feet and have diameters up to five feet. Associated with these two dominants are mountain hemlock, western red cedar, and Alaska cedar. Along the streams and the fringes of beaches are dense stands of red alder. On mountains high enough to have a timberline, the trees at the highest elevations are subalpine fir and Pacific silver fir. The most important shrubs beneath the Sitka spruces and western hemlocks include several species of blueberries, huckleberry, salal, copperbush, and the extraordinarily prickly devil's-club. All but the last are in the heath family. Black cottonwood dominates the floodplains, along with a good complement of willows.

Although approximately one-third of the Tongass National Forest is in congressionally designated wilderness areas, the average visitor still has an abundance of opportunities to enjoy the lush scenery, vegetation, and wildlife in this national forest. Here, even the wilderness experience is different from that in the lower 48 states. Most of the wilderness can be reached only by float planes or boats, including kayaks, canoes, motorboats, sailboats, charter boats, ferries, and cruise ships. Public recreation cabins are for rent from the Forest Service in much of the backcountry. Reservations for these rustic cabins are required. Do not expect indoor plumbing. Some campgrounds are available, and in general, you can pitch a tent in the middle of nowhere.

Although hunting, fishing, hiking, and nature study are primary activities in the Tongass, winter sports enthusiasts have plenty to do, including cross-country skiing, snowmobiling, and ice skating on lakes, ponds, and even in frozen-over sloughs.

For many visitors, Juneau is the starting point for an unparalleled adventure. If you are interested in glaciers—and being in Alaska is incentive enough—you can drive 15 miles north to the lower end of Mendenhall Glacier (pl. 3) to the marvelous Forest Service visitor center where you can learn all about glaciers. The Mendenhall Glacier is an active ice floe fed from the massive Juneau Ice Field high above. The ice field provides the moisture for at least 38 different glaciers. Although the Mendenhall Glacier is moving forward at the rate of two feet per day, enough melting and breaking off of large pieces, called calving, occurs that the face of the glacier is actually slowly retreating. Today Mendenhall Glacier is 12 miles long, and the visitor center is situated at its terminus. As glaciers melt and retreat, the land that becomes exposed begins its succession of plant communities. As you hike near the glaciers in the Tongass, you will encounter all stages of succession, from mosses and lichens on seemingly barren soil to wildflowers, then shrubs, and finally trees. Near the visitor center are areas of grooved, polished rocks where the glacier has scraped its way along. Along a 1.5-mile loop, the Moraine Ecology Trail, you can see the rocks and debris left by the glacier where it has retreated.

The Glacier Highway goes for 41 miles northwest from Juneau, with the first 33 miles paved, the next four unpaved, and the final four very rough. Along the way you get superb views of mountains, glaciers, and the sea. Several trails may be taken from this road. It also leads to picnic areas and Forest Service and state recreation cabins.

South of Juneau on the other side of Stephens Passage is Admiralty Island. Most of this large, beautiful island has been protected as the Admiralty Island National Monument under the management of the Tongass National Forest, and all but 18,000 acres at the north end of the island is in the Kootznoowoo Wilderness. Surrounded by Stephens Passage, Chatham Strait, and Frederick Sound, the interior of the island rises steeply to heights of more than 4,000 feet. A dense temperate rain forest of Sitka spruce, western hemlock, and Sitka alder clothes the mountain, penetrated only by occasional rocky spires. Several bays and inlets make deep incisions into the island, and these protected areas are perfect to moor your boat. The large Seymour Canal has a great density of nesting bald eagles, and the water in this canal is home to harbor seals, porpoises, sea lions, and humpback whales.

For canoers, a 26-mile Cross Island Canoe Route is just that—it crosses the island from the Seymour Canal on the east to the Tlingit Indian village of Angoon, with a population of 450, on the west. It is necessary to portage between the eight lakes along the route. Admiralty Island can be reached by canoe, kayak, chartered or private float plane, or ferry from Juneau to Angoon.

Along Pack Creek in the northeastern corner of Admiralty Island, the Forest Service has constructed an eight-foot by eight-foot platform about 20 feet above the ground for visitors to observe bears. This platform has an unobstructed view for 200 yards each way up and down the stream at the mouth of Pack Creek. Once you climb the iron ladder and emerge through a trapdoor in the platform, there are benches on three sides and a strong hand rail around the entire platform. A roof gives protection from rain. By anchoring your boat between the west end of Windfall Island and the shores of Windfall Harbor, you are only a short distance via a good foot trail to the observatory. During July and August, when the salmon run, as many as a dozen different bears come to the stream, some of them several times. You will see the bears deftly pin the salmon to the bottom of the stream in shallow riffles with their paws. Then, taking the fish in their mouths, they walk to nearby gravel bars or into dense brush to enjoy their catch. You need a permit, available from the Juneau Ranger District, to set foot in the area from June 1 to September 20. Visitation is limited and strictly enforced. It is best to book early when the salmon are running.

Admiralty Island is a perfect habitat for the large Alaskan coastal brown bear (fig. 3). With dense forests and an abundance of salmon streams, Admiralty supports one of the densest populations, with more than one bear for every nine square miles.

At the northernmost edge of the Tongass National Forest, at the upper end of Yakutat Bay, is the Russell Fiord Wilderness. This extremely narrow fiord, being pinched by encroaching glaciers, penetrates 35 miles inland. On the heavily forested mountain slopes and in the alpine meadows live the so-called blue, or glacier, bears, an unusual color form of the black bear. This wilderness also supports a large moose herd. Hiking in this uncharted wilderness will likely bring you into contact with bears (brown, black, and glacier), wolves, and mountain goats. As you travel up the fiord by boat, you will probably see harbor seals swimming along. Sea lions usually hang around the outer coast.

Another popular region for wilderness explorers is the Tracy Arm–Fords Terror Wilderness on the east side of Stephens Passage about midway between Juneau and Petersburg. Tracy Arm is popular for passengers in larger touring boats. Tracy Arm and Endicott Arm are two narrow fiords that ex-

Figure 3. Brown Bears.

tend 30 miles westward toward British Columbia's Coast Mountains. Near the beginning of Endicott Arm is an area of tall, sheer rock walls known as Fords Terror. As you travel deep into the fiords, you will see impressive waterfalls among the dense stands of Sitka spruce and western hemlock. Look for mountain goats in the higher elevations.

A short distance north of Tracy Arm and out of the wilderness area is the very narrow Limestone Inlet. Terrain on either side of the secluded little inlet rises to 3,000 feet, with a series of several lakes found at 1,400 feet.

Northwest of Juneau on the western side of the Lynn Canal are the Chilkat Mountains. One-third of this area is in the Endicott River Wilderness, dominated by 5,280-foot Mt. Young, which stands high above the rest of the wilderness. This is one of the best places in the northern Tongass to see exceptionally large Sitka spruces and western hemlocks. Several boggy openings in the forest, known as muskegs, are kept wet by an average of 92 inches of rainfall each year. Do not attempt to step out into these boggy areas, even if some beautiful wildflower is beckoning to you. This wilderness has no marked trails.

West of Admiralty Island are two islands of very different size: the larger Chichagof Island and the smaller and more northern Yakobi Island. These islands make up the West Chichagof–Yakobi Wilderness. They have rugged trails and three Forest Service cabins for rent. If you choose to go here, you will see all that Alaska has to offer: mountains covered with temperate rain forests, alpine tundra, muskegs, bays, coves, lagoons, reefs, tidal meadows, narrow passages, estuaries, and even evidence of former gold mining. Sitka black-tailed deer are plentiful, as are grizzly bears, sea lion rookeries, sea otters, and harbor seals, the latter seemingly sprawled out on every available rock.

Cape Fanshaw, a large promontory protruding into Stephens Passage, is a pleasant area where the small village of Fanshaw and a cannery used to be located. The mountain slopes behind the townsite used to have the finest stand of Alaska cedar in the world, but it was decimated by cedar die-back. South of Cape Fanshaw, at the northeastern corner of Kupreanof Island and just across the Wrangell Narrows from Petersburg, is the Petersburg Creek–Duncan Salt Chuck Wilderness. The Duncan Salt Chuck is a huge salt marsh influenced daily by tidal changes. It is one of the best places to see an Alaskan salt marsh. Petersburg and Salt Chuck, the two major streams in the wilderness area, have an abundance of steelhead trout and several species of salmon. Nearby mountain tops regularly have an accumulation of 200 inches of snow.

The southern end of Baranof Island below Sitka projects into the Gulf of Alaska. From the shoreline to the summit of 4,528-foot Mt. Ada is a distance of only three miles, emphasizing the steepness of the mountains in this region. The temperate rain forest trees are thicker here than anywhere else because the area receives as much as 400 inches of rain each year. The coastline from Whale Bay to Cape Ommaney is among the prettiest in the world.

Between Baranof Island and Kupreanof Island is Kuiu Island. On the secluded and protected western side of the island is Tebenkof Bay, with many attractive coves, inlets, and lakes. The mountainous area behind the shoreline is in the Tebenkof Bay Wilderness.

Southeast of Petersburg and north of Wrangell, the Stikine River penetrates deep into the mountains, finally crossing the Canadian border just below Elbow Mountain. Along the river is a fine riparian community of black cottonwood and several species of willows. Moose are very plentiful here. The slopes that rise steeply from either side of the river have typical spruce–hemlock forests. North of the Stikine River, the LeConte Glacier slopes down to Frederick Sound where icebergs have resulted from calving of the glacier. The tidal flats around the mouth of the Stikine are excellent for observing migrating waterfowl.

No trip to Alaska is complete without a visit to Misty Fiords National Monument east of Ketchikan. This beautiful area is administered by the Tongass National Forest. The national monument is part of the Coastal Mountains of southeastern Alaska and has extraordinarily long and deep fiords with steep cliffs rising several thousand feet. Near the eastern edge across from the Canadian border are several active glaciers. The Unuk River cuts across the northwestern corner of the national monument through extremely steep valleys. Glaciers hang above the valleys and, in sharp contrast, there are a few lava flows. Farther south in the monument, the Chickamin and LeDuc Rivers are lined by steep canyons before they empty into the Behm Canal that forms the western edge of the area. Big Eddystone Rock, a volcanic plug 200 feet tall, is a conspicuous landmark along the canal. The most scenic area of all is the Rudyerd Bay area. One fascinating point of interest is Punchbowl Lake, which can be reached by hiking a .9-mile trail from Punchbowl Cove. You have to arrive at the cove by boat. The trail begins on a boardwalk across a large muskeg. Growing from this watery, spongy substrate are several species of lichens and mosses and some flowering plants. Species you may be able to identify are bog rosemary, crowberry, bog kalmia, Labrador tea, rusty menziesia, bog cranberry, mountain cranberry, and bog blueberry. Low-growing and sprawling juniper, undergreen willow, and Barclay willow are present, as well as dwarf plants of lodgepole pine and Alaska cedar. The trail then begins a steep climb, with steps carved out of old fallen trees and a few switchbacks to ease the pain. About halfway to the lake is a breathtaking waterfall (if you have any breath left). The trail levels out as it follows a stream to Punchbowl Lake. The lake, surrounded by tall, granite cliffs, has many islands at its center.

One of the most popular areas in the Misty Fiords National Monument is Fish Creek Wildlife Observation Site. It is reached from the Canadian side via Highway 37A. From this highway take the gravelly Salmon River Road for 41 miles to the Alaskan community of Hyder, and then continue three miles north to the Fish Creek bridge. A trail leads to the observation site. From mid-July through September, chum and pink salmon come to Fish Creek to spawn in the clear shallow water. Grizzly and black bears are easily observed fishing for these fish. Because traffic congestion at the parking lot is common and the gravel road to it is long, visitors may hire a shuttle bus in either Stewart, British Columbia, or Hyder, Alaska, to take them to the observation site. If you have your own vehicle, we suggest continuing on the Salmon River Road as it climbs past the Salmon Glacier, ending at the abandoned Grand Duc Mine. On the way you pass alpine meadows full of wildflowers. Some of the wildflowers you might see are the yellow- and white-flowered mountain avens, alpine meadow bistort, prickly saxifrage, narrow-leaved chrysanthemum, northern yarrow, and yellow anemone. For the more energetic, there is a five-mile Titan Trail from the observation site to alpine ridges along the United States–Canada border. Before starting the rugged ascent, you have to wade across Fish Creek.

In the vicinity of Ketchikan are three very accessible and worthwhile trails.

Eight miles north of town, reached by a paved road, is the Ward Lake Day Use Area. A 1.3-mile gravel nature trail encircles Ward Lake, and the Perseverance and Connell Lake Trails are accessed here, as well. Farther north and not accessible by automobile is the Naha River National Recreation Trail. For 5.5 miles you can walk along the edge of the river and cast for fish the whole way.

Prince of Wales Island is the third largest island in the United States, after the large island of Hawaii and Kodiak Island. The island has steep forested mountains, U-shaped glacial valleys, bays carved by glacial ice, streams, lakes, and saltwater tidal flats. Once you reach the island by small aircraft or ferry, you will find nearly 1,500 miles of roads. From the community of Craig, start your journey by taking the Hydaburg Road. From the junction with the Hollis-Klawock Highway, it is two miles south on the Hydaburg Road to One Duck Trail, eight miles to Cable Creek Fish Pass, and 12 miles to the Soda Lake Trail. One Duck Trail, 1.25 miles long, climbs steeply for 1,400 feet, but the alpine vegetation at the summit and the views of Klawock Mountain are not to be beaten. At Cable Creek Fish Pass is a wheelchair-accessible boardwalk to a viewpoint overlooking the pass. Bears are often seen fishing for salmon here. The Soda Springs Trail goes through an extensive muskeg before coming to Soda Lake, 2.5 miles distant. If you are interested in the local culture, Klawock is a Tlingit village and Hydaburg is a Haida village.

In the northern section of Prince of Wales Island, Road 27 from Thorne Bay leads to two caves. From this road, walk 250 feet to an observation deck where you can look directly into the mouth of Cavern Lake Cave and the stream that gushes forth from it. You are not permitted to enter this cave. At another place along the road is a boardwalk and staircase to the entrance of El Capitan Cave. The cave is gated 200 feet inside. Rangers from the Thorne Bay District Station conduct guided walks into the cave, but you must register in advance.

Several islands west of Prince of Wales Island are in designated wilderness areas. Coronation Island Wilderness, Warren Island Wilderness, Maurelle Islands Wilderness, and South Prince of Wales Wilderness are all covered with lush temperate rain forests. The last wilderness area contains about 75 islands.

Finally, a tiny island not in a wilderness area is of botanical interest. Dog Island, only two miles long and 1.5 miles wide, at the very southern edge of the Tongass but a little north of Duke Island, contains the northernmost stand of Pacific yew. The yew is mixed with hemlock, western red cedar, Alaska cedar, Sitka spruce, and lodgepole pine. Growing in the understory are salal, blueberries, bunchberry, and rusty menziesia. In the very center of Dog Island is a raised muskeg with scattered, stunted plants of lodgepole pine and Alaska cedar.

Alpine Meadows and Tundra

During the most recent period of continental glaciation, from about 40,000 to 12,000 years ago, the southern and eastern portion of what is now the Tongass National Forest was largely covered by ice. Great glacial fields still lie in Alaska. The St. Elias Range, located east of the Tongass National Forest, is one of the highest mountain ranges in North America. In addition to glaciers and associated geological features, visitors to this wild area of the Tongass National Forest can see vast tracts of forest and, in summertime, spectacular mountain meadows.

Few roads penetrate the area, so exploration must be done almost exclusively on foot. Some areas have rock glaciers—huge sheets of rocks and boulders that were created by the action of ice. These rock glaciers, a type of ground moraine, were formed some 8,000 years ago when a small glacier lay higher up the mountain. Freezing and thawing fractured the underlying sedimentary rocks, and the fragments, lubricated by meltwater, were carried down the mountain by a slow-moving core of glacial ice. Eventually the ice melted, and the rock flow stopped, forming the rock glacier.

The area around Hyder is a good base from which to explore alpine meadows and tundra in the Tongass National Forest.

Meadows have a continuous cover of grasses and mosses, punctuated by colorful wildflowers and low shrubs. At this high altitude, the growing season is less than three months long, and many wildflowers are in bloom at the same time. White and yellow blossoms predominate, with a sprinkling of reds, pinks, blues, and purples.

Spruce forests contain an ample variety of low shrubs. Labrador tea and Lapland rose-bay have dry, inedible fruits, whereas common bearberry has mealy red fruits. More palatable are the shiny, purplish black fruits of black crowberry and the bright red fruits of mountain cranberry. Another common bush, shrubby cinquefoil, has five-parted leaves and bright yellow flowers. Common wildflowers are red baneberry, grass-of-Parnassus, bunchberry, twin-leaf, pink-flowering wintergreen, Arctic lupine, monkshood, and tall larkspur.

Tundra plants are usually dwarfed, although here and there are muskegs dominated by black spruce. Common shrubby species are bearberry, Labrador tea, alpine azalea, lowbush cranberry, black crowberry, and dwarf Arctic birch. Sphagnum and flat, leafy-looking lichens abound.

Black spruce muskegs are home to willows, dwarf blueberry, Beauverd spiraea, dwarf Arctic birch, Labrador tea, mountain cranberry, and black crowberry. Other plants are lady's-slipper, cotton grass, and a horsetail.

Mountain meadow vegetation is dominated by mosses and wildflowers, but various willows and other many-branched, wiry shrubs may also be seen. Wildflowers with white blossoms are alpine bistort, a small-flowered anemone, grass-of-Parnassus, dotted saxifrage, tufted saxifrage, and two kinds of mountain avens. Yellow flowers include milk vetch, heart-leaved arnica, and tundra groundsel. Pink or reddish flowers include meadow bistort, moss campion, and dwarf willow herb. The smallest blue flowers belong to a wild forget-me-not and an alpine speedwell. Among shrubs are bearberry, black crowberry, dwarf Arctic birch, Labrador tea, and mountain cranberry.

NATIONAL FORESTS IN ARIZONA

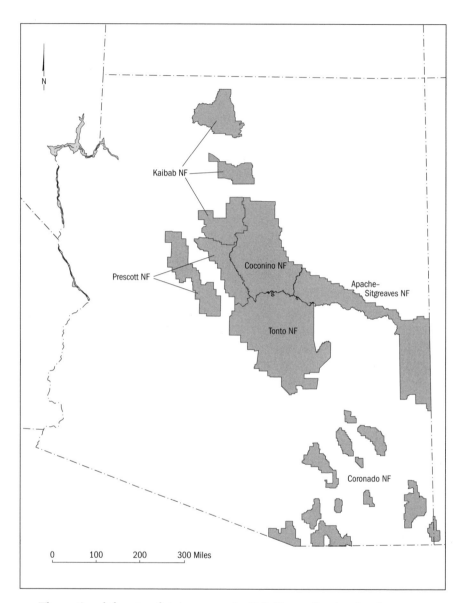

The national forests of Arizona are in U.S. Forest Service Southwestern Region 3. The regional office is in the Federal Building, 333 Broadway, SE, Albuquerque, NM 87102. The national forests of New Mexico are also in this region, and they will be discussed in a companion volume to this book. Arizona has seven national forests in six administrative units. They consist of 12.1 million acres and 36 wildernesses.

Apache-Sitgreaves National Forest

SIZE AND LOCATION: Approximately 2 million acres in eastern Arizona, with a small part in New Mexico. Major access roads are U.S. Highways 60, 180, and 191 and State Routes 87, 260, 273, and 277. District Ranger Stations: Alpine, Duncan, Lakeside, Overgaard, Springerville. Forest Supervisor's Office: Box 640, Springerville, AZ 85938, www.fs.fed.us/3r/asnf.

SPECIAL FACILITIES: Boat ramps; swimming beaches; winter sports areas.

SPECIAL ATTRACTIONS: Blue Range Primitive Area.

WILDERNESS AREAS: Mt. Baldy (7,079 acres); Escudilla (5,200 acres); Bear Wallow (11,080 acres).

The Apache-Sitgreaves National Forest is two forests in one. Although the two are contiguous, they are very different in character from each other. Nonetheless, in 1976, the Apache National Forest and the Sitgreaves National Forest were combined into one administrative unit.

The orientation of the Apache is north to south; the Sitgreaves east to west. The Apache is high mountain country to the north then drops dramatically over the Mogollon Rim (pronounced muggy-OWN) to the south. The Sitgreaves is entirely located in the Colorado Plateau on top of the Mogollon Rim. The Apache has major rivers, tremendous canyons, and high peaks; the Sitgreaves has numerous artificial lakes, relatively shallow canyons, and no peaks.

The White Mountains dominate the Apache side of the national forest, with Mt. Baldy topping out at 11,403 feet. The peak itself is in the Fort Apache Indian Reservation. All except the last mile of the trail to the summit on the East Fork Trail is in the Mt. Baldy Wilderness. The meadow around Phelps Cabin at the start of the trail teems with wildflowers from July to September. The East Fork of the Little Colorado River originates along this trail and is a fine place to see high-elevation riparian plants. State Route 273 accesses Crescent and Big Lakes, both picturesque lakes in a pretty mountain setting. The aspens above the lakes are in their golden glory in the autumn.

Greer is a small but interesting resort community in the White Mountains. One mile east of town is a mile-long trail that leads into Butler Canyon. The stream in the bottom of the cool canyon is lined with black alder and willows. River, Tunnell, and Bunch Reservoirs are in the national forest a short way northeast of Greer. To the west is the larger White Mountain Reservoir.

Almost all other activities in the Apache can be accessed off the scariest

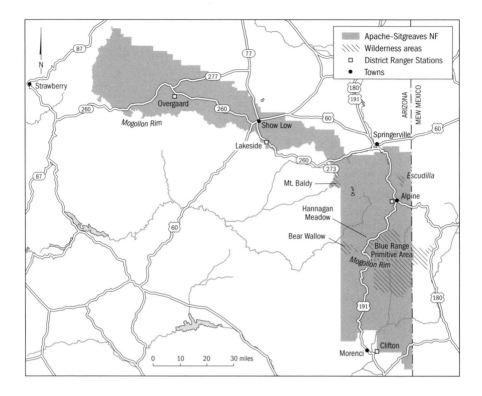

paved road we have ever taken. U.S. Highway 191, a scenic byway known as the Coronado Trail, connects Springerville and Alpine to Morenci in the south—an incredible 120-mile route, most of which is without a town or settlement in between. The highway goes through lush coniferous forests, past tranquil meadows, and along sparkling streams, and then descends the Mogollon Rim into a somewhat lower, but equally rugged, mountainous area. The highway reputedly follows the route that Francisco Coronado and his entourage of 1,100 men took in 1540 in their quest for the Seven Cities of Cibola. No documentation exists, however, to verify that this was the same route. After you have driven it, you will surely think that Coronado could have found an easier way. Robert Bates, former forest supervisor of the Apache National Forest, recounts the siting and actual construction of the road, which was completed in 1926. He recalls that as he was leaving a restaurant in Morenci, a wild-eyed tourist staggered in the door and exclaimed, "My God, they ought to wipe this highway off the map!" The highway is not for the fainthearted.

To the east of the Coronado Trail, between Springerville and Alpine, is Escudilla Mountain in the heart of the Escudilla Wilderness. The three-mile Escudilla National Recreation Trail to the summit is pleasant and not too strenuous for the average hiker. If you can spend an extra day in the Alpine area, a jagged, long road from Luna Lake zigzags past White Bluff and touches the New Mexico state line. From here, this gravel road follows a circuitous route along the scenic Blue River. At the Blue Crossing Campground, turn back to the northeast to rejoin the Coronado Trail, skirting the northern edge of the vast Blue Range Primitive Area.

From this junction with State Route 191, continue on the Coronado Trail to Hannagan Meadow, which is near the edge of the Mogollon Rim. The views from this open, wildflower area are breathtaking, as is the highway after it leaves Hannagan Meadow. It is advisable to stop frequently along the road, and, fortunately, there is no lack of things to do, particularly in the vicinity of the KP Cienega Campground. The campground is situated along scenic KP Creek, and nearby is the marshy cienega.

From the campground, turn off on a good gravel forest road that twists along the curvy northern edge of the Bear Wallow Wilderness to the Reno Lookout Tower. From the tower, the Bear Wallow Trail drops into the canyon along Bear Wallow Creek. At a trail junction you can take the left fork back to Double Cienega Creek on Forest Road 25 for a round trip of six miles, or you may take the right fork into Schell Canyon or proceed west before eventually coming to the San Carlos Indian Reservation.

For more ruggedness, take rough Forest Road 54 less than one mile south of the Reno Tower Road. In three miles, this dead-end road makes more turns than you can imagine. Near the end of the road is the northern terminus of the 29-mile Eagle National Recreation Trail, which drops 3,500 feet from the Mogollon Rim to a forest road on Eagle Creek. The trail passes through several canyons, including the highly scenic Chitty Canyon.

Back on the Coronado Trail, you continue curving toward the east, curving to the west, descending, and ascending. At Rose Peak there are fabulous views of the Blue Range Primitive Area to the east. Eventually you come to Grey's Peak Campground for additional scenic views and then Sardine Saddle. Be sure to hike the half-mile Big Tree Trail to an exceptionally large Arizona cypress that is nearly 100 feet tall with a circumference of 18 feet. The crown of the cypress has a spread of 41 feet. From the Sardine Saddle it is only five miles to the southern boundary of the Apache-Sitgreaves National Forest and 13 miles to Morenci.

The Sitgreaves part of the forest is on top of the Mogollon Rim and extends north into the Colorado Plateau as much as 30 miles in places. State Route 260 traverses most of the width of the forest from Show Low to where the highway plunges over the Mogollon Rim near Willow Springs Lake. As an alternate route, Forest Road 300 follows the edge of the rim through the entire forest. Be forewarned that this is a rough, sometimes narrow back road, but several splendid views into the desert country below make it worthwhile. Woods Canyon Lake and Bear Canyon Lake are both popular for family outings. Hikers may follow the entire length of the rim on the General Crook National Recreation Trail, which follows the military supply route used in the 1800s.

Another forest road to Chevelon Canyon, 30 miles to the north, passes through a beautiful ponderosa pine forest and eventually drops quickly to Chevelon Creek.

From the Canyon Point Campground a trail goes from the edge of a sinkhole bordered by ponderosa pines, Douglas firs, white firs, and quaking aspens to the bottom of the sinkhole. If you are there during July and August, you will see myriad wildflowers, including prickly poppy (pl. 4).

For a very easy hike, we recommend the mile-long Mogollon Rim Nature Trail located about two miles north of Lakeside, which provides a good idea of the common trees in this part of the Colorado Plateau. While you are in Lakeside, drive the short distance to the Porter Mountain Road. The Blue Ridge Cave Trail begins here and follows pretty Porter Creek. Many kinds of wildflowers grow beneath the ponderosa pines, alligator junipers, Utah junipers, and Gambel's oaks. After a little more than a mile you come to Blue Ridge Cave, a lava tube with a constant temperature of 52 degrees F.

For a change of pace, take the Pintail Wetland Interpretive Trail, which is about 3.5 miles north of Show Low. You will probably see various species of waterfowl if you look closely among the marshy vegetation.

Phelps Cabin

The high mountains of east-central Arizona are a world apart from the Sonoran Desert to the southwest and the arid mesas and plateaus of Navajo country to the north. These are the White Mountains, where winter temperatures often drop to –30 degrees F, snowfall averages six feet a year, and clear mountain streams ripple past towering spruces and Douglas firs. Dotted with small communities, such as Alpine, Eager, Springerville, and Greer, the region is divided between the Fort Apache Indian Reservation and the Apache-Sitgreaves National Forest. Among the biologically important locales on Forest Service land is a wetland and adjacent grassy meadow and forest that make up the 330-acre Phelps Cabin Research Natural Area.

The area takes its name from a log cabin on a former homestead, now unoccupied. The Research Natural Area is reached from the southern arm of the Mt. Baldy hiking trail, which loops west from Forest Road 113. The trail climbs the slopes of the massive mountain but passes short of the summit, which is sacred ground owned by the Apache tribe. A few dozen feet from the trailhead, the path bursts through a grove of blue spruces and quaking aspens into a meadow that flourishes during the growing season, which is from late April to September.

The meadow owes its presence to a combination of high elevation and southern exposure. Apart from some weedy brome grass that grows along the hiking trail, it is pristine. The native grass species include Idaho fescue, Idaho bent grass, bluejoint grass, hair grass, and Junegrass. All are palatable to grazing animals, but domestic livestock are now excluded. The grasses generally produce inconspicuous flowers, but a number of wildflowers also bloom in the meadow. Among them are a blue flax (pl. 5), with a slender stem and a large flower; a magenta wild geranium; American harebell with its solitary, blue, bell-shaped, nodding flower; and three yellow wildflowers—a goldenrod, a groundsel, and a mountain dandelion.

The grassy meadow slopes toward the East Fork of the Little Colorado River, a sparkling stream only a few feet wide. Several wetland shrubs that are good indicators of unpolluted streams line the banks or grow in the flowing water. They are the western elderberry, thin-leaf alder, and three willows—coyote, Scouler, and whiplash.

The marshy to boggy wetland that extends several feet back from either side of the creek, covering about 35 acres, is the focal point of the Research Natural Area. Botanists have recorded 61 species of flowering plants there, including numerous grasses, sedges, and rushes and showier species representing a wide range of plant families. Among the common wildflowers are Rocky Mountain iris, marsh marigold, monkshood, two monkey-flowers, a skullcap of the mint family, hemlock parsley, Fendler's cowbane, and a bog orchid. Corn lily, which stands up to five feet tall, has huge, round leaves that appear corrugated because of their elevated veins. Growing submerged in the water or with some floating leaves are the aquatic white-flowered buttercup, the little heart-leaved buttercup, and a waterwort. The yellow-flowered cinquefoil, a shrubby species nearly ubiquitous in North American wetlands, is scattered all across the wetland and advances upslope into the grassy meadows.

The Research Natural Area also includes 194 acres of the adjacent forest, made up of Engelmann spruce, blue spruce, and Douglas fir. Beneath these cone-bearing trees, which are more abundant the higher you climb up the slopes of Mt. Baldy, are rich layers of shrubs and wildflowers. The most common shrubs are the gooseberry currant and bearberry honeysuckle, whose twin, blue-black berries are nestled between a pair of lip-red leaves. The wildflowers include three kinds of wild orchids, northern bedstraw, and false Solomon's-seal. Forests such as this are characteristic of more northerly regions. In all likelihood, the coniferous tree species and their associated flora were driven southward by glaciation during the Ice Age. Then, as conditions warmed and the ice retreated, they were left isolated on the high, cool mountains.

Chitty Canyon

Roughly tracing the route traversed 450 years ago by Francisco Coronado in his search for the fabled Seven Cities of Gold, my student Rod Doolen and I drove south along U.S. Highway 666, a few miles west of the Arizona–New Mexico border. We soon came to the brink of the Mogollon Rim, which separates the cool, pineclad Colorado Plateau to the north from the hot, dry lowlands—and desert—to the south. More than 200 miles long, the Mogollon Rim is the highly irregular edge of the plateau that runs from northwestern Arizona southeastward to the vicinity of Strawberry and then more or less eastward across eastern Arizona and western New Mexico, eventually connecting with the Mogollon Mountains. We paused at an observation point where the terrain abruptly dropped nearly 3,000 feet to the valley below. After enjoying the view, we headed to nearby Chitty Canyon, managed by the Apache-Sitgreaves National Forest, to explore the vegetation below the edge of the plateau.

Taking a Forest Service road west from the highway, we arrived close enough to the upper end of Chitty Canyon to park our car and hike in. Whereas a forest dominated by ponderosa pines covers the slightly lower plateau north of the rim, here we stepped out into a fir–aspen zone at an elevation of about 9,500 feet. Closely crowded Douglas firs, white firs, Engelmann spruces, and quaking aspens are enshrouded by low-hanging clouds. The heavy shade of the trees inhibits the growth of understory vegetation, but a number of attractive wildflowers thrive in the moist soil. They include the delicate calypso orchid, the mottle-leaved rattlesnake plantain orchid, wintergreen, pipsissewa (a relative of wintergreen), and spring-flowering coralroot. The last is an orchid that lacks chlorophyll and must obtain its nutrients from the organic matter in the soil.

The dense forest of the fir–aspen zone extends down to about 8,000 feet, continuous except for small openings where the soil is too thin for trees to grow. These openings, all less than five acres in size, are high meadows where, because of overgrazing by domestic livestock, the native vegetation has been replaced by weedy species such as bracken fern and western sneezeweed. (A few miles to the north, pristine Hannagan Meadow presents a high-mountain meadow of native grasses and wildflowers.)

Descending by trail to about 8,000 feet, the forest begins to open up, and the Douglas firs, white firs, and Engelmann spruces are replaced by ponderosa pines and smaller trees such as Gambel's oak, gray oak, Rocky Mountain maple, and alligator juniper. The trees are more widely spaced, allowing sunlight to filter through to the forest floor, where the soil is hotter and drier than in the fir–aspen zone. Still, some Douglas firs and quaking aspens do well in rocky crevices and other protected areas, even below 8,000 feet.

Conditions are drier still on the steepest south- and west-facing slopes and on the ridgetops, at elevations between 7,000 and 8,000 feet. Gambel's oak and alligator juniper are smaller and more gnarled, and the shrubby, white-fruited snowberry is plentiful. Scattered wildflowers include beargrass, banana yucca, mescal, prickly pear cactus, and wild lotus.

Below 7,000 feet, down to the bottom of Chitty Canyon at 4,500 feet, the vegetation consists of scattered, round-topped, piñon pines and alligator junipers. Grasses fill in the understory, along with Indian paintbrushes and beardtongues.

Chitty Creek, which long ago carved the canyon, is fed by rainfall, snow melt, and groundwater from several springs. Lining the creek and its tributaries are box elder, Arizona walnut, lance-leaved cottonwood, and thick-leaved alder. Red osier dogwoods form occasional thickets, all the more conspicuous because of their scarlet twigs. Wildflowers abound, including a gorgeous yellow columbine, a tall blue larkspur, bright yellow buttercups, white-flowering violets, and two kinds of false Solomon's-seals.

Here and there between 8,000 and 8,200 feet are seeps, or places where groundwater oozes to the surface from springs. These soggy openings, free from woody plants except for an occasional Bebb willow, are filled with fall panic grass, bog orchid, yellow-eyed grass (a member of the iris family), golden-glow (a black-eyed Susan with a yellow center rather than a black one), and Macoun's buttercup.

Chitty Canyon is only one of hundreds of inviting areas along the Mogollon Rim, which extends into several national forests. (Because the region includes private holdings as well as part of the Apache Indian Reservation, a Forest Service map is essential for guidance.) There are many places to camp, both on top of the rim and in the lands below. Few roads, however, link the two areas. Where we stopped along the highway is one of only six places in more than 200 miles where a paved road goes from the high country to the low. In fact, only a limited number of hiking trails go up and down the face of the rim; each is difficult, often strewn with loose, crumbly rocks, and treacherously muddy after a thunderstorm.

Coconino National Forest

SIZE AND LOCATION: Approximately 1,800,700 acres in central Arizona, extending from about 25 miles north of Flagstaff to the Mogollon Rim. Major access routes are Interstates 17 and 40, U.S. Highways 89A and 180, and State Routes 87, 164, 179, 209, and 260. District Ranger Stations: Sedona, Happy Jack, Flagstaff. Forest Supervisor's Office: 1824 S. Thompson, Flagstaff, AZ 86004, www.fs.fed.us/r3/coconino.

SPECIAL FACILITIES: Winter sports areas; off-highway vehicle areas.

SPECIAL ATTRACTIONS: Arizona Snow Bowl; National Wild and Scenic Verde River.

WILDERNESS AREAS: Wet Beaver (6,155 acres); West Clear Creek (15,238 acres); Strawberry Crater (10,743 acres); Red Rock–Secret Mountain (47,194 acres); Munds Mountain (24,411 acres); Fossil Springs (11,550 acres); Kachina Peaks (18,616 acres); Kendrick Mountain (6,510 acres, partly in the Kaibab National Forest); Sycamore Canyon (55,937 acres, partly in the Kaibab and Prescott National Forests).

A 10,000-foot drop in elevation from the snow-capped San Francisco Peaks to the desert highlands along the Verde River passes through several life zones in the Coconino National Forest. Separating these two diverse areas is the spectacular Mogollon Rim. The forest is a diverse landscape of colorful rock canyons, cinder cones and lava flows, crystal-clear lakes, permanent flowing streams, and wide-open, flower-filled meadows.

Although the Coconino National Forest is one large, contiguous area, it has five very different geographical regions. North of Flagstaff and Interstate 40 are the Volcanic Highlands. South and east of Flagstaff is the Colorado Plateau Country, and the southern edge of the Coconino is the Mogollon Rim area. All around Sedona are the colorful escarpments of Red Rock Country. The fifth geographic region is Desert Canyon Country, located east of Interstate 17 and the Verde River and north of the community of Camp Verde.

The Volcanic Highlands area contains Arizona's highest mountains: the San Francisco Peaks. Although not readily noticeable, volcanic cones formed during Pleistocene times have been eroded so that no trace of the former craters remains. Above the 11,000-foot timberline on Humphreys Peak is a two-square-mile area of alpine tundra that is composed of 51 plant species that are disjunct from the tundra regions of the highest peaks in the Rocky Mountains 250 miles away in northeastern Colorado. The rarest plant of all is the San Francisco Peak groundsel, a dwarf, yellow-flowered species.

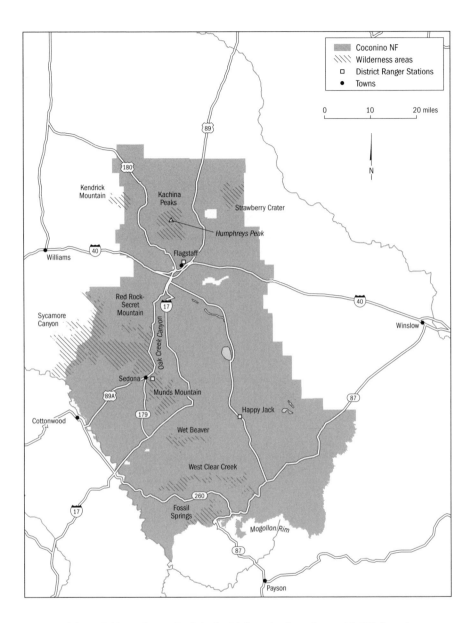

Although Humphreys Peak is the highest in elevation at 12,633 feet, Agassiz, Fremont, and Doyle Peaks all top 11,000 feet, forming a ring just west of the fantastic Arizona Snow Bowl. Several hiking trails are in the area.

A great orientation to the San Francisco Peaks is to drive a loop road that encircles these high mountains. Shortly after leaving Flagstaff, U.S. Highway 180 to the northwest enters the Coconino National Forest. After only three miles, the crooked, twisting road to the Arizona Snow Bowl splits off. If you decide to take this side trip, stop near Little LeRoux Springs and take a mile-long hike to a handsome, grassy meadow known as Freidlein Prairie. At the Snow Bowl is a ski lift to Agassiz Peak.

Back on the loop road is another forest road along the western side of the San Francisco Peaks, eventually passing Hockderffer Hills before it joins another forest road that was part of the old stage route that went to the Grand Canyon. If you choose to stay on U.S. 180, you come to access points for hiking into the Kendrick Mountain Wilderness. The forest road that follows the old stage route passes north of the San Francisco Peaks. The scenic part of the route passes several natural lakes and springs and crosses White House, Reese, and Jac Canyons before reaching U.S. Highway 89. Head south on this highway toward Flagstaff for three miles, then take the Schultz Pass back road, which rejoins U.S. Highway 180 two miles northwest of Flagstaff. This last leg of the journey crosses scenic Weathersford Canyon and skirts the eastern edge of the Kachina Peaks Wilderness Area, which includes Humphreys Peak.

The most striking evidence of volcanic activity in the highlands of the Coconino is east of U.S. Highway 89. Two of the better volcanic areas are in the Sunset Crater and Wupatki National Monuments. But the national forest also has its share of features. Strawberry Crater Wilderness just east of Deadman Mesa contains about 600 craters and cones dating back 50,000 to 100,000 years. At the northwest corner of the wilderness area is a huge lava flow, and low cinder cones are all over the south end. There are no marked trails here. A loop road provides access to many features. After going through Sunset Crater, the highway comes to the north end of the Cinder Valley Off-Highway Vehicle Area and then to the Strawberry Crater Wilderness. Along the highway is a vista with a scintillating view of the Painted Desert. The loop highway leaves the Coconino briefly and passes through Wupatki National Monument. Just below the highway into Sunset Crater, the forest service operates the Bonita Park Campground where there is a nature trail.

The Colorado Plateau Country could also be called the Forest and Lakes Country because the rolling surface of the plateau contains continuous forests interrupted by dozens of lakes and meadows. The forest at higher elevations is dominated by ponderosa pines. On lower slopes are the shorter and less densely crowded piñon pine and juniper community. Mormon Lake's 5,000 acres make it the largest natural lake in Arizona when it is full, but the level of the lake fluctuates with the amount of precipitation the area receives. During drier times, much of the lake becomes a marsh. For an overview of the Colorado Plateau Country, drive an 80-mile all-weather loop road. Start your adventure on Forest Highway 3 that begins just south of the junction of Interstates 17 and 40 at the south edge of Flagstaff. The forest highway parallels the long and narrow Upper Lake Mary for several miles and passes several campgrounds. It then skirts around Mormon Lake. The two campgrounds on the western side of Mormon Lake have short nature trails through the forest. Ten miles south of the lake, turn west onto another forest highway, which leads to Interstate 17. To complete the loop, drive north to Flagstaff.

One of the most truly dramatic areas in the west is the Mogollon Rim, the precipitous drop-off that separates the Colorado Plateau from the desert basin below. The rim extends for 200 miles east to west, with the elevation from the top of the rim to the desert a staggering 2,000 feet in places. Part of the rim forms the southeastern boundary of the Coconino National Forest. Just before State Route 87 begins its descent down the Mogollon Rim to the community of Strawberry, drive the forest road that comes dangerously close to the edge of the rim. Near the plaque describing the historic Battle of Big Dry Wash, turn north and cling to the top of Battleground Ridge, dead-ending at the actual battle site near a narrow arm of the scenic Blue Ridge Reservoir. The battle in 1871 was the last between the U.S. Army and the Apache Indians. Along the rim are numerous short hikes, but if you want to spend many days hiking, the 134-mile General George Crook National Recreation Trail is available. It follows the rim from the Clear Creek Campground all the way through the Coconino and Apache-Sitgreaves National Forests. General Crook used the trail as a supply route between Ft. Whipple near Prescott and Ft. Apache in the White Mountains.

The most spectacular colorful scenery to be found anywhere surrounds the resort village of Sedona. The region is known for the fabulous Oak Creek Canyon, but other canyons in the area are almost as fantastic. If you are coming to Oak Creek Canyon from Flagstaff via U.S. Highway 89A, you will drop over the rim south of Flagstaff and follow Oak Creek. The creek is lined by countless campgrounds and picnic areas. Or you may drop into Sedona on the white-knuckle Schnebly Hill gravel road from Interstate 17. This is a trip you will never forget. The road creeps along the northern edge of Munds Mountain Wilderness before coming to Sedona.

Northeast of Sedona is the equally colorful Red Rock–Secret Mountain Wilderness. A .9-mile trail just within the wilderness area leads to the unbelievable Devil's Natural Bridge. Other well-named features in the vicinity are Chimney Rock, Coffeepot Rock, and Steamboat Rock. Fifteen miles west of Sedona and parallel to Oak Creek Canyon is Sycamore Canyon, which is also in the Sycamore Canyon Wilderness. It is nearly as colorful, but the heavy vegetation obscures many of the rocky cliffs. The trail from Sycamore Pass Road drops down to Sycamore Creek, which is lined with willows, box elder, cottonwood, alder, and Arizona ash, as well as Knowlton's hop hornbeam.

The Desert Canyon Country is a northern outpost of the Sonoran Desert. Ocotillos, cat's-claw acacias, agaves, numerous cacti, and creosote bush are familiar desert species here. Despite the dryness of the region, Wet Beaver Creek has water the year round. The colorful steep-walled canyon carved by the creek is now part of the Wet Beaver Wilderness Area. Similarly, West Clear Creek, which has cut a very deep canyon less than two miles wide, has ample water. This canyon, extending for 20 miles with cliffs towering to 1,000 feet on either side, is part of the West Clear Creek Wilderness. Just below the Mogollon Rim and west of Strawberry are Fossil Springs, which produce 20,000 gallons of water per minute into Fossil Creek. The creek, the centerpiece of the Fossil Springs Wilderness, forms a 12-mile-long canyon before dividing at its northern end at Sandrock and Calf Pen Canyons.

Long before Yavapai and Apache Indians and pioneers lived in this area, the Sinagua prehistoric people occupied the land of the Coconino. They built dwellings of wood, mud, and stone around A.D. 600. By A.D. 1100, they were constructing multistoried buildings. Sinagua ruins can be seen at several places in the forest. These areas are fully protected. Just east of Flagstaff is the 70-room Elden Pueblo, which dates back to about A.D. 1130–1200. On a mesa overlooking the Verde River are the Clear Creek Ruins with a 50-room pueblo, a large courtyard, and a 150-foot-wide plaza. On Sacred Mountain near Beaver Creek is a pueblo with a ball court dating to A.D. 1300–1450. In the red cliffs at the southern edge of the Red Rock–Secret Mountain Wilderness are the Honanki and Palatki Ruins with Sinagua rock paintings more than 2,000 years old. Just off U.S. Highway 89 between Sunset Crater and Wupatki National Monuments is the stone Medicine Fort constructed from A.D. 900 to A.D. 1100 by a prehistoric group known as the Cohunoba.

San Francisco Peaks

Ten miles north of Flagstaff, Arizona, in the Coconino National Forest, a massive volcanic mountain rises nearly 6,000 feet above the Colorado Plateau. Culminating in Humphreys Peak (elevation 12,643 feet) and several lesser peaks (notably Agassiz and Fremont), the mountain is called the San Francisco Peaks. Here, a century ago, ecologist Clinton Hart Merriam observed the way the vegetation changed with increasing altitude, primarily in response to the lowered temperature and increased humidity. A similar pattern, he later noted, was evident as one traveled northward on the continent. On the basis of these observations, Merriam defined a series of biological life zones, which with some modification remain useful categories today.

The vegetational changes Merriam described are evident on the trip from Flagstaff to the San Francisco Peaks. Around Flagstaff, a desertlike community of grasses and shrubs abounds below 6,000 feet. This is replaced between 6,000 and 7,000 feet by a dry woodland habitat of widely spaced piñon pines and western junipers (fig. 4). Ponderosa pine dominates still higher terrain, characterizing most of the area from Flagstaff to the Snow Bowl, the winter resort area of the San Francisco Peaks. Merriam called these habitats the Sonoran, Piñon, and Pine Life Zones; today the terms Lower Sonoran, Upper Sonoran, and Transitional are preferred.

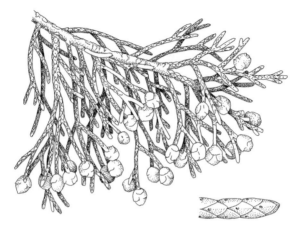

Figure 4. Western juniper.

From the Snow Bowl, hiking trails take the visitor through the remainder of Merriam's life zones. Corkbark fir and a sprinkling of Engelmann spruce trees populate the Canadian Life Zone, from about 8,500 to 10,000 feet, with the Engelmann spruce taking over in the Hudsonian Life Zone, 10,000 feet up to timberline. The timberline is marked by the appearance of gnarled, stunted bristlecone pines that rise only a few feet above sprawling shrubs of creeping junipers and gooseberry currants. Merriam termed the timberline a separate life zone, but nowadays it is simply considered part of the Hudsonian.

Finally, a bleak habitat known as tundra characterizes the Alpine Life Zone above timberline. The tundra occupies only two square miles, on Humphreys Peak and Agassiz Peak, but it is notable for its isolation from similar habitats. The nearest tundra vegetation grows on the San Juan Mountains, 250 miles away in southwestern Colorado.

The slopes above timberline are largely covered by angular boulders and an assortment of rocks, pebbles, and cinders. Wind, frost action, and erosion retard the formation and accumulation of soils. As a result, most of the tundra plant life consists of lichens and mosses that have managed to etch a foothold onto the rocks. Occasional tufts of grasses, sedges, and rushes, with their narrow leaves and inconspicuous green flowers, grow in pockets of soil that have filled crevices.

On east-facing slopes protected from bitter westerly winds, broad-leaved plants such as mountain avens, a member of the rose family, have been able to establish themselves. Once a seed of mountain avens lodges between rocks and germinates, it sends out creeping stems that form compact mats. As the older parts of the plant die, they become part of a gradually building soil layer. The soil, in turn, provides a niche where other flowering plants, such as Sibbald's cinquefoil, moss campion, and Parry's lousewort, become anchored.

The broad-leaved flowering plants are adapted to a short growing season that lasts from 101 to 113 days. All of them delay growth until June. In three weeks, mountain avens, Jacob's ladder, and mountain cress are in bloom, but most species do not flower until July or August. Two dwarf gentians and a dwarf goldenrod wait until September, when snow is imminent, before they produce flowers.

Rockslides on the steepest terrain create talus slopes, where some of the most interesting alpine wildflowers live. Parry's primrose grows tufted among the rocks; in July its leafless stalks bear a cluster of inch-wide pink flowers. The rock sandwort's compact, gray-haired leaves blend in so well with its stony surroundings that it is often difficult to spot without its white flowers. Most unusual is the San Francisco Peaks groundsel, whose small, yellow, daisylike flowering heads stand above a cluster of deeply cleft leaves. Discovered by botanist Edward L. Greene in 1884 and found nowhere else in the world, it is listed as a threatened species by the U.S. Fish and Wildlife Service.

Half of the flowering plants that grow above timberline on the San Francisco Peaks occur in similar habitats in the Rocky Mountains north of Arizona. Another 31 percent are circumpolar, that is, they are found north of the Arctic Circle in North America, Europe, and Asia, and 10 percent are confined to North American arctic regions from Greenland to Alaska. Only 8 percent of the plants are known to areas south of the San Francisco Peaks. Questions of interest to scientists are how these primarily northern plants got to the San Francisco Peaks, and how long they have been isolated from the nearest tundra zones, 250 miles away.

Geologist Henry Hollister Robinson estimated that the San Francisco Peaks began to form nearly 60 million years ago, when a volcano built up about 8,800 feet of lava. Subsequent uplifts that created the Colorado Plateau raised the elevation to nearly 16,000 feet. After the volcano became extinct, about 2 million years ago, erosion reduced the mountain more than 3,000 feet to its current height.

Fifty-five to 65 thousand years ago, when a much cooler climate prevailed, an advancing glacier moved into the area. Geologist R. P. Sharp believes the climate was cool enough at this time for tundra plants from the north to spread to the San Francisco Peaks. For nearly 50,000 years, the climate was such that the vegetation zones from Colorado to New Mexico and Arizona were shifted 4,000 feet lower than they are today. But as the climate began to moderate, about 24,000 years ago, the life zones were raised, and the highest peaks became isolated "islands in the sky."

Because of the southern latitude, the climate on the summits of the San Francisco Peaks has become too dry and too warm for most tundra species to survive and thrive. The amount and diversity of life on the San Francisco Peaks is less than on peaks of similar elevation in the Rocky Mountains. Botanist Thomas Moore predicts that the plants will become increasingly sparse and less vigorous if the current climatic trends continue.

Oak Creek Canyon

One of the most abrupt geological features in the nation is Mogollon Rim, a 75-mile-long cliff that separates the cool, moist high country of northern Arizona from the hot, dry deserts to the south. Alternate Highway 89 plunges over the 1,000-foot rim and twists down steeply, finally bottoming out along Oak Creek. The highway then continues south, through a corridor of populated private lands surrounded by the wilderness of Coconino National Forest. In a few miles, close to the postcard-pretty town of Sedona, Oak Creek Canyon's red rock formations burst into view. Traffic slows as each turn in the road reveals cliffs, buttes, and occasional arches tinted in pink, purple, and fire red. Finally, at Bell Rock, a few miles south of Sedona, the red rock country vanishes as suddenly as it appeared.

The canyon walls consist of five layers of rock, each formed by a different geological process. The bottom stratum, the Supai formation, is a series of brilliant red siltstones and shales, up to 1,800 feet thick, laid down by streams on an ancient flood plain. Above this is a 500-foot band of buff-colored Coconino sandstone formed by windblown sands, 200 feet of white Toroweap sandstone deposited by water, and a 300-foot layer of gray Kaibab limestone created from the shells and other remains of sea animals. Black lava from several volcanic flows caps the whole at certain places.

Sometime during the canyon's geological history, these rock strata slipped vertically, creating faults. Normally, when rock bends along a fault, the layers in the uplifted portion bend downward, and the layers that move down bend upward. In Oak Creek Canyon and some of its side canyons, an unusual phenomenon can be seen: the layers of the uplifted rock bend upward and the downward-dropped strata bend down. Geologist Brainerd Mears suggests that Oak Creek Canyon's rocks did not bend when the faults were first formed. Sometime later, he argues, horizontal compression buckled the layers at the faults to create the pattern we now see.

This feature, called reverse drag, is particularly evident along one of Oak Creek Canyon's side canyons, West Fork (pl. 7), whose steep cliffs and gentle benches were cut by water draining from Mogollon Rim. (Zane Grey lived here for a while and was inspired to write his *Call of the Canyon*.) West Fork supports an exceptional diversity of plants and plant communities for such a small area. To preserve the region for study, the U.S. Forest Service has designated a portion of West Fork as a Research Natural Area.

The West Fork stream runs through one of the richest riparian habitats in Arizona. Dense colonies of jointed-stemmed horsetails grow in the shade of box elders, Arizona walnuts, New Mexican alders, bigtooth maples, and Gambel's oaks. During the summer, patches of the showy, yellow monkeyflower and stiff gilia (pl. 6) alternate with the horsetails. On my first hike along the West Fork, I encountered a small, much-branched tree whose fruits resembled hops. The tree proved to be Knowlton's hop hornbeam, one of the rarest woody plants in the southwestern United States. This hornbeam, which was discovered in the Grand Canyon, also grows sparingly in southern Utah and again in the trans-Pecos area of Texas and New Mexico.

The box elder and Knowlton's hop hornbeam provide examples of how geographical distribution, ecological requirements, and external features— including the continuous or discontinuous gradation of characteristics—are important in determining whether a plant is a distinct species. The box elder, a member of the maple family, has one of the broadest geographical ranges of any woody plant in North America. Common in the eastern United States from Vermont to Florida, it is also found throughout the nation's heartland and as far west as central Montana, Nevada, and California. It even penetrates southern Canada and Mexico. Throughout all of its range, the tree displays similar compound leaves and pendulous, winged fruits. Although box elders along West Fork have leaflets that are a little thicker and a little hairier than those of box elders back east, in all other aspects the plants are the same, including the stream bank habitat. In addition, box elders from all across the United States show only a gradual transition in thickness and hairiness of the leaflets. For this reason, botanists confidently classify the box elder as single species.

On the other hand, botanists recognize two species of hop hornbeams in North America. The eastern hop hornbeam lives from Nova Scotia to Florida and west to the Dakotas and eastern Texas and into Mexico. This small tree usually inhabits moist but well-drained slopes and ridges. It has pointed leaves and two-inch-long fruits. Several hundred miles to the west of the westernmost location of eastern hop hornbeam, a second hop horn-beam with rounded leaves and fruits that barely reach one and one-half inches in length grows in isolated rocky canyons. This is Knowlton's hop hornbeam, considered a distinct species because of its geographical isolation from the eastern hop hornbeam and because of the marked variation in its leaves and fruits.

The remarkable biological feature of West Fork is that within just a few hundred feet of the wet, riparian habitat are two other, very different environments. Where the canyon runs almost due east and west, the slopes that face southward enjoy relatively dry conditions, and the branched, leather-leaved shrubs have developed into the brushy, dry plant community called chaparral, whose name is derived from the Spanish word *chaparro* meaning "dwarf evergreen oaks." Many areas of the chaparral are almost impenetrable because of the entangled growth of the pointed-leaved manzanita, mountain mahogany, silk-tassel bush, and shrubby live oak. Prickly pear cacti and century plants, found in openings in the chaparral, are reminders that the desert is not far away.

A short distance from such warm, dry areas are stands of Douglas fir and ponderosa pine, trees normally expected at higher elevations, such as at Mogollon Rim. These are able to grow where the canyon bottom is densely shaded. Such cool and moist conditions are especially evident in West Fork's narrow side canyons, where the steep walls block the sun's rays. The upper ends of these side canyons are natural channels for cool air.

Beneath these conifers is a contrasting mixture of the pink-flowered New Mexico locust and the Gambel's oak. The quiet observer may hear or see an Abert's squirrel, a golden-mantled ground squirrel, a western tanager, or a pygmy nuthatch as it scurries down a tree trunk.

Although Oak Creek and West Fork canyons can be visited at any season of the year, the traveler must be prepared for temperature fluctuations. Summer temperatures average 75 degrees F but occasionally climb to 100. In winter, the average temperature is between 35 and 40 degrees F, but it can fall below zero on the rim. Afternoon thundershowers can be expected during July and August, and there may be snow or light, steady rain during the winter.

Coronado National Forest

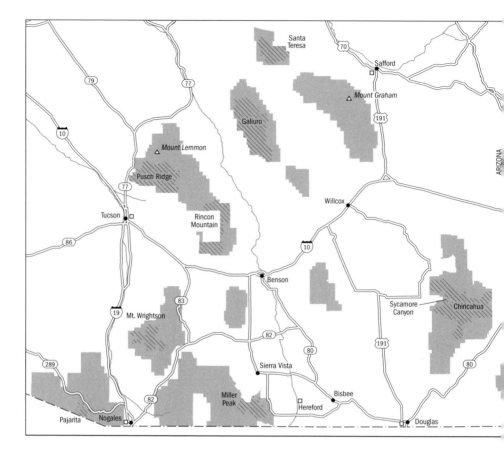

SIZE AND LOCATION: 1,780,000 acres in southeastern Arizona, with a small unit in southwestern New Mexico, extending from Winkelman and Safford in Arizona to the Mexican border and from New Mexico to west of Nogales. Major access routes are Interstates 10 and 19, U.S. Highway 191, and State Routes 77, 79, 82, 83, and 289. District Ranger Stations: Douglas, Nogales, Safford, Tucson, Hereford. Forest Supervisor's Office: 300 W. Congress, Tucson, AZ 85701, www.fs.fed.us/r3/coronado.

SPECIAL FACILITIES: Boat ramps; winter sports areas; visitor center.

SPECIAL ATTRACTIONS: Sky Island Scenic Byway; Sabino Canyon Recreation Area.

WILDERNESS AREAS: Mt. Wrightson (25,260 acres); Santa Teresa (26,780 acres); Rincon Mountain (38,590 acres); Pusch Ridge (56,933 acres); Pajarita (7,553 acres); Miller Peak (20,228 acres); Galiuro (76,317 acres); Chiricahua (87,700 acres).

An "Island in the Sky" is an isolated mountain range that stands dramatically above plains or desert that surround it. Because the summits of these mountains have been isolated from the summits of other mountains for millions of years, just as islands in an ocean are separated from other islands for long periods of time, several plants and even a few animals may be confined to a particular Island in the Sky and found nowhere else.

In southern Arizona, mostly between Tucson and Nogales and eastward into New Mexico, the Coronado National Forest encompasses 12 of these Islands in the Sky. Nearly all activities and points of interest in the national forest are associated with these islands. To appreciate fully the Coronado National Forest, you need to visit the 12 Islands in the Sky.

Because of their proximity to Tucson, the Santa Catalina Mountains and their satellite Rincon Mountains are the most visited as well as the most massive of the Islands in the Sky. Mount Lemmon, at 9,157 feet, towers above the northeastern part of Tucson. At the edge of Sabino Canyon Road from Tucson, just within the Coronado's northern boundary, is the extremely popular Sabino Canyon. At the visitor center you will learn about Sabino Lake, Thimble Peak, Rattlesnake Canyon, Bear Canyon, Seven Falls, and the sparkling streams that are in the area. A long hiking trail from the Lower Bear area up Bear Canyon past Seven Falls is a fine one if you are in good physical condition. No cars are allowed in Sabino Canyon. A shuttle is available, or you can walk in.

If you do not mind driving mountain roads, you must take the Santa Catalina Highway, also known as the Sky Island Scenic Byway (or General Hitchcock Highway as it is sometimes called) that goes from desert to Mt. Lemmon Ski Valley, passing through several life zones of vegetation. At the southeast corner of Mt. Lemmon is the heavily used Pusch Ridge Wilderness. From near the end of the paved Santa Catalina Highway is a hair-raising back road that twists over the northern end of the Santa Catalinas, ending mercifully at the town of Oracle.

Nearby are the Rincon Mountains, which pass directly through the eastern part of Saguaro National Park. That part of the Rincon Mountains in the Coronado National Forest on the north, east, and south sides of the national park are in the Rincon Mountain Wilderness.

South of the Santa Catalina and Rincon mountains, between Tucson and Nogales, are the attractive Santa Rita Mountains. You know you are there when the bright pink pads of the Santa Rita prickly pears begin to appear in the desert just before reaching the mountain foothills. Madera Canyon is a popular location with a lodge, a campground, picnic areas, and an abundance of hiking trails. This is one of the meccas for bird watchers, and a checklist of birds of the Santa Ritas may be picked up at the lodge. The area is usually congested, so you may want to branch off the road to Madera Canyon and take a rough back road through Agua Caliente Canyon. If you have the time and stamina, you may hike a trail through the canyon and then up to 9,453-foot Mt. Wrightson. If you make it that far, you are in the midst of the Mt. Wrightson Wilderness Area. The summit of the mountain has stands of ponderosa pines and Douglas firs, a sharp contrast to the desert vegetation encountered when you start the trail.

If you are into spelunking, or caving, three worthwhile, although vandalized, caves are in the Santa Rita Mountains. You need to check at the Nogales Ranger Station or the Forest Supervisor's Office and obtain a key to the gates and pay a key deposit to visit Cave of the Bells. Onyx Cave, at the end of an old road up Gardner Canyon, has several passageways and rooms. At the foot of the hill where the cave is located are the foundations of an old mill. To obtain a key to Onyx Cave, you will need to contact the Escabrosa Grotto Organization. Just north of Onyx Cave is Cave of the Bells in Sawmill Canyon. A passageway called Lake Tunnel leads to a permanent lake that is 80 feet lower than the entrance to the cave. Equally interesting is that the water in this underground cave is 76 degrees F, a full nine degrees warmer than the rest of the cave. The cave is closed from mid-August through September because of breeding activity of the lesser long-nosed bat. State Route 83, which follows the eastern side of the Santa Rita Mountains, provides several access points into the mountain.

At the crossroads community of Sonoita, State Route 83 joins State Route 82. On its way south to Patagonia and eventually Nogales, State Route 82 follows Sonoita Creek, which separates the Santa Rita Mountains to the west from the Patagonia Mountains to the east. Before exploring the Patagonia Mountains, drive to the historic Kentucky Camp where five adobe buildings built in 1904 remain from the Santa Rita Water and Mining Company complex. You may wander among the large "hotel" (which was reputedly the office), the small gold-processing building, a barn, and two cabins.

Although interesting and scenic trails are in the Patagonia Mountains, such as the trail along Harshaw Creek and up to Red Mountain, or the Paloma Trail to a spring and an old mine ruins, one of the best is in the lower Canelo Hills north of the mountains. Along the way is a nice marsh, or cienega, near Papago Springs where such wetland species as softstem bulrush, marsh violet, and a ladies'-tresses orchid grow. In May, an adjacent meadow glows yellow with *Ranunculus macranthus,* the buttercup with the largest flower of any buttercup in the world.

One of the most interesting adventures is to drive the dirt and gravel forest highway that begins four miles north of Nogales and terminates 55 miles later at the Coronado National Memorial. At no time does this scenic and historic road stray more than four miles north of the Mexican border. The road goes below the southern end of the Patagonia Mountains and ends around the southern end of the Huachuca Mountains. South of the Patagonia Mountains are several scenic canyons and bubbling springs. At one point is the largest velvet mesquite tree in the world. The road drops down to the dusty village of Lochiel, which has an actual border crossing into Mexico. You may wish to explore around the few old unpainted buildings that are there. You also can observe the Fray Marcos de Niza Historical Monument.

From Lochiel is a five-mile stretch of the forest highway that goes through private property, but as soon as you cross Parker Canyon, you are back in the national forest. A side road leads to lovely Parker Canyon Lake, where camping, boating, and fishing are popular. Back on the forest highway, you are now at the southern end of the Huachuca Mountains. The road crosses dozens of streams, most of them dry, but the canyons on either side of them are evidence of tumultuous water in times past. The road climbs to Montezuma Pass just before becoming paved as it enters Coronado National Memorial. To the north, in the national forest, is the very rugged Miller Peak Wilderness. A difficult trail exists from Montezuma Pass to the crest of 9,466-foot Miller Peak. You can also access the summit of Miller Peak by taking the three-mile-long Lutz Canyon Trail, which follows an old mining road. Along this trail, as you look eastward, are fine views of the San Pedro Valley and the Mule Mountains. The Huachuca Mountains are known for their hummingbirds, and 14 hummingbird species have been reported for this region.

West of Interstate 19 are the Tumacacori Mountains, which give way to the Pajarito Mountains on the Mexican border. The highlight of this region is Sycamore Canyon, a remarkably rough and scenic canyon that extends from Hank and Yank Springs on the gravel Ruby Road to the Mexican border. The significance of Sycamore Canyon is that it is the only place in the United States for more than a dozen species of plants that have come up from Mexico. Unusual animals such as the jaguarundi, the pencil-thin vine snake, and the elegant trogon occur here. After exploring Sycamore Canyon, you can relax around the 49-acre Pena Blanca Lake, which is located only five miles from Mexico. The lake is set among colorful bluffs dotted with oaks. You may hike the nearby Atascosa Trail to 6,255-foot Atascosa Peak. The steep trail, helped by several switchbacks, ends with breathtaking views of Sycamore Canyon to the south, the white buildings of Kitt's Peak National Observatory to the northwest, and the Santa Rita Mountains to the northeast.

Least impressive of the 12 Islands in the Sky are the Winchester Mountains, which lie west of Willcox. Although Ruby Peak rises to 7,631 feet, the only good trail goes up Rockhouse Canyon for a short distance to the Rock House.

Northwest of the Winchester Range are the rocky but often brushy Galiuro Mountains that top out at 7,663 feet on Bassett Peak. Most of this range is in the Galiuro Wilderness, but the trails are difficult to follow. You might try the trail up Ash Creek to Bassett Peak. The trailhead is at the end of a four-wheel-drive road, although there is a trail from the Deer Creek Work Station to Holder Spring that is scenic but challenging. On all sides of the national forest are desert plains. In the wilderness is wreckage of a B-24 bomber that crashed during a training mission in January 1943.

The remote and little known Santa Teresa Mountains are adjacent to the San Carlos Indian Reservation. These are extremely rugged mountains that only the experienced hiker should attempt. Much of the range is in the Santa Teresa Wilderness. Chaparral covers the lower third of the mountains, with forests of ponderosa pine and Douglas fir at the higher elevations. Black Rock Dome has a treeless summit, but it is not high enough to be above timberline. Black Rock Canyon Trail bisects the wilderness area from east to west and passes Black Rock Dome.

Southwest of Benson is the small Whetstone Mountain Range dominated by Whetstone Peak on the eastern side and Apache Peak on the west. Numerous springs are in the area, but few trails lead to them.

The Dragoon Mountains are known more for their role in western history than for their scenic beauty. Somewhere within the range is reputedly where the famed Apache leader Cochise is buried. The Cochise Stronghold Campground is supposed to be situated in the approximate locale. From this campground are interesting short trails up both East Stronghold Canyon and West Stronghold Canyon. The lightly used forest highway from Pearce to Tombstone crosses national forest land where it passes over Middlemarch Pass.

The Chiricahua Mountains are one of the best and most scenic of the Islands in the Sky. The most incredible rock formations have been protected by the Chiricahua National Monument, but the vast remainder of the mountain range is part of the Coronado National Forest. The central part of the Chiricahuas are the most rugged and are included in the Chiricahua Wilderness. You can get all the scenery and nature you want by driving the Onion

Saddle Road from the community of Portal on the east side of the mountain range to the Chiricahua National Monument. The paved part of the scenic road leaves Portal and immediately enters Cave Creek Canyon (pl. 8), which is touted as the "Yosemite of Arizona." The canyon not only has spectacular scenery but also is a mecca for wildlife viewers. Birds you might see are the elegant trogon, Strickland's woodpecker, zone-tailed hawk, painted redstart, flammulated owl, and a dozen different kinds of hummingbirds. What you will not see, unfortunately, is the broad-billed parrot, which was extirpated from the Chiricahua Mountains in the 1920s and fell to predators when it was twice reintroduced since the 1970s. The Chiricahua Mountains were its only location in the United States. At the American Museum of Natural History's Southwestern Research Station at the western end of the canyon, the road becomes dirt and gravel and very narrow in places as it winds its way over Onion Saddle and into Pinery Canyon.

You may also enter the western side of the Chiricahua Mountains by using the Turkey Creek Road through Pole Bridge Canyon to the end of the road at Morse Canyon. The Pole Bridge Research Natural Area contains the Apache pine, Chihuahua pine, Mexican piñon pine, southwestern white pine, netleaf oak, and silverleaf oak, among others. A forest road up Rucker Canyon crosses the wild and woolly Devil's Canyon. It is an unforgettable excursion.

Although the Peloncillo Mountains are mostly in extreme southwestern New Mexico, they are included in the Coronado National Forest. Skeleton Canyon crosses the central part of the Peloncillos, extending from New Mexico to Arizona. At the mouth of Skeleton Canyon in a corner of the Coronado National Forest on the Arizona side is the place where Apache leader Geronimo and Chief Naiche surrendered to General Nelson A. Miles on September 5, 1886, thus ending the Apache War. The site is marked.

Mount Graham is the dominant feature of the Pinaleno Mountains. The Swift Trail is a paved road that climbs into the Pinalenos for 42 miles from the town of Safford to Riggs Flat Lake. The first four miles are in chaparral dominated by mesquite, creosote bush, and burroweed. As the road proceeds up the Penalino foothills, ocotillos, sotols, and prickly pears become common. After 15 miles, the Swift Trail has climbed into a piñon pine–alligator juniper–scrub oak woodland. The Wet Canyon Picnic Area is at 6,000 feet. Streamside vegetation at the small picnic area includes black alder, Arizona walnut, velvet ash, and Wright's sycamore. The pine at the campground is the Chihuahua pine.

By the time the Swift Trail enters Turkey Flat at 7,500 feet, the forest is composed of ponderosa pine, Douglas fir, white fir, and quaking aspen. A two-mile side road leads to historic Heliograph Peak. The peak is at an elevation of 10,638 feet and was used as a U.S. Army heliograph station. The army was able to communicate from station to station across southern Arizona on sunny days by signals given by means of a mirror and sunlight.

Beyond the trail to Heliograph Peak is Graham Peak, home of the very rare Mt. Graham spruce squirrel and now the site of a controversial observatory. Near the terminus of Swift Trail is Riggs Flat Lake Recreation Area, which has a very popular campground and fishing lake.

Mt. Lemmon

The Santa Catalina Mountains loom high above the north and northeastern boundaries of the expanding desert city of Tucson. Dominated by 9,153-foot Mt. Lemmon, the range covers an area of about 200 square miles, most of which is under the jurisdiction of the Coronado National Forest. The Forest Service has provided many campgrounds, hiking trails, and other amenities to serve the nearby metropolis, including paved General Hitchcock Highway, which ascends all the way to the ski area near the top of Mt. Lemmon.

The 35-mile drive from Tucson to Mt. Lemmon (less than 20 miles as the crow flies) is often equated to driving from Arizona to Canada. The Sonoran Desert, at 2,500 feet above sea level, contains a great variety of cacti and spiny shrubs, such as ocotillo (pl. 9) and cat's-claw acacia (fig. 5). This community grades into a desert grassland with many grasses and several species of large century plants. Above the grassland is the open oak woodland, dominated by Emory oak, Mexican blue oak, and Arizona rosewood. This shades into

Figure 5.
Cat's-claw acacia.

pygmy conifer–oak scrub, whose common species are piñon pine, alligator juniper, Arizona oak, Emory oak, and two species of manzanita. Then, almost imperceptibly, a pine–oak woodland of ponderosa pine, Chihuahua pine, silverleaf oak, and Arizona oak takes over. By 7,300 feet only the ponderosa pine and silverleaf remain, and above 7,600 feet the forest consists entirely of conifers—ponderosa pine, southwestern white pine, and, on the summit, a montane fir forest of white fir and Douglas fir.

The differences in vegetation at different elevations are due, in large part, to the amount of moisture available and the position on the mountain; that is, whether the slope is north-facing or south-facing. The Desert Zone, for instance, receives only three to 15 inches of rainfall each year and therefore supports mainly cacti and various spiny shrubs and small trees that are adapted to arid conditions. This zone extends roughly from 2,500 feet to 4,000 feet on the hotter, drier south slope of Mt. Lemmon but only to 3,200 feet on the cooler, moister north slope. Similarly, the desert grassland community lies between 4,000 and 6,300 feet on the south side but between 3,200 and 4,400 feet on the north side.

The desert is an open community, but higher up the mountain, as moisture becomes increasingly abundant, the habitats become more and more crowded with plants until, in the fir forests at the highest elevations, the trees grow very close together. Many desert plants are leafless or have had their leaves reduced to spines, minimizing water loss. The sequence progresses upward to grasses; to shrubs with narrow but not needlelike leaves; to plants whose elongated, thick leaves are densely crowded at the plant base; to broad-leaved evergreen shrubs; to broad-leaved evergreen trees; and finally to needle-leaved evergreen trees.

A century ago, naturalist Clinton Hart Merriam called attention to the parallel between moving higher in elevation and higher in latitude. At higher elevations, more of the plants belong to species found mainly in the northern United States and Canada, with fewer from the southwestern United States. For example, Sonoran species are common in the desert, whereas grasses from the Great Plains are found in the desert grassland. In the lowest forested areas, the species that prevail are southwestern and Madrean (from Mexico's Sierra Madre), whereas at higher elevations are found such northern species as harebell, lady fern, and golden corydalis.

About halfway to the top along the General Hitchcock Highway, the Forest Service has erected a large, informative sign summarizing the biological life zones that can be observed, from Lower Sonoran to Upper Sonoran to Transition to Canadian. But the astute observer will note that these communities are not completely distinct and that others may be identified. Between 6,000 and 7,000 feet, for example, a mosaic of plants with various geographic affinities is found—typical Rocky Mountain and northern species such as male fern, small-leaved pussy-toes, green pyrola, and snowberry; southwestern species, including Engelmann prickly pear and Arizona wheatgrass; and Madrean species, among them Wilcox's barberry, powderpuff bush, sotol, alumroot, and a century plant.

As a result of erosion and other geological forces, Mt. Lemmon is more isolated today than it was millions of years ago, when a continuous chain of mountains, extending from Canada to Mexico, permitted many species of plants to migrate north and south. Although New Mexican and southwestern species continue to colonize the lower, drier regions of the mountain, adding to the diversity of plants, northern species are no longer able to bridge the gaps to reach the high peak. As a result, Mt. Lemmon and similar isolated peaks are commonly referred to in Arizona as mountain islands.

Agua Caliente Canyon

Rising above the Sonoran Desert, the Santa Rita Mountains extend north to south, situated roughly midway between Tucson, Arizona, and Nogales, Mexico. The highest peaks, including 9,453-foot Mt. Wrightson, are topped off with verdant forests of Engelmann spruce and ponderosa pine, whereas cacti and other desert plants grow at the base of the mountains, at about 3,000 feet, and penetrate dry canyons up to 4,000 feet. The range is a favorite among naturalists because of its varied plant and animal life and the scattering of natural springs, but only a few routes provide easy access into the rugged terrain. The most popular destination, reached by a highway that pierces the north side of the region, is a camping and recreational facility at the head of Madera Canyon; my favorite locale, however, is Agua Caliente Canyon, best reached from the west. Both canyons are under the jurisdiction of the Nogales Ranger District of the Coronado National Forest.

Lying at 3,600 to 4,200 feet, Agua Caliente Canyon harbors mostly desert grassland and isolated stands of oak, whereas piñon pine and alligator juniper are scattered on the surrounding dry, rocky slopes. Natural springs help feed some streams that flow intermittently in response to rainfall; other streambeds are washes, entirely dry except after heavy rains. One reason I am attracted to this canyon is that it is one of the few places in the United States where you can find two species of *Amoreuxia,* which are essentially tropical wildflowers. The bird life, too, is alluring. Several species of hummingbirds flit about the blooms of the Arizona trumpet and the desert honeysuckle, and the rat-tat-tat of five different kinds of woodpeckers may be heard. Occasionally the elegant trogon, more common in Mexico, will fly overhead. Mammals are also plentiful. Although rarely seen, mountain lions and bobcats periodically cross the canyon. Likelier to be encountered are javelinas (collared peccaries) and coatis (large, ring-tailed cousins of the raccoon).

To get to Agua Caliente Canyon, exit Interstate 19 at Canoa Road and follow the east frontage road south before turning eastward into the canyon on Elephant Head Road. Eventually you will connect with a road that climbs along the south side of the canyon, heading toward the Smithsonian Institution's Fred Lawrence Whipple Observatory. Tours of that facility, located on the 8,585-foot summit of Mt. Hopkins, depart from a visitor's center and are available by reservation. Before you reach the visitor's center, located at the base of Mt. Hopkins, you will find a rough, unpaved road that leads off to the left for a quarter mile, down into the canyon, to Agua Caliente Spring. As its name indicates, the water from this spring is hot. Heated underground through geothermal processes, the water feeds an intermittent stream.

From the spring, you can pick up any of several hiking trails. One that heads toward Mt. Hopkins passes two rocky knolls that top off at 4,962 feet and at 6,012 feet. Together they are known as the Devils Cash Box, named for a 1909 gold strike. Intrepid hikers can also hike onward to the Madera Canyon campground or to the summit of Mt. Wrightson.

Desert grassland plants include many cacti, the most spectacular of which are the saguaro, with its thick, armlike branches, and a number of large chollas, such as the tree cholla, the jumping cholla, and the cane cholla. Also prominent are various prickly pears, with their flat, padlike stems, and the fiercely spined Arizona barrel cactus, which may grow to a foot in diameter and more than three feet tall. Smaller species include the Arizona rainbow cactus, sporting multicolored spines; claret cup, named for its flower; pancake cactus, which grows up to four inches in diameter and lies flat on the ground; pincushion cactus, which stands about two inches tall; and two- to five-inch-tall nipple cactus, whose stems end in small, rounded tips.

A plant with cactuslike spines that is actually related to Mexico's boojum tree is ocotillo. After a rain, its barren, gray black stems change overnight to green as small leaves emerge from buds covering the plant. After a few days of dry weather, the leaves fall off and the plant resumes its dormant appearance. Whereas cactus flowers have many petals, the ocotillo's red flowers have only five, united at their base to form a tube.

Stands of oak may appear in shallow depressions and other places that have slightly more moisture. The dominant (and often only) species is Gambel's oak. Dry, rocky slopes support a few kinds of stunted trees, among them Gambel's oak, silverleaf oak, piñon pine, and alligator juniper. Most do not grow more than 20 feet tall under the arid conditions. Typical shrubs include squawbush (a type of sumac); coral bean and fairy duster (both members of the pea family); two types of hibiscus, or rose mallow; and cliff Fendler bush. Snapdragon vine, a plant with arrowhead-shaped leaves and dark red or bluish flowers with a yellow center, scrambles over some of the shrubs. Prickly pear cacti found here are the clock-face prickly pear, Engelmann prickly pear, cliff prickly pear, and purple prickly pear. Wildflowers that grow scattered in rocky soil include pine-needle milkweed, Arizona trumpet, long-leaved phlox, and a true wild cotton. Two uncommon species with three-inch-wide orange flowers are *Amoreuxia palmatifida* (yellow show) and *A. gonzalezii,* the only two members of the tropical Cochlospermaceae family that are found in the United States.

On some dry slopes—perhaps because they are not quite so rocky—shrubby species include wild quinine bush, silk tassel bush, desert sumac, mescat acacia (white thorn), and wait-a-minute bush (a type of mimosa). Wildflowers that grow here are dogweed, springleaf zinnia, wedgeleaf scurf-pea, and trailing windmills (a creeping plant whose flowers are grouped by threes into what appear to be single blossoms).

Streams, although they may flow only intermittently, help support the growth of several trees, among them blue palo verde, mesquite, velvet ash, small-leaved mulberry, netleaf hackberry, and soapberry. A common shrub is seep willow, a member of the aster family whose narrow leaves resemble those of willows. Wildflowers near streams include the bright yellow-flow-ered monkey-flower; a Mecardonia with tiny purple snapdragon-like blos-soms; a thoroughwort with lavender flower heads; and Abert's dome, a plant with yellow flower heads and each head having a central, raised dome.

Washes support only plants that can survive long stretches of extremely arid conditions. Two shrubs here are desert broom and burro bush, and wildflowers include desert honeysuckle, with long, tubular, brick red flow-ers, and woolly feverfew, with white or greenish flowers.

Sycamore Canyon (Coronado National Forest)

The air temperature had soared above 112 degrees F, and the Fremont cot-tonwoods were stilled by the absence of any wind. The pool of water in the creek was a welcome sight, and I dropped eagerly to my knees to press my face into it. A flash of movement in the water brought my attention to a small fish that I later identified as the Sonoran chub. This chub is found occasion-ally in Mexico, but in the United States it is confined to this one special place, Coronado National Forest's Sycamore Creek, a stream that flows out of southwestern Arizona and into Mexico. Sycamore Canyon, which has been carved by the creek, harbors a variety of plants and animals that are at the edge of their geographic ranges or that have unexplained gaps in their dis-tribution. The Sonoran chub is just one of these species.

Sycamore Canyon extends south for five miles to the Mexican border. Steep slopes and wild, rugged cliffs, dominated by rhyolites, shale, and sand-stone, frame the creek and its inviting tributaries. Occasional spirelike pin-nacles rise nearly 100 feet. Near the border, the canyon walls diminish in size, and the creek dissipates less than a mile into Mexico. True to the region's name, several beautifully arched, white-barked specimens of the Arizona sycamore line the creek, and growing on some of the rocky slopes is the rare Goodding ash, discovered here in 1936 by Leslie N. Goodding, one of the first botanists to explore the southern part of Arizona.

To reach the canyon, my family and I had traveled about 15 miles via paved road west from Nogales, Arizona, to Pena Blanca Lake. We then covered another 12 miles on a graveled, often narrow, and at times difficult road until we crossed Sycamore Creek a few miles east of the nearly abandoned mining town of Ruby. After parking under a spreading Emory oak at the nearby campground, we hiked to the canyon's entrance, passing the ruins of the old Bartlett Ranch.

Legend has it that the once prosperous Bartlett cattle ranch was attacked by Apaches on April 28, 1886. Although two men were wounded, further tragedy was averted when two 10-year-old boys were able to warn neighboring ranchers. Few such signs of early habitation exist in the canyon. Leslie Goodding noted that the terrain is so rugged that "both cows and man alike have been discouraged from violating the sanctum." The area, like most of the other ranges in southern Arizona, became a part of the Coronado National Forest around the beginning of the twentieth century.

It was a mid-August day when we visited. By late afternoon, the thunderclouds would likely darken the skies, for it was the rainy season in southern Arizona. During much of July and August, thunderstorms come almost daily, giving temporary relief from the 100-degree heat, usually dropping the temperature to the mid-60s by nightfall. A cooling trend in the canyon begins during September; by February, the afternoon highs reach only the lower 60s, and at night the temperature hovers around freezing. Snow falls occasionally during winter but rarely lasts more than a day on the ground.

A short distance into the canyon we were delighted by the sight of a large, purple, sweetpea-shaped flower on a stem that trailed over a rocky slope. We recognized it as a butterfly pea. Like several flowering plants in the canyon, the butterfly pea has a disjunct, or discontinuous, range. Common along the East Coast and west to Missouri and eastern Texas, it skips the rest of Texas and all of New Mexico, turning up next in the Chiricahua Mountains in southeastern Arizona. Finally, after another 150-mile gap, it pops up in Sycamore Canyon.

Such discontinuous ranges are puzzling and provide biogeographers with material for speculation. One of the most remarkable ranges is that of *Asplenium exiquum,* a spleenwort fern. Plentiful in the Himalayas, this fern was inexplicably found in Mexico in 1865. In 1882 it was noted by a botanist exploring Conservatory Canyon, a few miles east of Sycamore canyon, and in 1937, Goodding discovered it here.

A number of flowering plants that are primarily Mexican occur in the United States only in the canyon. Most of them have a continuous distribution, but a trailing rhynchosia, also of the pea family, grows in Chihuahua, Mexico, and in Sycamore Canyon—some 400 miles apart—but apparently not in between. The short and stiff-branched whisk fern, discovered in rocky terrain in Sycamore Canyon less than four decades ago, is not found again for 300 miles to the south in Sonora, Mexico, and for 1,200 miles to the east in Texas. A few of the disjunct plants in Sycamore Canyon range many miles to the north. The Utah shadbush, common in the Grand Canyon as well as in Colorado and Utah, skips most of Arizona before it grows again on the canyon's rocky slopes.

No simple explanation exists regarding why some plants have discontinuous ranges. The spleenwort and the whisk fern, which reproduce by exceedingly lightweight spores rather than by seeds, may be dispersed to distant and widely separated areas by the wind. The butterfly pea and the rhynchosia, on the other hand, produce seeds that are many times heavier than spores. That their disjunct distribution can be explained by the effects of wind dispersal seems unlikely. Perhaps both had continuous distributions at one time, but some populations were wiped out, leaving gaps in their ranges. Drastic climatic changes, past geological disturbances, or biological factors such as disease and predation could cause such disruptions.

Although the distribution of plants in Sycamore Canyon has received the greatest amount of attention, there are animals in the canyon that have equally fascinating ranges. Vying for top billing among the birds are the coppery-tailed trogon, the rose-throated becard, and the five-striped sparrow. The trogon is an elegant, robin-sized bird with a hooked yellow bill and a long, square-cut, copper-colored tail. The male is dark emerald green with crimson underparts. This Central American species enters the United States in Sycamore Canyon and in a few other areas in southern Arizona. Even though its clear, cuckoolike call is distinctive, the trogon often goes unnoticed because it conceals itself in the middle layer of the forest canopy. The coppery-tailed trogon is a close relative of the colorful Central American quetzals, which were birds of importance to the Maya emperor-gods.

The rose-throated becard is a flycatcher that measures about six inches long. Although rather common in Mexico, in the United States it is found only in southern Arizona and in the lower Rio Grande Valley of Texas. It lives in the lower canopy layer of the forest. Another Mexican bird, the five-striped sparrow, has recently been seen in Sycamore Canyon.

We encountered several mammals or signs of them in Sycamore Canyon. Raccoon and deer tracks were abundant in the soft earth next to the creek, and a variety of small prints traceable to some of the four kinds of skunks that live in the canyon were evident. As we wandered up one of the side canyons, a rustling of vegetation near the base of a rocky slope drew our attention. The creature announcing its presence was a coati-mundi—the first I had ever spotted in the wild. We stared at each other for half a minute or so until I decided to back off, not wishing to disturb this animal further in its natural habitat. Jaguars probably live here too, and there are earlier reports of the jaguarundi, a small member of the cat family that is on the federal list of endangered species. At the present time the only place in the United States where the jaguarundi is known to occur with certainty is Texas.

One of the most unusual of all snakes in the United States, and one of the rarest as well, is the Mexican vine snake. Leslie N. Goodding found a vine snake in Sycamore Canyon several years ago. This grayish snake usually grows to a length of four feet yet is only the diameter of a pencil. Its slender head, which may reach six inches in length, is usually raised a few inches off the ground as the snake glides rapidly along in search of lizards.

The pools in Sycamore Creek are home not only to the Sonoran chub but also to the tarahumara frog, a four-inch-long amphibian whose ranges are primarily in Mexico and which rarely crosses into the United States.

Much of Sycamore Canyon is steep, rugged, and difficult to traverse. The canyon has numerous habitats, from the rich riparian areas adjacent to the creek to the dry, Sonoran desert-scrub. Oak woodland communities are common, as are areas of grassland and mesquite. Several succulents, including at least 14 different kinds of cacti, have been recorded.

Biogeographers continue to seek an explanation for the presence in Sycamore Canyon of plants and animals that have disjunct ranges or that are at the edge of their ranges. No doubt the permanent water supply plays a major role in the distribution of organisms, as does the cool air that penetrates into the bottom of the canyon after sunset. Combined with the geological history of the area, these factors have made Sycamore Canyon a naturalist's paradise.

Mt. Graham

"This is a great Pleistocene museum," exclaimed Tom Waddell as we neared the 10,720-foot summit of Mt. Graham in southeastern Arizona's Coronado National Forest. A seasoned wildlife biologist with the Arizona Fish and Wildlife Department, Tom has hiked and ridden horseback over most of the surrounding Pinaleno Mountains during the past several decades. He was referring, for one thing, to the glacial till that has been found on the north and east sides of the mountain, the southernmost location of such features in the United States.

The Pinalenos became isolated from other mountains when the Ice Age ended, about 11,000 years ago. The Canadian type of spruce–fir forest that formerly occupied the surrounding valleys was stranded on the steep mountain peaks as the glaciers retreated and the Sonoran Desert advanced into lowland areas. Various small mammals, reptiles, birds, insects, and mollusks that could not tolerate the desert heat and aridity were also confined to the top of Mt. Graham and other Pinaleno peaks. These included some, such as the Clark nutcracker and the Arizona black rattlesnake, that had originally migrated down from the north across the Colorado Plateau and others, such as the twin-spotted rattlesnake and Yarrow's spiny lizard, that had moved up from Mexico's Sierra Madre Occidental.

Through the subsequent period of isolation, Rusby's mountain fleabane and a number of other plants have developed into distinct species, and animals such as the Mt. Graham spruce squirrel, the white-bellied vole, and the Mt. Graham pocket gopher have become recognizable subspecies. The fleabane, squirrel, vole, and pocket gopher live only above 9,400 feet in the Pinalenos and nowhere else in the world.

Thought to be extinct in 1965, the Mt. Graham spruce squirrel was rediscovered a few years later and now numbers less than than 200 individuals. It is the smallest and southernmost subspecies of the red squirrel, a common animal that ranges all the way into boreal Canada and Alaska. Unlike other red squirrels, which chatter incessantly, the Mt. Graham spruce squirrel is mute except in the most life-threatening situations. Some zoologists believe that the 11,000-year isolation of this subspecies in the absence of mammalian predators—namely, the pine marten—has eliminated its need for being vocal.

During the first three decades of the twentieth century, the spruce squirrel ranged down the mountain slopes to about 6,500 feet, where at least some spruces and Douglas firs were available to provide seed cones for food. But intensive logging, which began in the 1930s, confined the squirrel to higher elevations. When, in the 1940s, the Abert's squirrel was introduced into the adjacent elevations of ponderosa pine for hunting purposes, the spruce squirrel was driven even farther upland.

Today the Mt. Graham spruce squirrel lives only within the closed forest canopy above 10,000 feet on Mt. Graham and on some nearby peaks. Here the squirrel selects a tree with a cavity for its nest and begins to compete with bandtailed pigeons and crossbills for spruce cones on the trees and with chipmunks for cones on the ground. The squirrel will perch on a branch with a cone, devouring the seeds and dropping the inedible cone debris to the ground. In time, deep layers of this debris build up, and the squirrel uses this midden as a cache in which to bury fresh cones and even mushrooms for future consumption. The squirrel builds the midden in dense forest stands, sometimes in hollow logs and stumps, where sunlight cannot penetrate and dry out the cache of food.

Although the Forest Service is not now logging the spruce–fir forest, and the spruce squirrel seems to have adapted to the occasional intrusion of the

Abert's squirrel, its existence is still threatened. Some biologists believe that any organism reduced to less than 200 individuals may lack the genetic diversity needed for survival. The University of Arizona's Steward Observatory has constructed a multimillion-dollar complex with 18 sophisticated telescopes on a 3,500-acre area that encompasses most of the terrain above 9,400 feet on Mt. Graham and other Pinaleno peaks. Many of the remaining spruce squirrels live in this high-impact area, and the area is no longer accessible to visitors.

Kaibab National Forest

SIZE AND LOCATION: 1.5 million acres in northern Arizona, on both sides of Grand Canyon National Park. Major access routes are U.S. Highway 180 and State Routes 64 and 67. District Ranger Stations: Fredonia, Grand Canyon, Williams. Forest Supervisor's Office: 800 S. 6th Street, Williams, AZ 86046, www.fs.fed.us/r3/kai.

SPECIAL FACILITIES: Winter sports area; horse trails; boat ramps.

SPECIAL ATTRACTIONS: Kaibab Plateau Scenic Byway; Beall Wagon Road.

WILDERNESS AREAS: Sycamore Canyon (55,937 acres, partly in the Prescott and Coconino National Forests); Kendrick Mountain (6,510 acres, partly in the Coconino National Forest); Kanab Creek (70,460 acres, partly on Bureau of Land Management land); Saddle Mountain (40,539 acres).

The Kaibab National Forest has three distinct units. North of the Grand Canyon is the Kaibab Plateau unit, and the national forest covers all of it down to the North Rim of the Grand Canyon. South of Grand Canyon National Park, the Tusayan District of the Kaibab National Forest occupies the Coconino Plateau. The third unit is west of Flagstaff where Kendrick Mountain is located and south of Williams where Bill Williams Mountain is found.

On most of the mountain slopes in the Kaibab are extensive woodlands dominated by piñon pine, Utah juniper, one-seeded juniper, and Rocky Mountain juniper. Alligator juniper occurs only on the south side of the Grand Canyon area. Because these trees rarely grow tall, local residents refer to these woodlands as pygmy forests. As elevation increases, ponderosa pines form continuous stands. Gambel's oak and quaking aspen are also common in the forest. High above the stands of ponderosa pine and quaking aspen, two additional species, Douglas fir and white fir, become common. The

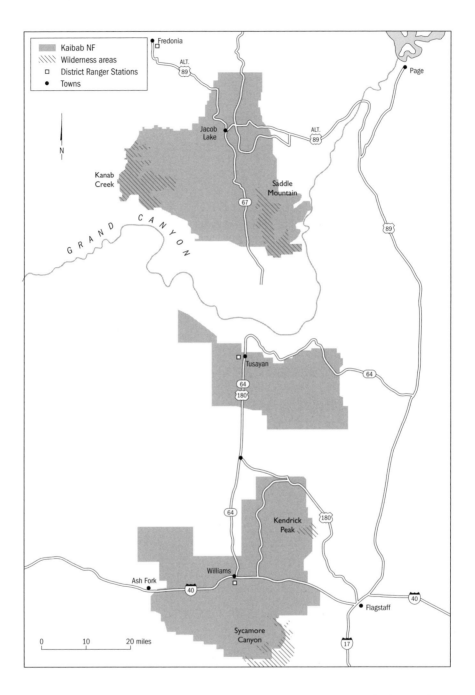

higher north-facing summits of most of the mountains consist of a forest of subalpine fir, Engelmann spruce, and blue spruce. In the Kanab Creek drainage below 5,000 feet at the western edge of the Kaibab Plateau is an extensive desert-scrub community.

State Route 67 from the community of Jacob's Lake to the north entrance to the Grand Canyon National Park bisects the northern unit of the Kaibab. At the extreme western edge of the national forest is the rugged Kanab Creek Wilderness. At Snake Gulch in the wilderness are excellent rock paintings on nearby cliff walls that date back to the Archaic and may have been produced by ancestors to the Hopi and Kaibab Paiute Indians. At the extreme eastern edge is the rough-and-tumble Saddle Mountain Wilderness Area. Between the two and adjacent to State Route 67 is serene DeMotte Park, a broad, tranquil meadow where wildflowers bloom during the day and wildlife come to reach water in the evening. If you wander around DeMotte Park, you come across such wildflowers as starry false Solomon's-seal, death camas, fairy-slipper orchid, columbine, larkspur, golden corydalis, and at least two kinds of lupines.

Westside Road is a long, loop route over much of the western side of the Kaibab Plateau. Leave DeMotte Park on this road and drive up Lookout Canyon. Then join a forest road that passes through Warm Springs Canyon before turning south for your return to DeMotte Park. The large area that you encircled has hundreds of springs and dozens of side canyons. One very rough side road off of the Westside Loop ends at the Indian Hollow Campground on the north rim of the Grand Canyon. From here you may hike the Thunder River Trail into the national park. If you see a black squirrel with a white tail, it is the Kaibab squirrel, known essentially only from the north rim area. It is replaced on the south side of the Grand Canyon by the Abert squirrel.

Another forest highway goes east from DeMotte Park to the very edge of the Grand Canyon and follows the north rim for three miles before turning abruptly to the north and ending at Marble Viewpoint for one of the most remarkable views around. Just before the viewpoint is the Marble Sinkhole.

Seven miles north of the northern entrance to the Grand Canyon National Park is part of the old Beall Wagon Road that brought emigrants from Ft. Smith, Arkansas, to the Colorado River. Twenty-three miles of the old road runs through the Kaibab National Forest.

The Coconino Plateau on the south side of the Grand Canyon is a beautiful area. To get a general idea of the plateau, drive a forest highway diagonally from the Tusayan Ranger Station to the southeast corner of the national forest. For a spectacular treat, drive to Grandview Lookout at the northern edge of the Coconino Plateau, at which lookout is a super view into the Grand Canyon. Once you have savored the view, drive the very rough Coconino River Road from the lookout to the eastern edge of the national forest.

The summit of Kendrick Mountain is 10,418 feet, and only here in the Kaibab National Forest can you find corkbark fir, southwestern white pine, and limber pine. Kendrick Mountain, the centerpiece of the wilderness of the same name, is an old volcanic plug. The trail to the summit begins at the end of the Pumpkin Center Road.

Bill Williams Mountain lies south of the community of Williams and is encircled by the Bill Williams Loop Road. Be sure and drive this loop. Two side roads off the loop are worth taking. A short trip is down the Bixler Saddle Road on the western slopes of Bill Williams Mountain. Some fantastic rock formations occur along the road. A longer side road leads to White Horse Lake, a most pleasant surprise in a nice mountain setting. From the lake is a short drive through desert scrub to the road's end at Sycamore Point. From this precipitous point, you can look straight down into the colorful Sycamore Canyon Wilderness (fig. 6).

Sycamore Canyon (Kaibab National Forest)

When I was researching this field guide to the national forests, a resident of Arizona encouraged me to inspect Sycamore Canyon, a small-scale Grand Canyon about 20 miles southwest of Flagstaff. The route on my first approach soon narrowed to a one-lane gravel road, passing scenic White Horse Lake and meandering across the top of the Colorado Plateau through ponderosa pine forest. It ended at Sycamore Point, which is an overlook perched on the Mogollon Rim, the southern edge of the plateau. Three thousand four hundred feet below, Sycamore Creek shimmered at the bottom of its canyon like a silver thread.

Sycamore Canyon is a designated wilderness area in parts of the Coconino, Kaibab, and Prescott National Forests. Its 55,937 acres lack roads, so the only way to get a firsthand look at the canyon is by trail. Six entrances exist along the eastern and southern boundaries of the wilderness. I entered at Packard Place, the southernmost point, near where Sycamore Creek joins the Verde River.

The trail, which parallels Sycamore Creek for much of its way, is dwarfed by the colorful rock walls that tower above. In places, all seven of the canyon's major geological formations are visible. Lowest, and hence oldest, is a thick layer of gray limestone deposited by a sea that covered the area some 400 million years ago. Occasional pinkish beds of rounded grains of quartz are intermixed in this gray rock.

After withdrawal of the sea more than 300 million years ago, the brilliant red sandy shale and sandstone of the Supai formation were formed, in some places nearly 1,000 feet thick. This red rock is a prominent feature of Oak Creek Canyon, a few miles to the east. Then, in succession, thick layers of buff-colored Coconino sandstone, white Toroweap sandstone, gray Kaibab limestone, and red Moenkopi sandstone were deposited on the Supai. Later, gravels washed down from the higher Bradshaw Mountains to the south and covered much of the Moenkopi sandstone.

Finally, 60 million years ago, volcanic eruptions followed by major uplifting of the area triggered abrupt changes. Lava flowed from the volcanoes and formed a hard cap of dark basalt over the region, and faulting drastically altered the configuration of the Colorado Plateau. During the Pleistocene, which began 1.8 million years ago, Sycamore Creek and its tributaries began the relentless cutting action that exposed all these formations.

Figure 6. Sycamore Canyon Wilderness.

Sycamore Canyon is host to a variety of vegetation communities. There is a difference between the west side, where conditions are cooler and more moist, and the east side, which receives the direct rays of the afternoon sun. High on the west side are forests of white fir, Douglas fir, and ponderosa pine. On the east side of the canyon, nearly pure stands of ponderosa pine dominate north- and west-facing slopes, and a shrubby community known as chaparral appears on the southern slopes. Deer, elk, black bear, mountain lion, and wild turkey frequent all these forested slopes.

Along Sycamore Creek where water is more plentiful, a forest of deciduous hardwood trees has developed. The type and density of the woody plants vary with the width of the canyon. Arizona ash, often growing in dense stands, dominates the narrowest parts of the canyon. Beneath the ash, next to the creek, are hundreds of plants of Arizona alder. Other trees include netleaf hackberry, box elder, Arizona walnut, and Arizona sycamore, from which the canyon takes its name. Cottonwoods and willows, common species along river banks, are rare because the narrow parts of the canyon receive less sunlight and are subjected to cold air that accumulates in the canyon bottom.

In areas along Sycamore Creek where the canyon broadens, increased sunlight permits the vigorous growth of mesquite, cat's-claw acacia, and shrub live oak; and cottonwoods and willows turn up with a little more regularity. Finally, in the broadest sections, which provide good growing conditions for a wide diversity of species, Goodding's willow and Fremont's cottonwood dominate, the latter sometimes growing nearly 100 feet tall. Red bats and pipistrelles use both trees for their summer roosting sites, and summer tanagers and zone-tailed hawks nest among the upper branches. This habitat is also home to Arizona gray squirrels and small, long-tailed, raccoonlike animals called ring-tail cats (pl. 10).

Although only 21 miles long and seven miles across at its widest point, Sycamore Canyon does resemble the mighty Grand Canyon, which is nearly 75 miles to the north. The vegetation may be denser, but the rocks are just as colorful and represent a very similar geological history, with the same sequence of formation. One thing Sycamore Canyon does lack are the crowds that descend on the Grand Canyon. Fewer than a thousand people enter the wilderness during the entire year.

Prescott National Forest

SIZE AND LOCATION: 1,238,154 acres in central Arizona, from Seligman and Williams on the north to Crown King and Arcosanti on the south. The Verde River forms the eastern boundary. Major access routes are Interstates 17 and 40, U.S. Highways 89 and 89A, and State Route 169. District Ranger Stations: Chino Valley, Camp Verde, Prescott. Forest Supervisor's Office: 344 S. Cortez, Prescott, AZ 86303, www.fs.fed.us/r3/prescott.

SPECIAL FACILITIES: Winter sports area; off-highway vehicle areas; boat ramps.

SPECIAL ATTRACTIONS: Verde Wild and Scenic River.

WILDERNESS AREAS: Apache Creek (5,666 acres); Castle Creek (25,215 acres); Cedar Bench (14,950 acres); Granite Mountain (9,762 acres); Juniper Mesa (7,406 acres); Pine Mountain (20,061 acres, partly in the Tonto National Forest); Sycamore Canyon (55,937 acres, partly in the Coconino and Kaibab National Forests); Woodchute (5,833 acres).

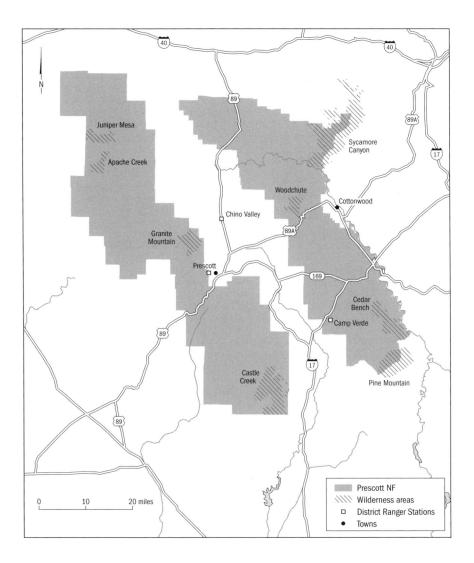

Cool mountains above 7,500 feet in elevation to arid Sonoran Desert communities epitomize the range of diversity to be found in the Prescott National Forest. History can be found throughout the forest, from gold-mining along Lynx Creek to a gunbattle at Battles Flat to near–ghost towns such as Cleator and Crown King. Horsethief Basin is an area of former cattle rustling.

The eastern and western units of the Prescott National Forest are separated by the broad Chino and Lonesome valleys. Most of the southeastern boundary of the national forest is the Wild and Scenic Verde River, Arizona's only river with such a designation. Whitewater rafting is at its best in the spring when there is maximum water flow. Most prominent in the eastern unit are the Mingus Mountains, the Black Hills, and Black Mesa. The western part of the Prescott is dominated by the Bradshaw Mountains and the lesser Juniper, Santa Maria, and Sierra Prieta Mountains.

You may want to start your exploration of the Prescott National Forest in the western section by driving a very scenic mountain road through the Bradshaw Mountains. These rugged mountains are rich in history, primarily because of the minerals they contain. Gold was discovered in 1864 along Lynx Creek, and a few people still live in the picturesque mining community of Crown King. Leave Interstate 17 at Cordes Junction and drive through chaparral to the nearly abandoned town of Cleator. From Cleator, the narrow road climbs via switchbacks to Crown King, winding around the northern end of Castle Creek Wilderness. The granite peaks in this wilderness are on the eastern flank of the Bradshaws. A forest trail follows Castle Creek from the Kentuck Campground in Horsethief Basin and crosses the wilderness just below Twin Peaks, staying at an elevation around 6,800 feet. However, if you take the forest trail along Boulder Creek from Horsethief Lake to the south edge of the Prescott, you drop into desert where saguaros, mesquite, jojoba trees, cat's-claw acacias, and palo verdes dominate. On the way you go through a good chaparral community of mountain mahogany, scrub live oak, and manzanita. Parry's penstemon (pl. 11) is a beautiful wild flower.

After looking around Crown King, drive north through the heart of the Bradshaws on the Senator Highway. Just past the ghost town site of Goodwin is the Palace Station Stage Stop (fig. 7), one of the few remaining stagecoach stations left in the West. This two-story log building was built in 1875 as the midpoint between Prescott and the Peak Mine. It consists of two rooms and a kitchen downstairs and a sleeping loft above. Between Palace Station and Prescott, the Senator Highway climbs past the dense ponderosa pine forests on Mt. Union and Spruce Mountain. Patches of aspen, light green during the summer and golden in the autumn, interrupt the darker green cover of the pines. Just after Spruce Mountain the road drops down to Groom Creek where the old wooden Groom Creek Schoolhouse stood from 1896 until it was rebuilt of stone in the 1930s. The Prescott National Forest has used the building as an administrative site since the closing of the school in 1952. A picnic area, amphitheater, and nature trail through the ponderosa pine forest are nearby. Eventually you make your way to the southern end of the city of Prescott. Along the way are several inviting-looking campgrounds.

The Granite Mountain Wilderness is a few miles northwest of Prescott. The trail from Granite Basin Lake at the southeast corner of the wilderness to Blair Pass and then through Long Canyon passes gigantic granite boulders.

At the extreme northwestern corner of the Prescott National Forest are Juniper Mesa and Apache Creek Wildernesses, each covering less than 7,500 acres. Juniper Mesa is a flat-topped mesa that runs east to west at the south-

Figure 7. Palace Station, one of the few remaining stagecoach stops.

ern end of Juniper Mountain. A hiking trail traverses the entire length of the mesa. The drier, south slopes of Juniper Mesa are covered by a piñon–juniper forest, whereas the northern side has fine stands of ponderosa pine. Apache Creek flows diagonally through the Apache Creek Wilderness, but the region is so remote that there are no marked trails. It is at the northern end of the Santa Maria Mountains.

On another day, drive east of Prescott on State Route 169 for about five miles, then take a forest road to Lynx Creek where you may do a little panning for gold. Nearby Lynx Lake has a boat ramp.

The lower half of the vast Sycamore Canyon Wilderness is in the Prescott National Forest and is perhaps the prettiest part of the forest. The rocky outcrops in the canyon are reminiscent of the Grand Canyon, but the presence of more vegetation gives the canyon a totally different look. The Verde River passes by the southern boundary of the wilderness, and the Packard Mesa Trailhead begins in this vicinity. The trail stays to the west of Sycamore Creek, which has carved the canyon, and the trail may be followed nearly the entire length of the wilderness area.

The Woodchute Wilderness is just west of Jerome, a little town that is perched on the edge of the plateau above Deception Gulch. The 2.5-mile trail to the summit of Woodchute Mountain provides magnificent views of the San Francisco Peaks far to the northeast. If you prefer shorter trails, try those in the Mingus Mountain area, which is not in a wilderness, south of Woodchute and Jerome.

At the far southeastern corner of the Prescott National Forest are Cedar Bench and Pine Mountain Wildernesses. Cedar Bench is a ridge that runs diagonally across the wilderness area and allows for fine views of the Verde River. You can drive a rough road from the old Dryer Mine and up Tank Canyon to Chalk Lake at the edge of the wilderness area. From here there is a half-mile hike to picturesque Ox Bow Lake. The Verde River passes through the heart of the Pine Mountain Wilderness and 6,814-foot Pine Mountain. Ponderosa pine and Douglas fir grow on the top of the rim above the Verde River.

Bradshaw Mountains

The rugged Bradshaw Mountains, in the southern reaches of Arizona's Prescott National Forest, were once prime gold country. The territory's principal back road follows the abandoned bed of an ore-hauling railroad, the Crown King branch of the Prescott and Eastern. Rocky and narrow, the road meanders from the near–ghost town of Cleator, in the eastern foothills, west to Crown King and then north to the city of Prescott.

Cleator (elevation nearly 4,000 feet) lies along the northern edge of the Sonoran Desert. The surrounding scrubland is dominated by the foothill palo verde, a small tree with spiny branches that belongs to the legume family, but saguaros occasionally stand as emblems of the desert, along with clumps of other cacti—Fendler's hedgehog, Engelmann prickly pear, and teddy bear cholla. Other desert plants include the feather duster calliandra, ocotillo, and jojoba, all small trees with little leaves or with leaves modified into spines, adaptations that reduce water loss in the arid climate.

From Cleator the road begins a 2,000-foot climb to Crown King, leaving the saguaros and other desert-floor plants below. In their place appear round-topped woody plants that are about six to eight feet in height. Except for bare patches of rocky soil, the uniform vegetation gives the landscape a velvety appearance. This chaparral is a plant community dominated by the shrub live oak.

Although elevation is a factor in determining where chaparral occurs (most chaparral is found at elevations from 3,000 to 5,800 feet), biologists attribute its presence to a Mediterranean-type climate characterized by cool, damp winters and hot summers. Similar brushy habitats, referred to as garigue, occur along the Mediterranean Sea; Corsica has a distinctive type known as maquis.

In the United States, chaparral is found across central Arizona and again in southern California. Although these two habitats look similar, they differ in species composition and climate. Arizona has two major rainy periods—

one from late November through December and one from July to September—and most chaparral growth takes place during the summer. In California, the rainfall is concentrated in the winter, and that is the growing season.

Daniel Axelrod, a botanist at the University of California at Davis, believes that the two chaparrals had a common origin but diverged more than 2 million years ago as a result of topographical changes. Newly formed mountain ranges in California confined summer precipitation to the coast, giving competitive advantages to those plants that were suited for long, dry summers. In Arizona, the climate was less modified: the plant community continued to receive summer rain and remained similar to its original form.

The chief species in the Arizona chaparral are the shrub live oak, mountain mahogany, pointleaf manzanita, desert ceanothus, skunkbush sumac, and yellow silk-tassel (fig. 8). Most of these are evergreen shrubs with small, leathery leaves; dense, compact crowns; and deeply penetrating roots. Most rely on root sprouting as their major means of propagation. These modifications enable the species to survive the dryness of the chaparral habitat and the periodic fires that are fed by the tinder of fallen leaves.

In the chaparral, the rounded crowns of adjacent shrubs usually touch, forming a canopy that provides at least 60 percent cover. Lower branches often die for lack of sunlight, with the result that living branches enlarge and stems become gnarled and contorted. The canopy of many a shrub is supported by only a narrow strip of living tissue.

Figure 8. Yellow silk-tassel.

Root systems are extensive, penetrating quickly to great depths and spreading out laterally to capture all available surface moisture. One biologist measured a three-month-old desert ceanothus and found that its roots were three feet deep and two feet across.

Chaparral leaves are usually two or three times thicker than those of non-chaparral species and are covered with a thick cuticle, or wax layer. In addition, the small openings in the leaves, called stomates, are confined to their lower surface and are often embedded in deep crevices. These characteristics inhibit water loss. The leaves of most chaparral shrubs are also small, minimizing the danger of overheating during the hot summer.

An important part of plant adaptation in the chaparral is the ability to reproduce successfully despite the fires that lightning touches off once or twice each decade, killing many plants and burning others back. Those few species that depend on seeds for propagation produce seeds that will not germinate until scarred by fire. The seeds may lie dormant in the soil for many years before a fire activates them. The seedlings then start their lives at the most opportune time, when cover is reduced and the danger of further fire is minimized.

Most chaparral species produce few seeds, however, reproducing instead by root sprouting. A swollen root crown, or burl, develops at the base of the plant, beginning in the seedling stage and enlarging each year until it is far thicker than any of the main stems of the shrub. The burl contains numerous buds that produce vigorous sprouts when activated by fire. A chaparral shrub capable of sprouting after a fire has several advantages over those that propagate through seeds, as pointed out by California botanist Ted Hanes. One is that the plant retains its established place in the habitat; another is that nutrients and water in the root system are immediately available, with the result that the first year's growth is much more rapid than that of seedlings.

Because of these adaptations, chaparral regains its normal appearance much more quickly after a fire than other community types. Characteristic shrub density in the chaparral returns within five years, and crown cover is established in seven years.

In the Bradshaw Mountains, the chaparral is gradually replaced by a ponderosa pine community at about 6,000 feet. The increase in moisture and cooler climate are readily apparent on the road from Cleator as it nears Crown King. At Crown King itself, a few weathered frame buildings nestle beneath the swaying pines. They are all that remains of a mining community that had its heyday around the beginning of the twentieth century.

Tonto National Forest

SIZE AND LOCATION: Approximately 3.6 million acres in central Arizona, between the Mogollon Rim on the north and the eastern Phoenix suburbs, Globe, and Miami on the south. Major access roads: Interstate 17, U.S. Highway 60, and State Routes 87, 88, 188, 260, 288. District Ranger Stations: Scottsdale, Mesa, Roosevelt, Payson, Globe. Forest Supervisor's Office: 2324 E. McDowell Road, Phoenix, AZ 85006, www.fs.fed.us/r3/tonto.

SPECIAL FACILITIES: Boat ramps; swimming beaches; Roosevelt Visitor Center; Blue Point River Access Point.

SPECIAL ATTRACTIONS: Apache Trail National Scenic Byway; From the Desert to the Tall Pines National Scenic Byway; Highline National Recreation Trail; National Wild and Scenic Verde River.

WILDERNESS AREAS: Pine Mountain (20,061 acres, partly in the Prescott National Forest); Superstition (160,285 acres); Sierra Ancha (20,850 acres); Salt River Canyon (32,101 acres); Salome (18,531 acres); Mazatzal (250,517 acres); Hell's Gate (37,440 acres); Four Peaks (60,743 acres).

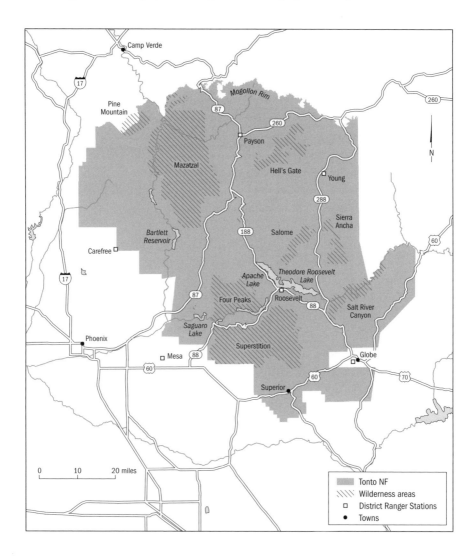

One of the largest national forests in size, the Tonto is also one of the most frequently visited because of its proximity to Phoenix and the Valley of the Sun. There is enough Sonoran Desert habitat to please everyone, and there are also mountains whose summits are covered by forests of ponderosa pine and Douglas fir. In between are brushy chaparral communities above the desert and piñon pine–juniper–oak woodlands on the mountain slopes.

Water enthusiasts, including boaters, flock to Theodore Roosevelt Lake, Apache Lake, Saguaro Lake, Canyon Lake, Bartlett Reservoir, and Horseshoe Reservoir. Intermittent and perennial streams, often bordered by steep, vertical cliffs, provide sensational scenery.

Almost all features of the Tonto may be accessed from State Route 87 (the Beeline Highway) from Mesa to Payson, from the Apache Trail National Scenic Byway (State Route 88), from The Desert to the Tall Pines National Scenic Byway, and from U.S. Highway 60 from Apache Junction to Globe.

From Mesa to Payson, the highway gradually climbs from desert through mountains to the Mogollon Rim. Not far from Mesa you may look to the right to see the curved narrow peak of Weaver's Needle in the distance. Here is a good place to see many plants typical of the Sonoran Desert.

From State Route 87, the Bush Highway makes a loop over to Saguaro Lake where there are excellent boating facilities. A few miles farther north on State Route 87 is a side road (Forest Road 143) that snakes its way toward the Four Peaks Wilderness. Three trailheads along this road provide an opportunity to explore the western side of the wilderness.

At Payson, State Route 87 heads toward Pine and Strawberry below the Mogollon Rim. From Pine you can hike a scenic trail up Pine Canyon to the top of the rim, or you can take a forest road over Hardscrabble Mountain or a four-wheel-drive road over Deadman Mesa. Both lead to trailheads from which you can hike the northern part of the Mazatzal Wilderness.

For less strenuous activity, drive Forest Highway 64 where it branches east off State Route 87 just past Tonto Natural Bridge State Park. Called the Control Road, this paved route sits on a plateau below the Mogollon Rim but above the desert. You will be rewarded by magnificent views and interesting side canyons. After crossing the East Verde River, where there is the nice Verde Glen Campground, you should keep going to the best lookout in the forest. Here, at Diamond Point, named for the sparkling quartz crystals in the soil, the view south to the desert is as far as you can see. The Control Road eventually joins State Route 260 west of Kohl's Ranch. At Kohl's Ranch, take the road to the Tonto Fish Hatchery, the whole way paralleling the exceedingly rocky Tonto Creek. On my first visit in the 1970s, I enjoyed seeing Zane Grey's cabin at the end of the road and beneath the Mogollon Rim. This is the area he wrote about in *Under the Tonto Rim*. When the deadly Dude Fire roared through this part of the forest in 1991, Zane Grey's cabin was one of the casualties.

If you want a close look at the area along the Control Road, you can hike the Highline National Recreation Trail No. 31 from State Route 87 just south of Pine to the Two Sixty Trailhead on State Route 260 about 12 miles east of Kohl's Ranch.

From the town of Apache Junction, State Route 88, known as the Apache Trail National Scenic Byway, branches off U.S. Highway 60. This is one of the most enjoyable driving routes anywhere. The scenic byway first winds its way through desert terrain to Canyon Lake where there are fine areas for picnickers and boaters. You can even take a scenic cruise on this beautiful lake on the Canyon Queen. Canyon Lake is all that separates the Four Peaks Wilderness Area to the north and the Superstition Wilderness Area to the south. The road goes through picturesque Tortilla Flats where you may enjoy some refreshments before continuing. After a very twisting stretch of road, there is a short side road to Apache Lake—a boater's paradise. As you look across the lake to the north shore, you will see the fantastic Painted Cliffs.

Seven miles farther on is Burnt Corral Recreation Site, another area with a boat ramp. From here it is only a short distance to the largest of them all, Theodore Roosevelt Lake. State Route 88 ends at the lake, but you can take State Route 188 either to the north end of the lake and eventually to State Route 87 or to the south end of the lake and eventually to Globe. Three miles south you should stop for a worthwhile visit to Tonto National Monument. A few miles southeast of the monument, The Desert to the Tall Pines National Scenic Byway leaves State Route 188 and heads north on State Route 288. This gravel and dirt road is the wildest scenic route of them all. For the first four miles this scenic byway follows the western edge of the Salt River Wilderness, but there are no marked trails in this large, wild region. Where the road cuts back sharply, you pass by Diversion Dam at the extreme east end of Theodore Roosevelt Lake. Two miles beyond the dam is the rough-and-tumble Cherry Creek Road, which comes to an abrupt end at the Flying H Ranch where there is a trailhead for an unforgettable hike into the Sierra Ancha Wilderness. Along State Route 288, the Young Highway, is a short drive along Workman Creek to a small parking lot. From there you may hike to see Workman Creek Falls. Yes, there is a waterfall in Arizona! After recovering from the shock of this beautiful site, continue on for 18 miles to the village of Young. Between the Workman Creek Falls (Forest Road 487) and Young, the scenic byway (State Route 288) passes by the Reynolds Creek Campground, which is located on a narrow passage between the Salome Wilderness to the west and the Sierra Ancha Wilderness to the east. Trails to both wilderness areas begin near the campground. Group reservations are necessary to use this campground. The trail to Hell's Hole in the Salome is rough and rugged but well worth the effort. Should you choose to hike in the Sierra Ancha Wilderness, be on the lookout for cliff dwellings from the Salado Indians. A few rough roads in the wilderness that are now closed to vehicular traffic were used in past uranium mine prospecting.

For several miles on either side of Young, the scenic byway crosses private land in Pleasant Valley, only to begin its grueling final ascent to the Mogollon Rim in the Apache-Sitgreaves National Forest.

One of the most famous areas in the Tonto National Forest is the Superstition Mountains, where legend has it that the Lost Dutchman Gold Mine is somewhere in "them thar hills." As a result, the wilderness has been severely damaged by thoughtless prospectors. Weaver's Needle, a volcanic plug that pokes its pointed summit high into the sky, has been a landmark since the first pioneers came to the area. The trailhead at Peralta Canyon is in the heart of Sonoran Desert country. The Superstition Mountains are in the Superstition Mountain Wilderness.

If you continue eastward on U.S. Highway 60 past Peralta Canyon, you will come to the town of Superior. Outside town is Apache Leap where, according to legend, a band of trapped Apaches leaped to their deaths rather than be captured during an uprising. The shiny black pieces of obsidian in the canyon are called Apache Tears and are said to be the hardened tears of those who loved the ill-fated warriors.

U.S. Highway 60 passes through rocky and scenic Devil's Canyon where a pull-out on the south side of the highway is directly across from a small mesa on which the rare yellow-spined barrel cactus lives. Near the settlement of Top of the World, a side road to Five Point Mountain brings you to the heart of the habitat for the Arizona red claret cactus.

At Globe, in the most southeastern part of the Tonto National Forest, a forest highway goes into the Pinal Mountains, climbing to Pinal Peak after passing Russell Kellner Canyon. Although there are radio towers on the summit, the view from here is outstanding.

North of Scottsdale and Cave Creek, Forest Road 205 (Horseshoe Dam Road) provides access to the trailheads on the west side of the Mazatzal Wilderness and to Bartlett and Horseshoe Reservoirs. A fine nature trail extends from the Jojoba Campground around part of the Bartlett. The forest highway passes along the eastern edge of the New River Mountains, home of the rarest species of agave in Arizona.

Desert Vista

As one drives approximately 40 miles northeast of Phoenix on the Bush Highway (State Route 89) toward Payson, there are panoramic views to the east across the Sonoran Desert. The pointed top of Weaver's Needle may be seen in the distance. Extending from southwestern Arizona and southeastern California into Baja California, the Sonoran Desert encompasses 119,000 square miles, including 2,360 miles of seacoast along the Gulf of California. Three other great deserts lie nearby: the Chihuahuan Desert to the east and the Mojave and Great Basin Deserts to the north. Of the four, the Sonoran Desert has been dry for the longest time and has acquired the most complex assortment of plants: a subtropical type of vegetation midway between temperate and tropical floras.

In their chapter on the Sonoran Desert in *The Reference Handbook on the Deserts of North America,* biologists Frank and Carol Crosswhite have traced the drying out of the Sonoran Desert to 40 million years ago, when, they believe, a semipermanent area of atmospheric high pressure arose off the Pacific Coast (such a system exists today, blocking the moist air masses that form over the ocean). Then, 12 million years ago, tectonic movements separated Baja California almost entirely from the rest of the continent. Stretching, thinning, and faulting of the earth's crust resulted in the formation of a huge basin, which became hotter as the elevation fell. Although the basin subsequently partly filled with seawater, creating the present Gulf of California, the Sonoran Desert has remained extremely dry for a number of reasons. The Crosswhites point to the high-pressure system that repeatedly forms off the Pacific Coast; the cold north-south ocean currents off the coast of California, which encourage any moist air heading toward the continent to shed its moisture into the ocean; and the rain shadow, or blocking of moisture, caused by the mountains that rise above the desert.

The Sonoran Desert receives some rain in both summer and winter, ranging from an average of 13 inches annually in the east to little or no rain in the west. The late desert authority Forrest Shreve recognized seven regional divisions of the Sonoran Desert based on the type of vegetation, which is influenced by latitude, elevation, and topography. This area north of Phoenix falls into the Arizona Upland Division, where hot and dry conditions for most of the year are interrupted by seasonal precipitation, especially violent summer thunderstorms with heavy runoff. Because of the intermittent rain pattern, the vegetation is sometimes dense, sometimes sparse, and usually consists of several layers of plants, from dwarf cactuses that hug the ground to giant saguaro cactuses that may stand more than 40 feet tall.

In the flatter regions of the desert, where the soil is deepest, creosote bush, which may make up as much as one-fifth of the vegetation, is mixed with white-thorn acacia, cat's-claw acacia, and several kinds of cholla cactuses. On hillier terrain, the plants increase in diversity, if not in numbers, because of the variation in conditions. Trees and large cactuses dominate: foothill palo verde, mesquite, desert hackberry, ocotillo, large barrel cactuses, and the incomparable saguaro. Small perennials that frequently grow here are the Christmas cholla cactus, desert zinnia, and brittlebush. Washes—ephemeral rivers whose beds are dry for much of the year—are lined with desert willow and blue palo verde.

All plants that live in the desert must be able to endure high temperatures and fierce solar radiation, coupled with a limited and irregular supply of water. A great number of desert species are annual wildflowers that spend most of their life cycle in a dormant stage. Known as ephemerals, they only spring to active life after a rainy period, blooming and forming their seeds in a short time. The ocotillo, a perennial that typifies another category of desert-dwelling plant, forms leaves after a rainy period, only to drop them once droughty conditions return.

Drought-enduring evergreens, instead, are shrubs or small trees that keep their leaves throughout the year. These plants all have some device to prevent excessive loss of water from the leaves. Desert hackberry has a thick wax coating, whereas the abundant creosote bush has resinous leaves (the resin also deters browsing animals). Desert marigold has leaves covered by dense mats of hairs, which not only slow down water loss but also reflect sunlight. The cat's-claw acacia has minute leaves that reduce the amount of leaf surface exposed to drying forces. In jatropha, the stomata—the openings in leaves that permit the necessary exchange of carbon dioxide and oxygen—are recessed.

In cactuses, the water loss is minimal because the leaves have evolved into spines, and photosynthesis is carried out in the thick, green stems. Saguaros have active chlorophyll-bearing cells buried as much as three inches deep in their stems.

To collect moisture, some desert plants, especially those that live along the washes, have long root systems. Creosote bush has not only a deep root system but also many roots just beneath the surface of the soil that absorb water effectively during and after a rainfall. Many desert species, including the myriad cactuses, are succulents: fleshy plants equipped with special cells for storing water. Most succulents have shallow roots that take up water quickly when it rains.

Recently, scientists have discovered that desert succulents usually have a special mechanism for photosynthesis. Most plants open their stomata during the day, when photosynthesis takes place, to permit the exchange of carbon dioxide and oxygen. But to prevent water loss when the sun is most intense, these succulents close their stomata. Instead, they take in carbon dioxide at night and sequester it in the form of organic acids. Then, during the day, the organic acids release the carbon dioxide needed for photosynthesis.

Diamond Point

Extending more than 200 miles, from northwestern Arizona into New Mexico, the Mogollon Rim is the irregular southern edge of the 9,000-foot-high Colorado Plateau. Along the rim, the terrain may drop rapidly down to 6,000 feet before beginning a much more gradual descent down to 3,000 feet. A moist, ponderosa pine forest prevails on the plateau, whereas the region below is dry, supporting scrub forests and desert plants.

Much of the Mogollon Rim area falls within four national forests that offer visitors a variety of campgrounds, picnic spots, fishing lakes, and scenic trails. One natural attraction is Diamond Point (pl. 12), a protuberance that is part of the rim topography, but which lies 10 miles south of the plateau, rising from a 6,000-foot base to 6,384 feet. Diamond Point receives its name from the quartz crystals scattered over the rocky terrain and included in many of the rocks. In the sun, Diamond Point truly glistens.

Part of the route to Diamond Point—along the Control Road, or Forest Highway 64—parallels a stream. New Mexico locust, Lowell ash, and Arizona walnut line the streambank, and grasses cover much of the forest floor. Here and there, a touch of color may be added by the blooms of lupine, scurf pea, beardtongue, milfoil, wild geranium, or the shrubby wild Arizona rose. But after the final turnoff onto Forest Highway 65 toward Diamond Point, this lush vegetation falls behind. As the road begins to climb above 6,000 feet, the soil becomes extremely dry and rocky, and the plant life gives way to a uniquely Western habitat known as chaparral.

As discussed earlier, the word "chaparral" derives from the Spanish for dwarf evergreen oak; in Arizona it refers to a habitat dominated by shrubs and a stunted species of live oak, with relatively little ground cover provided by wildflowers and grasses. In general, Arizona's chaparral plant communities develop at between 3,000 and 5,500 feet, but in some places, such as Diamond Point, these limits differ as a result of slope exposure, soil type, and climate. Below 3,000 feet, chaparral tends to merge into desert scrub or desert grassland. Above 5,500 feet, chaparral usually gives way to forests of small (30-foot) piñon pines, one-seeded junipers, and alligator junipers— woodlands that are sometimes considered part of the chaparral but are more accurately termed Madrean woodland for their association with plant communities of the Sierra Madre.

Most of the plants in the true chaparral are trees and shrubs eight to 10 feet tall. These plants usually have a multitude of branches; broad, leathery, evergreen leaves; a dense, compact crown; and a very deep root system from which new growth sprouts readily. In the drier sites, chaparral shrubs and trees cover only about 40 percent of the ground, whereas in more moist areas, they may provide twice that cover.

The most abundant plant is a particular species of live oak that grows only about 10 feet high. It has blue-green, spine-tipped, hollylike leaves. Almost as common is the birch-leaf mountain mahogany, a shrub that—like so many other species with dense wood—is sometimes called ironwood. It is a member of the rose family, but its flowers lack petals. Mule deer browse on it, and its seeds and leaves are a staple in the diet of grouse. Another abundant shrub is desert ceanothus, a five-foot plant with small, thick leaves arranged in opposing pairs on the branchlets. Its small clusters of sweet-scented, white flowers bloom during the summer.

The chaparral contains two types of sumac. Sugar sumac has simple, evergreen leaves; the leaves of the other sumac, called skunkbush, are divided into three leaflets. Skunkbush is one of very few plants in the chaparral that shed their leaves in winter, and as its name suggests, its leaves emit an unpleasant odor when crushed.

One striking shrub is manzanita, whose bright green, thick, oval leaves contrast markedly with its twisted, mahogany red stems, which glisten in the sunlight. The leaves stand vertically with the edges toward the sun, an adaptation that helps reduce water loss.

Datil, a nonwoody, succulent species, is a type of yucca. The leaves arise from ground level, are about two inches wide, and have fibrous threads that hang from the edges. Datil's white flowers are borne on a stalk up to six feet tall.

One of the most conspicuous chaparral plants, at Diamond Point and elsewhere, is the century plant. This species of agave has clusters of succulent leaves that often are nearly two feet long and up to six inches wide, with spiny teeth along the edges. After 40 to 60 years (not 100, as the common name implies), a thick green stalk begins to grow upward from the leaves. When the stalk is about 15 feet tall, it puts forth clusters of attractive yellow flowers. After producing seeds, a process that takes about six months, the plant withers and dies.

The chaparral's fallen leaves are thick and leathery and do not decay readily, creating a fire hazard. In June 1990, a conflagration in Tonto National Forest came within a few miles of Diamond Point. Known as the Dude Fire (named for Dude Creek), it consumed chaparral and woodland over an area three miles wide and 14 miles long, killed several people, and destroyed 51 cabins, including one that belonged to Zane Grey, the legendary writer of Western novels.

Most of the chaparral's woody species are well adapted to survive fires, resprouting rapidly from their massive root systems. Most also begin to produce seeds during the first five years of life, rather than later, which is common in many woody plants. The seeds are often fire resistant, and those of some species will germinate only after being subjected to fire. Plants lacking fire-resistant seeds usually produce a very large number of seeds.

Arizona chaparral and Madrean woodlands grow where winters are mild and summers are hot. Precipitation at Diamond Point, averaging about 18 inches annually, falls primarily during two seasons. About 55 percent falls during the winter (from November through April), usually in the form of gentle rains, and another 35 percent drops during intense thunderstorms in July, August, and September. May and June, the early part of the growing season, are usually dry. Where similar conditions prevail, as in the Mazatzal Mountains 20 miles southwest of Diamond Point, a similar selection of chaparral species appears. But a different chaparral community arises in California, 200 miles to the west, where rain falls mostly in winter.

Workman Creek Falls

The lure of a waterfall within 75 miles of the arid metropolitan area of Phoenix seemed too good to be true, so when I first saw the words "Workman Creek Falls" on my map of the Tonto National Forest, I had to investigate. The falls are in the Sierra Ancha, a 2.5-hour drive northeast from Phoenix through typical desert terrain. Below 5,000 feet, cacti and tiny-leaved trees and shrubs are dominant. Columnar saguaros grow in great numbers in some places, elsewhere as isolated specimens. Much shorter are several kinds of prickly pear cacti and the jumping cholla, whose spine-covered branchlets seem to jump at you as you brush past. Other species, with leaves reduced to prevent excessive water loss, include mesquite, cat's-claw acacia, Christ's-thorn, and the green-trunked palo verde. Be warned that the water in Workman Creek may contain high amounts of radioactive material.

As it approaches the Sierra Ancha, whose Aztec Peak looms at 7,748 feet, the road (State Route 288, the Young Highway) turns to gravel, and the desert is gradually replaced by a community of grasses and other plant species. Agaves, yuccas, bear grass, and sotol—all clump-forming perennials—are abundant. One curious plant is the ocotillo, which looks like a thin cactus until after one of the region's infrequent rains, when its wandlike branches become briefly clothed with leaves.

At 6,000 feet, rolling hills with steep south-facing slopes are evident. In drier areas, a shrubby chaparral community predominates—a sometimes nearly impenetrable thicket of mountain mahogany, manzanita, hollyleaf buckthorn, and other species. Apache plume (pl. 13), whose fruits are covered with featherlike tufts of hair, is common. Interspersed with the chaparral are moister areas supporting woodlands of alligator juniper, two kinds of piñon pine, Arizona cypress, and a few species of oak. Brightly colored penstemons and Indian paintbrush appear in late summer.

Short of the village of Young, a rough track leads east of the main road and follows Workman Creek upstream. When in Arizona, I usually consult my cumbersome and now field-worn *Arizona Flora* for almost every plant I see, and walking along this trail to the falls was no exception. I stopped to identify painted alumroot, Griffith's stonecrop, Rusby's echeveria, and several others. Many nonnative weeds, however, have invaded along the disturbed trail: white and yellow sweet clovers, woolly mullein, shepherd's purse, curly dock, and orchard grass.

After about three miles, the trail reaches an overlook with a view of the falls, a narrow cascade of silver water plummeting over a precipitous ledge to a catch pool 250 feet below. By scrambling down the rocky slope, visitors can glimpse the falls from the edge of the pool.

The trail continues along Workman Creek in the direction of its source. The creek bank, kept moist by the cool, flowing waters is an ideal habitat for trees, whose shade favors herbaceous vegetation. Here I was pleasantly surprised to see several native species familiar in my home state of Illinois, 1,400 miles distant. I put aside my *Arizona Flora* because these were plants I knew: box elder, golden glow coneflower, northern white violet, fragile fern, spotted coralroot orchid, cardinal flower, and false Solomon's-seal. On one moss-covered boulder at the edge of the creek was a handsome American spikenard, a perennial that grows in similar habitats in the eastern United States. Poison ivy and woodbine, two vines common eastward, were also much at home.

According to Roy Johnson, who has written a thesis on the flora of the Sierra Ancha, a quarter of the species are also found in Illinois or other parts of the Midwest and East. Although a number of these are nonnative European introductions that grow in any suitable, disturbed area, most are natives. They are found primarily along streams or other moist habitats in the Sierra Ancha; far fewer occur in the drier areas. In fact, the desert or semidesert habitats do not contain a single native species in common with the Midwest or East. Plant species may need to be more specialized to become established in the drier habitats, especially to survive over the long term, when they will inevitably be subjected to occasional periods of extreme drought.

About one mile upstream from Workman Creek Falls is the western boundary of the Sierra Ancha Wilderness, a 21,000-acre, roadless area highlighted by 2,500-foot cliffs, precipitous box canyons, mesas, and Indian ruins. Here, geologists have found rocks dating back 1.2 billion years. Rocks as old as these can be found elsewhere in Arizona only near the bottom of the Grand Canyon. In 1982, geologist Paul Knauth of Arizona State University and Mark Beeunas, a graduate student, found samples of these ancient rocks whose chemistry suggests that at that time the exposed land surface was the site of organic activity. They thus concluded that some type of primitive green land plants existed 1.2 billion years ago, three times earlier than has generally been assumed.

In talking with Paul Knauth I learned another arresting fact: the water of Workman Creek Falls sometimes contains a significant amount of radioactive material. Uranium deposits are in the surrounding area, and in fact, many of these deposits have been disturbed by prospecting and mining. Anyone planning to explore any of the abandoned mine shafts should be aware of possible radon gas hazards.

Peralta Canyon

East of Phoenix and in Tonto National Forest, the arid Superstition Mountains rise above the northeastern edge of the Sonoran Desert. One of the landmarks of this volcanic range is Weaver's Needle, a pinnacle-like remnant of eroded lava. The rumor that the Lost Dutchman Gold Mine lies hidden somewhere up in the hills makes the region even more intriguing. One recent autumn, with visions of being the one to rediscover the gold, I hiked to Weaver's Needle along a well-traveled route up Peralta Canyon.

The trailhead is located in the desert, some 25 miles east of Phoenix. It starts at 2,400 feet, in a basin at the foot of the Superstition Mountains (pl. 14). Jagged cliffs of a glassy quartz called dacite frame the area. A prominent feature of the landscape is the saguaro, the giant, sometimes comical-looking columnar cactus that may grow to 35 feet, although its spreading root system rarely penetrates deeper than three feet.

For the first 1.5 miles, the trail climbs 1,200 feet, rising through different vegetation zones. At first, small round-topped foothills paloverde trees, with tiny yellow green leaflets, are scattered across the desert floor. Teddy bear cholla, a yellow-spined cactus that looks misleadingly soft, is also common. But as the trail climbs up the creek bed of Peralta Canyon, a diversity of short trees and shrubs appears, including desert hackberries, ironwoods, velvet mesquites, Emory oaks, and cat's-claw acacias. Although the Sonoran Desert has a lot of bare spaces and appears to be very open, most of the trees and shrubs are surrounded at their base by small perennials that are able to live in the shade. The saguaros, on the other hand, usually stand alone, as they cast little shade.

The trail occasionally crisscrosses the canyon's rocky streambed. Where pockets of moisture accumulate, small ferns, mosses, liverworts, and clubmosses sometimes grow beneath deep-green-leaved sugar sumac trees. Finally, after a series of switchbacks, the trail reaches a 3,890-foot summit known as Fremont Saddle. At this elevation, several grasses dominate the vegetation. Short-statured Arizona rosewood and jojoba trees are the prevailing woody plants, beneath which grow agaves, hedgehog cacti, and beargrass. The Superstition Mountains are visible in their full splendor from this vantage point; Weaver's Needle looms mightily about one mile ahead.

The Sonoran Desert is one of four deserts occupying parts of western North America. Forrest Shreve, who in the first half of the twentieth century spent his botanical career studying North American deserts, defined their characteristics. They are regions of low and unevenly distributed rainfall, low humidity, high air temperatures with great daily and seasonal ranges, and very high surface soil temperatures. Desert streams flow only sporadically, and the soil, low in organic content and high in mineral salt content, is subject to violent erosion by water and wind.

Most northern of the Western deserts is the Great Basin Desert, centered in northern Arizona, northern New Mexico, and Utah but with tongues that extend as far as Washington and Oregon. This desert lies mostly above 4,000 feet, occupying the valley floors between numerous mountain ranges. Precipitation is between four and 11 inches annually, usually in the winter, frequently in the form of snow. The Chihuahuan Desert, mainly in central Mexico but penetrating into southern New Mexico and extreme southeastern Arizona, is somewhat warmer because of its slightly lower elevation and more southerly position. Rain falls there mostly between mid-June and mid-September. In southern Nevada and adjacent California, the smaller Mojave Desert receives only two to five inches of rain, nearly all in late winter. It occupies terrain from sea level to 1,400 feet. Plant life in the Great Basin, Chihuahuan, and Mojave Deserts appears similar because of the many low shrubs.

The 120,000-square-mile Sonoran Desert covers much of central and western Arizona, southern California, and Baja California, with a strip on the east side of the Gulf of California. Although no single plant community ranges unchanged over such a large area, the Sonoran Desert looks different from the other deserts because of its tree-sized cacti and succulents. Two periods of rain occur each year, from December to mid-April and again from July to early September, with intervening periods of drought. Total annual rainfall averages from three inches in the driest parts to more than 20 inches. The aridity of the Sonoran Desert is maintained by high-pressure cells that prevail between 23° and 30° north latitude. In addition, mountains west of the Sonoran desert intercept much of the moisture in the air coming from the Pacific Ocean.

Until a decade ago, most scientists believed that the North American deserts had been around for many millions of years, but botanist Daniel Axelrod of the University of California at Davis, after studying the fossil floras in these regions, has concluded that in their current form, the deserts have existed for only about 10,000 years. Citing geological changes that have been brought about in part by movement along the San Andreas Fault and the opening of the Sea of Cortez, Axelrod has shown that the topography of the Sonoran Desert today is far different from what it was during most of the past 80 million years. Fossil evidence indicates that the vegetation went from savanna to dry tropical forest to short-tree forest to woodland chaparral to thorn forest before becoming the desert that it is today. These changes in plant life resulted from a progressive drying out of the area and a strengthening of the high-pressure system, which increased periods of drought.

The ancestors of many of the plants that live in the Sonoran Desert originally grew in cooler and moister climates. Their descendants apparently persisted by adapting to increasingly drier conditions. The succulents store large amounts of water; other plants have compact leaves or no leaves at all, thus reducing water loss. Some plants have extensive root systems that spread laterally beneath the plant. And some short-lived herbs sprout within days; following heavy rains in summer or winter. Their rapid growth, followed by early flowering and seed maturation, completes the life cycle within six to eight weeks, before the next drought begins. These ephemerals often spring up overnight in great masses, prompting aficionados to rush to the desert to see it in bloom. To look for these flowers at any other time is like hoping to stumble across the Lost Dutchman Gold Mine!

NATIONAL FORESTS IN CALIFORNIA

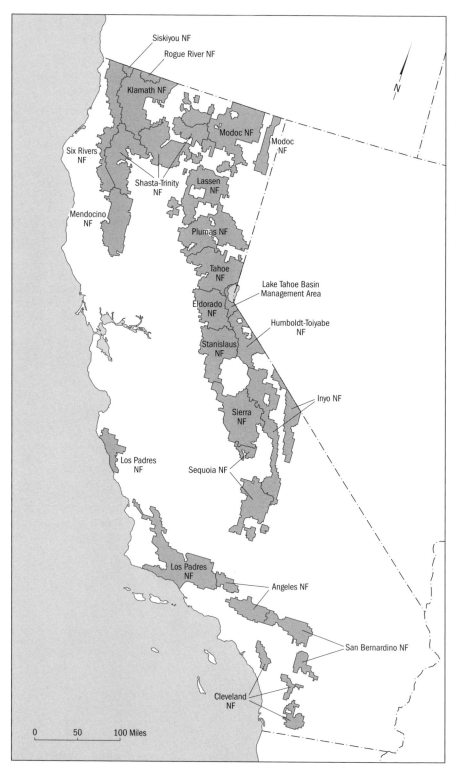

Siskiyou NF

Rogue River NF

Klamath NF

Modoc NF

Modoc NF

Six Rivers NF

Shasta-Trinity NF

Lassen NF

Mendocino NF

Plumas NF

Tahoe NF

Lake Tahoe Basin Management Area

Eldorado NF

Humboldt-Toiyabe NF

Stanislaus NF

Inyo NF

Sierra NF

Los Padres NF

Sequoia NF

Los Padres NF

Angeles NF

San Bernardino NF

Cleveland NF

N

0   50   100 Miles

The national forests of California are in U.S. Forest Service Pacific Southwestern Region 5; that office is located at 1323 Club Drive, Vallejo, CA 94592. California has 17 national forests and the Lake Tahoe Basin Management Area. The forests occupy 18.6 million acres and have 49 wildernesses.

Angeles National Forest

SIZE AND LOCATION: 654,000 acres in southern California, between Interstates 5 and 15; State Route 14 bisects the forest. Other access highways are Interstate 210 and State Routes 2 and 39. District Ranger Stations: Arcadia, San Fernando, Glendora, Saugus. Forest Supervisor's Office: 701 N. Santa Anita Avenue, Arcadia, CA 91006, www.r5.fs.fed.us/angeles.

SPECIAL FACILITIES: Boat ramps; swimming beaches; winter sports areas; off-highway vehicle areas.

SPECIAL ATTRACTIONS: Mt. Lowe Railway site; Crystal Lake Recreation Area.

WILDERNESS AREAS: San Gabriel (36,118 acres); Cucamonga (17,781 acres, partly in the San Bernardino National Forest); Sespe (219,700 acres, partly in the Los Padres National Forest); Sheep Mountain (41,883 acres, partly in the San Bernardino and Cleveland National Forests).

State Route 14 bisects the Angeles National Forest into two sections. The larger eastern part extends from the western part of the San Gabriel Mountains to State Route 14 and from Mt. Emma on the north to Interstate 210 on the south. It includes the San Gabriel Wilderness and part of the Cucamonga Wilderness. The main activities of the western section of the Angeles center around the large Bouquet Reservoir.

In the eastern portion of the forest, the Angeles Highway (State Route 2) wiggles and winds its way for more than 75 miles through the heart of the national forest. If you enter the highway at Sunland, you immediately find yourself in scenic Tujunga Canyon. The highway forks just beyond Big Tujunga Lake. The left fork, known as the Angeles Forest Highway, follows Mill Creek to the Mill Creek Summit Guard Station and then past Mt. Emma before exiting the forest. The right fork becomes the Angeles Crest Highway (still State Route 2) and snakes its way east through the San Gabriel Mountains. A side road leads to the Mt. Wilson Observatory. When you reach the Chilao Visitor Center, walk behind the building to observe the old but no longer used West Fork Ranger Station, the oldest standing ranger station in

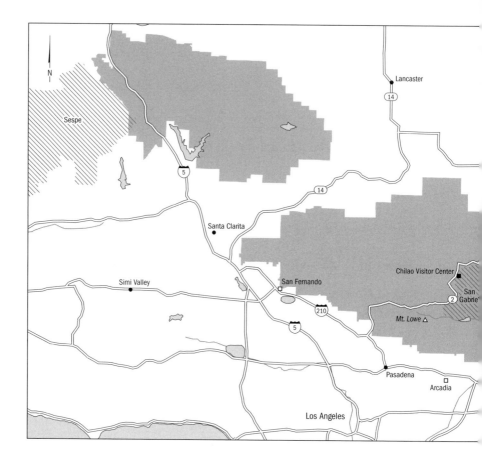

California. This tiny cabin was built in 1900 from alder and cedar logs for a cost of $70.00! The cabin originally stood 5.5 miles to the east but was moved to Chilao to protect it against vandalism.

Beyond Chilao the highway skirts the northern boundary of the San Gabriel Wilderness, passing such landmarks as Mt. Waterman and Kratka Ridge, both big ski areas. A pull-out provides a remarkable view from Dawson Saddle at the edge of Mt. Lewis. Five miles farther east, at Vincent Gap, take a rather hair-raising trail with multiple switchbacks to a forest of ancient limber pines on the north flank of Mt. Baden-Powell. Some of these pines are thought to be at least 2,000 years old. The Angeles Crest Highway continues for another five miles until reaching the boundary of the San Bernardino National Forest.

From near the Manker Flats Campground in the southeastern corner of the Angeles is a mile-long trail to the 500-foot, roaring San Antonio Falls. After admiring the falls, drive to Icehouse Lodge at the foot of Manker Flats for a longer hike along the Chapman Trail into Icehouse Canyon. Go as far

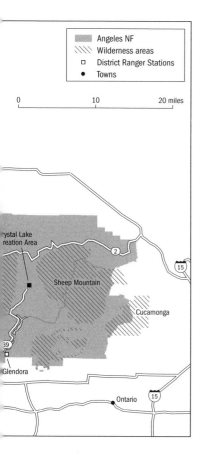

Angeles NF
Wilderness areas
□ District Ranger Stations
● Towns

0   10   20 miles

as the fork in the trail, then take the left fork that goes through a majestic stand of incense cedars in Cedar Canyon. You can smell the intoxicating scent of the cedars long before you reach the grove.

In the city of Pasadena, drive Lake Avenue north to Loma Alta. From here, walk the Sam Merrill Trail for two miles to the historic summit of Echo Mountain. During the 1880s, T. S. C. Lowe built two hotels on the mountaintop and then had an electric railway constructed to transport people to and from the Los Angeles area. Today you can hike along the 3.5-mile route of the railroad bed from Echo Mountain to Mt. Lowe.

One of the most popular long-distance trails is the 28-mile Gabrielmo National Recreation Trail. From Arroyo Seco Canyon just above the town of La Canada to the Chantry Flat Forest Station above the community of Sierra Madre, this rough and wild trail circles north of Mt. Disappointment and San Gabriel Peak, follows the West Fork of the San Gabriel River, and climbs over Newcomb Pass before passing to the west of Mt. Zion.

A great place for families is the Crystal Lake Recreation Area where there are several easy nature trails from which to choose. Rangers offer naturalist programs in the summer at the Yerba Santa Amphitheater.

Cleveland National Forest

SIZE AND LOCATION: 460,000 acres in southwestern California, from San Diego north to Corona and other suburbs of Orange and Riverside counties. Major access roads are Interstates 5, 8, and 15 and State Routes 74, 76, 78, and 79. District Ranger Stations: Alpine, Ramona, Corona. Forest Supervisor's Office: 10845 Rancho Bernardo Road, Suite 200, San Diego, CA 92127, www .r5.fs.fed.us/cleveland.

SPECIAL FACILITIES: Horse trails; off-road vehicle areas.

SPECIAL ATTRACTIONS: Sunrise National Scenic Byway.

WILDERNESS AREAS: Agua Tibia (15,933 acres); San Mateo (38,484 acres); Pine Creek (13,480 acres); Hauser (7,547 acres).

In contrast to most of California's national forests, which are found in the high mountain regions of the state, the Cleveland National Forest, for the most part, occupies much lower terrain. Monument Peak, at 6,271 feet, is the highest elevation in the forest. Chaparral is the common habitat type in the foothills and lower elevations in the Cleveland, with chamise, red shank, manzanitas, and madrones seemingly everywhere. As elevation increases, woodlands of coast live oak and California black oak surround meadows full

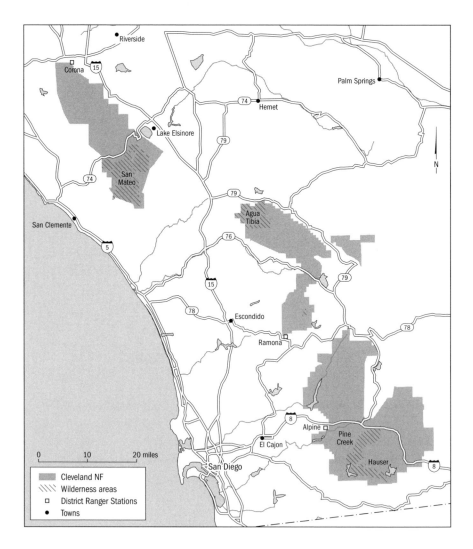

of wildflowers during the spring and early summer. At the highest elevations, Jeffrey pine is occasionally joined by the incomparable Coulter pine (fig. 9), whose foot-long, prickly and heavy cones are the largest of any pine in the world. The local name of "Widow maker" for this pine is appropriate if you were unfortunate enough to be hit by a falling cone. In addition to the typical plants of the area, the Cleveland National Forest is home to 60 sensitive plant and animal species and 22 endangered ones.

The forest comprises three major disjunct units. The Descanso District is

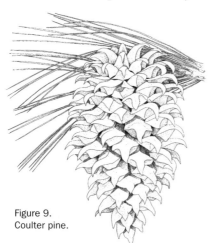

Figure 9.
Coulter pine.

just a few miles east of San Diego; the central Palomar District includes the famous Palomar Observatory (not a forest site); the Trabuco District goes north nearly to Corona and Fullerton. Each district has its own unique features.

For an all-inclusive introduction to the Descanso District, take the Sunrise National Scenic Byway, which meanders for 24 miles through the forest. From Pine Valley, near that community's exit on Interstate 8, follow State Route S1, the Sunrise Highway, north, eventually climbing over the Laguna Mountains. After driving through rolling countryside, you can stretch your legs on the short and easy Woodland Hill Nature Trail. Your reward is a nice assemblage of wildflowers beneath the trees. Next, drive up to the Burnt Rancheria Campground for one of the best trails in the Cleveland. The 1.2-mile Desert View Nature Trail at one point perches at the edge of the Laguna Mountains for a sensational view of the desert more than 1,000 feet below. On a clear day it is said you can see the Salton Sea from here. The highway skirts around one of the largest meadows we have ever seen. Laguna Meadow, once occupied by the Coast and Desert Indians, encompasses 900 acres. More than two dozen archaeological sites have been discovered and inventoried. You may wish to take the nearby Indian Hills Nature Trail. A short side road to the El Prado Campground brings you to the historic El Prado Cabin, a one-room log structure built in 1912 as the first ranger station in the forest. As you approach the northern end of the scenic byway, you come to the Noble Canyon Campground. From the campground is a 10-mile trail along the banks of Noble Creek that passes remnants of bygone gold-rush days. You may see what is left of a flume and several arrastras (old mills for pulverizing ore).

The centerpiece of the Palomar District is the Agua Tibia Wilderness. For the person who does not want to spend days hiking, there is a seven-mile Agua Tibia Trail from the Dripping Springs Campground on State Route 79 that leads to Agua Tibia Mountain. The trail goes through chaparral until it reaches the upper elevations of the mountain where there are forests of Jeffrey pine, red fir, and California black oak. From the top of Agua Tibia Mountain is the Palomar-McGee Trail that descends to Crossley Saddle before climbing to Eagle Crag, the highest point in the wilderness.

Several canyons mark the Trabuco District. The very rough and remote San Mateo Canyon and the area around it compose the San Mateo Wilderness. North of the wilderness is Juan Canyon, which is much easier to traverse. The 11-mile San Juan Loop Trail circles the canyon with a side trip to nifty Chiquito Falls. Because this area only receives about 20 inches of rainfall annually, the vegetation is mostly dry and desertlike with chaparral and coastal sagebrush common.

Eldorado National Forest

SIZE AND LOCATION: 676,780 acres in central California, east of Sacramento and southwest of Lake Tahoe. U.S. Highway 50 and State Route 89 serve the forest. District Ranger Stations: Pioneer, Georgetown, Pollock Pines, Camino. Forest Supervisor's Office: 100 Forni Road, Placerville, CA 95667, www.r5.fs.fed.us/eldorado.

SPECIAL FACILITIES: Boat ramps; swimming beaches; winter sports areas; off-highway vehicle staging areas.

SPECIAL ATTRACTIONS: Carson Pass Scenic Highway; Caldor Auto Tour.

WILDERNESS AREAS: Desolation (63,475 acres); Mokelumne (91,161 acres), partly in the Stanislaus and Toiyabe National Forests).

This is gold country!! After James W. Marshall discovered gold near Coloma in 1849, 50,000 wealth seekers rushed to the area now occupied by the Eldorado National Forest. Remnants of gold days abound, from abandoned mines to ghost towns to cemeteries nearly in ruins. The Eldorado National Forest is located between the foothills of the central Sierra Nevada and Lake Tahoe, taking in a range of elevations from 3,382 to 9,983 feet.

Hikers converge on the Desolation Wilderness for superb hiking. Because of its location between Reno, Lake Tahoe, and Sacramento, this wilderness area is the most visited per acre of any in the country. Although only 12.5

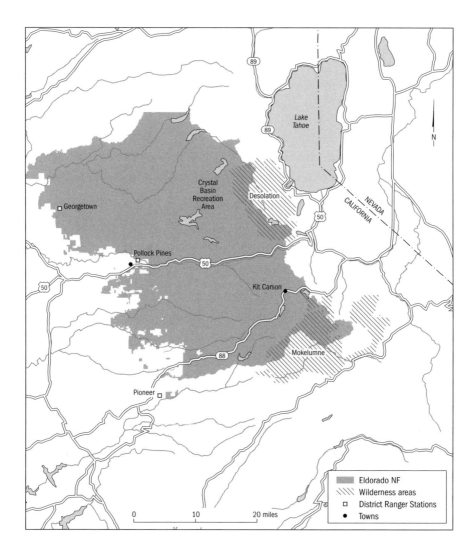

miles long and 8 miles wide, the wilderness contains alpine areas with stunted plants, mixed coniferous forests, and brushy chaparral. Numerous streams crisscross the wilderness, lined by button willow, sandbar willow, white alder, and cottonwood. About 130 lakes, many of them dotted with colorful water lilies and all of them full of fish, are in the wilderness area. Seventeen miles of the Pacific Crest Trail are laid out north to south through the wilderness, at one point climbing to Dicks Peak at 9,380 feet. The most popular trailheads are Wrights, Echo, Eagle Falls, and Fallen Leaf. Wrights Trail near the southwestern corner of the wilderness climbs over Rockbound Pass to China Flats. One nice trail begins from the Bayview Campground along State Route 89. From here you end up later at Sugar Pine Point State Park.

Along the way you pass Maggies Peak, the Velma Lakes, and climb over Phipps Pass.

Only a part of the Mokelumne Wilderness is in the Eldorado National Forest, but it is worthy of a visit. This wilderness is in a remote mountainous area that is bisected by the Mokelumne River. Mokelumme Peak at 9,337 feet is the tallest of several volcanic peaks in this wild area. Twenty miles of the Pacific Crest Trail wind through the wilderness. You might wish to take the 3.5-mile hike from the Woods Lake Campground to Fourth of July Lake. For the more rugged hiker, the deep and rugged terrain of Mokelumne Canyon is challenging.

A more accessible part and heavily visited region in the Eldorado is the Crystal Basin Recreation Area. The granite peaks covered by white and red firs and numerous pines provide ample opportunity to explore mountains. On the other hand, it is possible to just sit and relax around such crystal-clear lakes as Loon Lake, Upper Valley Reservoir, and Ice House Reservoir. All have campgrounds, picnic areas, and boat ramps.

Near Placerville, Traverse Creek flows through an area of serpentine rock. Although many plants find serpentine rock to be toxic, others seem to grow only on serpentine. The area along Traverse Creek is an easy place to study the vegetation.

The Caldor Auto Tour follows the old railroad bed of the California Door Company, which was in existence from 1884 to 1964. You can enter this 30-mile, one-way route over unpaved roads about 18 miles southeast of Placerville via the Pleasant Valley and Grizzly Flat Roads.

From its junction with U.S. Highway 50, State Route 89 southbound becomes the Carson Pass Scenic Highway. Near the pass at 8,850 feet are ski resorts and the pretty Silver Lake Campground. On the descent from Carson Pass, Mokelumne Peak can be seen off to the south.

Traverse Creek

Originating in the foothills of California's Sierra Nevada, rocky-bottomed Traverse Creek descends south for more than a mile through a tree-covered canyon and then passes gently through the middle of a shallow basin. The basin, carved out by the creek in ages past, is only slightly lower than the neighboring flat terrain but is easily distinguished by what grows there. Surrounded by a dense green forest of ponderosa pine, Douglas fir, and incense cedar, the basin itself contains only shrubs and a scattering of digger pines — often multitrunked trees with large, heavy cones and long, pendulous, gray-green needles. Because of its unusual vegetation, the basin's 200 acres are managed as a botanical special interest area by the Eldorado National Forest.

As I learned from forest botanists Mike Foster and Mark Williams, digger pine grows in the basin, and ponderosa pine, Douglas fir, and incense cedar do not because Traverse Creek is adjacent to deposits and rock outcrops made of the mineral serpentine. Geologists believe that serpentine rock, or serpentinite (named for its undulating, layered texture and mottled coloring), was first exposed in California about 150 million years ago. Today it is common enough to be California's state rock, covering many discontinuous areas for a total of about 1,100 square miles. It is most common in the South Coast Range, the North Coast Range, the Bay Area, and the western foothills of the Sierra Nevada.

The soil that forms when serpentine rock weathers is so low in some of the elements plants depend on—calcium, potassium, and even the molybdenum needed in trace amounts—that most plants cannnot survive in it. In addition, the soil is unusually high in nickel, cobalt, chromium, and magnesium, which are toxic to most plants. As a result, serpentine soils usually have a sparse cover of plants that can extract the minerals they need while coping with the toxic chemicals. For example, a wild mustard known as milkwort jewelflower, which grows only near Traverse Creek, can take up nickel in excess of 1,000 parts per million without any apparent harm. Other serpentine species of jewelflower and many other serpentine-tolerant plants take up nickel in modest amounts or exclude it altogether. Plants not found on serpentine soil, including some species of jewelflower, may die in soils containing only a few parts per million of nickel.

Arthur Kruckeberg, an authority on the botany of serpentine areas, notes that the vegetation in such relatively arid locales as Traverse Creek is made up of chaparral with a sprinkling of digger pines. In the Traverse Creek basin, the chaparral consists of four- to eight-foot-tall bushy shrubs, including four species in the buckthorn family—buckbrush, deerbrush, California coffeebush, and red inkberry—as well as leather oak and white manzanita. Most of these shrubs bloom in May. The manzanita is notable for bearing its little white, bell-shaped flowers at the tips of very sticky stalks. The stalks impede ants that might crawl to the flower in search of a pollen meal: instead, the pollen is reserved for the flying insects that pollinate the plants.

Because chaparral plants are adapted to arid terrain, all these shrubs have water-saving adaptations, such as small leaves, leathery leaves, or leaves with a whitish, waxy coating or hairy surface. Sometimes the microscopic openings, or stomata, in the leaves are sunk deep in the leaf tissue to further reduce evaporation. The leaves of the white manzanita, which are relatively broad, stand upright so that the rays of the midday sun fall obliquely on their surface.

Between April and June, many wildflowers bloom in scattered openings in the Traverse Creek chaparral. These colorful "serpentine flower fields" consist of low-growing species that are tolerant of serpentine soil, although many grow elsewhere as well. Most of these wildflowers are also drought tolerant; among them are a dwarf sedum with succulent leaves, a wiry buckwheat with a three-pronged flowering cluster, Sanford's wild onion, Congdon's lomatium, and the brilliant, rose-pink bitterroot. One species found only at Traverse Creek is the rare Layne's groundsel.

Some moisture-loving plants inhabit shallow depressions that accumulate water when it rains. Among them are yellow monkey-flower, which has five bright-yellow petals; bicolored monkey-flower, with two white petals and three yellow petals; pink-flowered whisker brush, with five pink petals and a rosy center above a tuft of short, slender, green leaves; a yellow violet; a two-inch-tall wild white clover; and an equally small native plantain.

Seeming anomalies at Traverse Creek, not far from the visitors' parking area, are a few large ponderosa pines and an incense cedar. According to the forest botanists, enough nonserpentine soil has washed down from higher terrain to create a foothold for these conifers.

Inyo National Forest

SIZE AND LOCATION: 2,000,000 acres east of the Sierra Nevada Crest in central California. U.S. Highway 395 divides the western side of the forest from the eastern side. State Routes 120, 158, 168, and 203 also service the forest. District Ranger Stations: Lee Vining, Mammoth Lakes, Bishop, Lone Pine. Forest Supervisor's Office: 351 Pacu Lane, Bishop, CA 93514, www.r5.fs.fed.us/inyo. Mono Basin Visitor Center: P. O. Box 429, Lee Vining, CA 93541, (760) 647-3044.

SPECIAL FACILITIES: Boat ramps; visitor centers; ski areas.

SPECIAL ATTRACTIONS: Bristlecone National Scenic Byway; Mono Basin National Forest Scenic Area; Mt. Whitney; Lee Vining National Scenic Byway.

WILDERNESS AREAS: Hoover (48,611 acres, partly in the Humboldt-Toiyabe National Forest); Ansel Adams (280,258 acres, partly in the Sierra National Forest); John Muir (580,293 acres, partly in the Sierra National Forest); South Sierra (60,084 acres, partly in the Sequoia National Forest); Golden Trout (303,511 acres, partly in the Sequoia National Forest); Boundary Peak (10,000 acres); Inyo Mountains (205,000 acres, partly on Bureau of Land Management land).

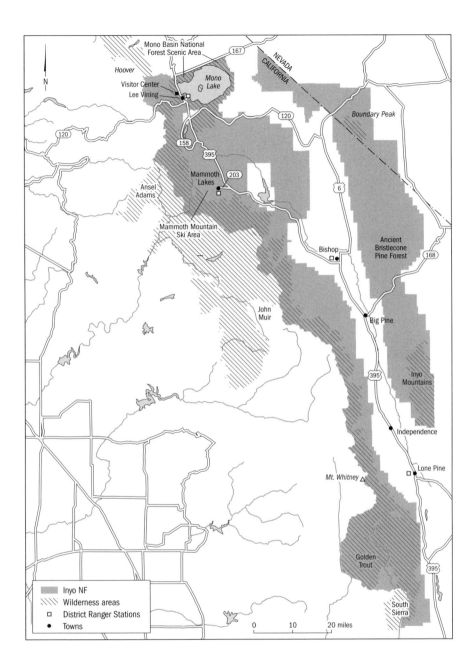

After studying a map of the Inyo National Forest, the quandary is where to go first. The forest is adjacent to Death Valley National Park, the lowest elevation in the country, and extends to Mt. Whitney, the highest elevation in the lower 48 states. The forest has numerous natural features, from the remains of volcanic activity to the oldest living trees in the world to California's greatest concentration of bighorn sheep (pl. 15) to Mono Basin. Historical sites in the forest range from earlier Indian occupation to remnants of past mining and grazing operations.

You may wish to start your adventure near the southern end of the forest by taking State Route 168 from Lone Pine and heading east across the broad Owens Valley. After about eight miles you enter the Inyo National Forest and begin to climb into the White Mountains. At Cedar Flat, branch off State Route 168 and onto the Bristlecone National Scenic Byway that leads to the Forest of the Ancient Bristlecone Pines. After a brief stop at Sierra View for an outstanding panorama of the Sierra Nevada Crest, including Mt. Whitney, you come to the Schulman Grove at 10,000 feet elevation. Although it may be cold and blustery, hike the Methuselah Loop Trail, which takes you through a forest of these gnarly, ancient wonders, including Methuselah, reputedly the oldest of these bristlecone pines, estimated to be almost 4,700 years old. If you want to see more, drive another 12 miles north to the Patriarch Grove, at 11,000 feet, and take another loop trail. You will be rewarded by seeing the Patriarch, biggest of them all with many trunks and a total overall circumference of nearly 38 feet, although only 40 feet tall.

After coming back down to earth, take the County Road (Big Pine Creek Road) west out of Big Pine and follow Big Pine Creek to Glacier Lodge at the edge of the John Muir Wilderness. A steep two-mile trail leads from the lodge and goes just inside the wilderness boundary to a gabled cabin made out of granite rocks. This was the summer home of one of Hollywood's scariest actors, Lon Chaney. The cabin appeared in the movie about Chaney's life, entitled *Man of a Thousand Faces.*

Next, drive to Bishop and take the western part of State Route 168. You essentially follow Bishop Creek and skirt Table Mountain to the road's end at the beautiful Lake Sabrina on the border of the John Muir Wilderness. On your return to Bishop, take a side road that follows the course of the South Fork of Bishop Creek to a lovely high mountain lake called South Lake, which lies at the base of the Inconsolable Range.

West of Independence is a Forest Service road that passes the wildflower-laden Grays Meadow before making its way to Onion Valley Campground at the foot of Kearsarge Peak. The most hardy persons may take the trail to the peak, which is in the John Muir Wilderness and in the heart of the largest concentration of California bighorn sheep in the state.

At Lone Pine, the most popular County Road (Whitney Portal Road) crosses the rocky Alabama Hills (not in the forest proper)—used in many Western movie sets—on its way to Whitney Portal, the jumping-off place for a hike to the top of 14,495-foot Mt. Whitney. If you do not wish to start the climb to Mt. Whitmey, just enjoy the picnic grounds and listen to the gushing sound of Lone Pine Creek as it plunges over granite boulders on its way from the flanks of Mt. Whitney to lower destinations. Persons climbing to the summit of Mt. Whitney (pl. 16) and back can make it in two days, usually stopping to camp in the vicinity of Mirror Lake, but you need a permit and reservations to make the climb. The first half-mile of the trail is steep enough to discourage many would-be mountain climbers.

Follow U.S. Highway 395 north from Bishop to Hot Creek Geological Site. Along Hot Creek, boiling water bubbles up from the creek bed, and there are fumaroles and geysers, as a chamber of hot magma lies about three miles below the surface of the earth here. After seeing this fascinating geological area, travel a few miles on State Route 203 to the resort community of Mammoth Lakes. Between Hot Creek and Mammoth Lakes, near the Sherman Creek Campground, is the three-chambered Mammoth Creek Cave carved out of olivine basalt. There are petroglyphs on the back wall of the cave probably carved by ancestors to the Paiute Indians. The cave is accessible by ranger-led tours.

After regrouping, take a drive south of Mammoth Lakes on the Lake Mary Road. The Panorama Dome Trail to the east is worth taking. The road comes to several sparkling mountain lakes, each with a campground. In addition to the large Lake Mary, there is Lake George and the smaller Crystal Lake farther south. Crystal Lake lies at the foot of 10,377-foot Crystal Crag. A nice hiking trail leads from the parking area above Lake George to Crystal Lake. If you continue to the end of the road, there are hiking trails to Heart Lake and Emerald Lake and an interpretive trail to the Mammoth Consolidated Mine, a gold mining camp from the 1920s.

After returning to Mammoth Lakes, continue west on State Route 203. Your first stop may be at a crack in the rock known as Earthquake Fault. This is not really a fault but a huge cooling crack 10 feet wide and 60 feet deep. A trail around the crack goes through a majestic forest of Jeffrey pine and red fir. Then drive to the Mammoth Mountain Ski Area, where you have to leave your car. At this point you need to board a shuttle bus to the Minaret Summit Vista, Agnew Meadows, Pumice Flat, Minaret Falls, Sotcher Lake, and Rainbow Falls. Devil's Postpile National Monument is along the way. At the Minaret Summit Vista, you may hike a short trail to get spectacular views of the breathtaking serrulated peaks of the Minarets to the west. All around are wildflowers covering the meadow. A little farther on, you will cross the Pacific Crest Trail near Agnew Meadows Campground. Anyone wanting a face-to-face view of the Minarets will have to hike to them from here. Agnew Meadows has a nice wildflower walk. Beyond Agnew Meadows, the shuttle takes you south through an area of pumice at Pumice Flat. From Pumice Flat is a short trail to Minaret Falls. Nearby is the Devil's Postpile National Monument (administered by the National Park Service). While still in the Inyo National Forest, just east of the Devil's Postpile, you may wish to hike the pleasant 1.25-mile trail that completely encircles Sotcher Lake. At the end of the shuttle bus route is the 1.25-mile trail to spectacular Rainbow Falls. The falls drops 101 feet into a gorge from the San Joaquin River.

From Mammoth Lakes north to Lee Vining, via U.S. Highway 395, you will be able to enjoy one geological phenomenon after another. Pumice starts showing up everywhere. A short side road will bring you to Inyo Craters where blue water has filled one depression and milky green water a second one. Just north of Inyo Craters off to the east of U.S. Highway 395 is a short road to Lookout Mountain, a rhyolite dome. During the eruption of the Long Valley Caldera about 760,000 years ago, obsidian glass flows covered the sides of Lookout Mountain, and the obsidian was quarried from here for many years. Then another rough side road off of U.S. Highway 395 will bring you to the fabulous Obsidian Dome where you will ogle the shiny black, glasslike obsidian. A little farther north is another crater called the Devil's Punchbowl.

The ash-gray summits of the Mono Craters, relatively modern volcanoes, loom up to the east of the highway. The roads leading toward the craters are through fields of pumice, and it is illegal to drive your vehicle off the roadway.

At Lee Vining, State Route 120 becomes the Lee Vining Canyon National Scenic Byway as it follows a circuitous route to Yosemite National Park, going between Hoover Wilderness (pl.17) to the north and Ansel Adams Wilderness to the south. Before getting to Yosemite, stop and take the short, interpretive nature trail to Tioga Tarns. You can see glacial ponds (tarns) and the vegetation that surrounds them.

Just before reaching Tioga Pass and the entrance to Yosemite, take the mile-long trail to the Bennettville Silver Mining Town on the crest of the Sierra Nevada. Although silver has not been mined here since 1889, several foundations of buildings may be seen, along with a large barn, the reconstructed assay office, various pieces of mining machinery, and the entrance to the Great Silver Tunnel.

If you are exploring the Inyo from south to north, you will come to one of the highlights of the forest. For years visitors have been attracted to Mono Lake to see the tufa formations that rise up out of the water and along the shoreline. In 1984, the area was designated the Mono Basin National Forest Scenic Area, a designation that gave additional protection to the geology and ecology of the area. Before starting on your exploration of these not-to-be-forgotten phenomena, stop at the modern Visitor Center just north of Lee Vining for orientation and a chance to purchase books and maps. If time is limited, drive to the South Tufa area and take the easy mile-long interpretive trail through dry tufa towers and past other tufa towers that are in the water. The black ring at the edge of Mono Lake is due to millions of alkali flies, but they never land on humans. This is a fee area.

While you are in the Mono Lake area, visit the mystical Piagi Park located just off of Forest Road 1S05 (Bald Mountain Lookout Road). This was an important site of the Mono Lake Paiute Indians. Here the Indians collected larvae of the Pandora moth, called piagi. The larvae were collected by digging trenches around the bases of Jeffrey pines. The piagi were then roasted and eaten. Nearby is a branch and brush structure known as a wikiup, which was used to store the piagi.

Ancient Bristlecone Pine Forest

During each growing season, a tree adds a layer of wood to its trunk and branches, creating the annual rings we see in a cross section. The new wood cells arise from the living tissue that lines the bark. In the spring, when it is moist, the tree devotes most of its energy to growth; at this time, the new cells are relatively large, but as summer progresses, they get smaller and smaller. In the fall, the wood cells die, and no new cells are added until the following spring. The contrast between the last layer of small cells from one year and the first layer of large cells the next year is usually strong enough to make ring counting rather simple.

Early in the twentieth century, A. E. Douglas of the University of Arizona observed that a wide ring, at least in some species of trees, indicated that the wood was formed during a wet year, whereas a narrow ring denoted a dry year. Archeologists, botanists, and foresters used this discovery to develop the science of dendrochronology—the dating of past events or climatic changes by the study of growth rings. Climate patterns in specific geographic areas can be ascertained by examining ancient living trees, and if suitable fragments of wood that overlap in age are found, the procedure can be extended back in time. Samples of unknown age can then be dated by matching their growth patterns with the established sequence of tree ring growth. For example, dendrochronologists have dated Pueblo Bonito, a prehistoric Indian settlement in New Mexico, back to the seventh century by analyzing the wood used in its construction.

In the 1950s, hoping to determine exactly when ancient droughts occurred in the Southwest, wood anatomist Edmund Schulman of the University of Arizona began a systematic search for the region's oldest trees. He knew that the oldest California redwoods were about 2,200 years old and that some western junipers in Yosemite were nearly 2,500 years old. Some giant sequoias have been living for more than 3,000 years; the famous General Sherman Tree, for example, may be as much as 3,500 years old. But Schulman wanted to find still older trees. In 1953, while exploring the White Mountains on the California-Nevada border with Frits Went of the California Institute of Technology, Schulman stumbled onto a promising grove of weathered, gnarled bristlecone pines (pl. 18), most of which were clinging to life by a mere ribbon of living bark.

For the next several years, Schulman studied the growth rings of these trees, which looked like living driftwood. Using a boring tool, he removed thin cylinders of wood from the trunks and counted their rings. (Although this procedure leaves a small opening, the pine's capacity to exude resin quickly into the wound prevents subsequent decay.) The rings were minuscule, and counting them was no easy matter. Some of the trunk tissues had eroded away, and those rings that were left usually were not in neat concentric circles. The band of living tissue in an old, eroded stem sometimes grew at right angles to its original rings. Schulman had to match the rings of one tree with those of others that had the missing section or that were undistorted. His efforts were amply rewarded, however, as the remarkable ages of some of these trees began to emerge. The first tree Schulman found that was more than 4,000 years old he called Pine Alpha; later he identified a 4,600-year-old denizen and christened it Methuselah. (During the 1960s, forester Donald Currey discovered a bristlecone pine near Wheeler Peak in eastern Nevada that is more than 4,900 years old!)

Several years ago I went to the Ancient Bristlecone Pine Forest, part of California's Inyo National Forest, to see Schulman's trees for myself. A Forest Service parking area is provided for the well-kept hiking trail known as Methuselah Walk, which circles past many of the oldest trees, including Pine Alpha and Methuselah. (I found that a leisurely pace was advisable to avoid exhaustion at the 10,000-foot elevation.) At the time I thought that any organism that lived longer than the norm had to have optimal conditions going for it. For plants, these conditions would be adequate moisture, moderate temperature, continuous sources of nutrients, and protection from severe weather. When I stood looking at Methuselah, I knew I had been wrong. The cold was bone chilling (even in July); the patchy pockets of soil showed little evidence of moisture, and the wind battered my body with repeated gusts.

The bristlecones, along with the similar limber pines, dominate a subalpine coniferous forest that extends from about 9,500 feet up to timberline (about 11,800 feet), mostly within the western Rocky Mountains. The White Mountains, where Schulman found his trees, lie in the "rain shadow" of the towering Sierra Nevada, which are to their west. Cyclones, moving inland from the Pacific Coast, shed their moisture before reaching the west slopes of the White Mountains, leaving only strong, dry winds that whip across the mountainsides. Rainfall is less than 12 inches per year.

The greatest concentration of very ancient trees is near the crest of a ridge, in exposed patches of dry, rocky, dolomitic limestone. The limestone is relatively cool because the light gray or even white ground surface reflects more solar radiation than nearby outcroppings of sandstone or granite. In addition, nutrients are readily leached from the porous dolomite. Yet under these harsh conditions, the trees here survive to extreme age. Mature pines that grow on sandstone or in protected areas are neither as misshapen nor as old.

The longevity of the pines appears to depend on their ability to develop and maintain a special growth habit. Trees 1,500 years old or older usually have only a narrow strip of living bark on the trunk and only a few living twigs and needles on one remaining live branch. The reduction of living tissue to a single strip results partly from weathering and abrasion by soil particles that remove the bark from the upper surface of the roots and partly from repeated loss of foliage and twigs on the windward side of the tree. These old trees grow exceptionally slowly, even during their very first year. Most of them produce only an inch or less of growth rings in a century. Schulman found one gnarled tree that was only three feet tall and had a trunk diameter of three inches; it was 700 years old.

Bristlecone pines that receive more moisture, more nutrients, and greater protection show comparatively fast growth during their early decades. These "normal" trees produce an increasing volume of new wood each year to maintain a constant annual increase in thickness around their symmetrical trunks. Like most trees, they become less efficient at absorbing and metabolizing carbon dioxide as they age. Consequently, they become uniformly sickly and die. But although most of an old bristlecone pine has already died over the centuries, a small living part of it remains healthy. Such a tree is able to maintain a constant ratio of living to nonliving tissue almost indefinitely with little expenditure of energy. During years with severe weather, minimal metabolic activity is required to keep the tree alive.

The durability of the exposed, dead wood contributes to the tree's longevity. In most trees, dead tissue ultimately rots and the tree falls, but in the old bristlecones, the nonliving wood remains sound throughout the life of the tree. This resistance may be due to the (as yet unexplained) greater than usual concentration of resin canals in the trunk and to the absence of wood-destroying fungi, which apparently cannot tolerate the cool, dry environment where old bristlecones grow. Fast-growing bristlecone pines in less rigorous habitats tend to develop heart rot after reaching a diameter of 15 inches or so.

Other favorable conditions may also help explain the longevity of the bristlecone pines. In the oldest groves, the trees are widely spaced, reducing or eliminating competition of tree roots and crowns. And the dolomite retains more moisture than either sandstone or granite. Consequently, the timberline on dolomite does not begin until about 11,800 feet, whereas the growth of trees on sandstone and granite ceases at 11,000 feet. Finally, at high elevations the dolomite that underlies the bristlecones supports very little understory vegetation, thus reducing the possibility of ground fire.

A few miles north of Methuselah Walk is another impressive group of bristlecone pines—Schulman Grove. The Patriarch, the largest bristlecone pine in the world, grows here. Thirty-eight feet wide and 40 feet high, the Patriarch is a tree to behold, yet scientists say it is a mere youngster, only about 1,500 years old.

At this point in history, the ancient bristlecones are a phenomenon, but experts predict that most of the old pines will succumb in a few centuries. Trees like the General Sherman sequoia, which, unlike the bristlecones, live under apparently optimal conditions, seem destined to become the longevity champions. At about 3,500 years, General Sherman is in the prime of life, showing little sign of deterioration. If conditions remain favorable, it may survive for another 3,000 years.

Mono Basin

As we drove over the crest at Conway Summit on U.S. Highway 395 a few miles north of Lee Vining, California, a large silver-looking lake filled the valley before us. As we got closer to the lake, grotesque, gray formations appeared in the lake and along the edges of it. In the background were gray-topped mountains that looked like they had just been dusted by snow. In actuality, the gray was ash and pumice from fairly recent volcanic activity, at least in geological time. We had come to the Mono Basin. The deep depression in the center of the basin was filled by the water of Mono Lake. The ashy mountains in the distance were the Mono Craters. The mystic figures in and around the lake were tufa towers (pl. 19), incredible formations of calcium carbonate.

This area has been one of fascination from the time that Paiute Indians lived here to the time of the California emigrants of the nineteenth century to the eager tourists of today. Because of the shrinking of the lake and the deterioration of the tufa towers, the federal government designated the lake and its surroundings as the Mono Lake Scenic Area, under the care and management of the U.S. Forest Service. A modern visitor center overlooking the west side of the lake is a good starting point to learn about the region and purchase maps and books necessary for in-depth exploration.

Mono Lake is salty and alkaline—2.5 times saltier than seawater and 80 times more alkaline. Because the region receives on the average only 10 inches of precipitation each year, mostly in the form of snow, the water in the lake comes from streams that originate high up in the Sierra Nevada to the west. These mountain streams contain salts and minerals that enter the lake. Because Mono Lake has no natural outlet, the water eventually evaporates, and the salts and minerals are left behind. The water of the lake has a very soapy feeling because of the high alkalinity and is so strong that soiled clothing easily becomes clean in it.

Three or four million years ago, as the Sierra Nevada was uplifting, the western floor of the basin tilted downward, and the southern and northern parts tilted inward, forming a large, bathtublike depression. As water drained from the mountains into the basin, Mono Lake was formed. About 12,000 years ago, at the end of the last Ice Age as the glaciers melted, Mono Lake filled to overflowing. Eventually the water of the lake covered 358 square miles and had a depth of 900 feet.

Calcium in the water from freshwater springs beneath the lake began to mix with carbonates from the salts that had been precipitated during evaportaion. The result was the formation of solid structures of calcium carbonate, or tufa, many of these below the surface of the lake but some of them protruding above. The tufa towers formed around the mouth of each spring. As the level of the lake lowers, more towers become exposed, and others are left stranded on the shore. Out of the water, the tufa towers and spires cease to grow. Some of the tufa towers on Mono Lake's most ancient shores are estimated to be more than 10,000 years old. Those you see today in the mysterious South Tufa area are 200 to 900 years old.

On the north shore of Mono Lake, an area known as Black Point began to erupt beneath the lake more than 10,000 years ago and protruded above the surface of the water. As its top cooled and contracted, it developed narrow crevices more than 50 feet deep. Today you can walk among the fissures of Black Point, with the cliff walls towering above you. Tufa towers are in the crevices that remain from a time when the tip of Black Point was still under water.

In 1941, the city of Los Angeles, in an effort to obtain a sufficient water supply for its burgeoning population, began to divert the water from four of the five major streams that enter Mono Lake. Because of this, Mono Lake dropped 40 feet and doubled in salinity. Its surface area was reduced to 60 square miles. After considerable and often heated debate between Los Angeles officials and people concerned about the future of Mono Lake, a measure was passed on September 28, 1994, that protects the lake and the streams that feed it. It is thought that within 20 years or so, the lake will rise approximately 17 feet.

Although sometimes referred to as a Dead Sea, Mono Lake is far from that. Trillions of organisms are in the lake, although most of them belong to three species. Most plentiful are one-celled green algae. In winter, when the algae reproduce, the lake turns a pea soup green color. The algae are devoured by the other two organisms in the lake, a brine shrimp and a brine fly. The brine shrimp are about half-an-inch long, and it is estimated that 4 trillion of them live in the lake from April to October. During the summer, the brine shrimp lay their eggs, which then overwinter on the lake bottom. In spring, the eggs hatch and the cycle begins again. In the meantime, brine flies land on an air bubble in the lake and float to tufa where they lay their eggs. Before becoming an adult, the brine fly goes through the larva and pupa stages. The Paiute Indians who lived in the area used the pupa stage for food. When we visited Mono Lake, we saw a ring of black at the edge of the shoreline. These are the brine flies.

During spring and summer, the shrimp and flies are a major source of food for 80 species of migrating birds. Each year, an estimated 800,000 eared grebes, 150,000 Wilson's phalaropes, 50,000 California gulls, and 400 snowy plovers visit Mono Lake. Red-necked phalaropes stop off at Mono Lake on their way to their winter home in South America.

The soil immediately around the lake is very salty and alkaline, containing great amounts of calcium, magnesium, and carbonates. Only plants that can tolerate these conditions can grow here. Most of the salt-tolerant species have deep tap roots to reach fresh water far below. Other plants have succulent leaves or glands that excrete salt. Trees cannot grow here, but two shrubs, rabbitbrush and greasewood, are common. Saltgrass is a mat-forming species, but protruding above the mat are smotherweed or bassia, stinkweed, tansy mustard, and Russian thistle. The attractive desert buttercup is also here.

In other areas, particularly around Panum Crater on the south shore, is a sagebrush scrub community. The plants in this community grow in pumice and volcanic ash. They also have long taproots to reach water far underground. The shrubs that grow here are often spaced some distance apart and include big sagebrush, bitterbrush, shrubby ambrosia, spiny hopsage, and desert peach. The peach, which does not have an edible fruit for humans, has attractive pink blossoms.

A couple of wetlands around the lake have freshwater substrate. These marshes are home to a variety of sedges and rushes as well as cattails. Growing among the tufa in the marshes are hoary nettle, which often surrounds each tower, willow herb, Missouri iris, horsetails, docks, groundsels, marsh paintbrush, and an aquatic speedwell. The marshes are excellent areas to see Virginia rails, swallows, red-winged blackbirds, and yellow-headed blackbirds.

Along the streams that enter Mono Lake are black cottonwood, quaking aspen, coyote willow, and lodgepole pine, and the shrubby wood rose, red osier dogwood, buffalo berry, and mugwort. The night-blooming Hooker's evening primrose is a common wildflower.

On the mountains around the lake are piñon pine–juniper woodlands. Associated with these dominant species are curl-leaf mountain mahogany, Mormon tea, serviceberry, and desert ceanothus.

Klamath National Forest

SIZE AND LOCATION: 1,726,000 acres in northern California, with a tiny extension into Oregon. California's State Route 96 meanders diagonally through the forest; California' State Route 299 provides access to the southern end of the forest. District Ranger Stations: Fort Jones, Macdoel, Happy Camp. Supervisor's Office: 1312 Fairlane Road, Yreka, CA 96097, www.r5.fs.fed.us/klamath.

SPECIAL FACILITIES: Boat ramps; winter sports areas.

SPECIAL ATTRACTIONS: Four Wild and Scenic Rivers.

WILDERNESS AREAS: Russian (12,000 acres); Marble Mountain (242,500 acres); Siskiyou (152,680 acres, partly in the Six Rivers National Forest); Trinity Alps (502,764 acres, partly in the Shasta-Trinity and Six Rivers National Forests and 4,623 acres on Bureau of Land Management land).

High mountains, narrow ridges, steep-sided canyons, turbulent rivers, and tranquil lakes highlight the Klamath National Forest. The mountains are clothed by continuous stands of ponderosa pine, Jeffrey pine, Douglas fir, red fir, and incense cedar. White fir grows in higher elevations. The Forest Service estimates more than 175,000 acres of virgin forest are in the Klamath. Much of the forest is designated wilderness, but there is still plenty of room for nonwilderness activities. The Klamath, Salmon, Trinity, and Scott Rivers drain the forest and provide some of the best fishing for steelhead, salmon, and rainbow trout. One hundred fifty-two miles of these rivers have been designated as Wild and Scenic Rivers.

Begin your exploration of the forest by driving twisting State Route 96 diagonally across the forest, always staying near the Klamath River. Start from the eastern side of the forest at Tree of Heaven Campground and leave the forest at the tiny community of Somes Bar. At Somes Bar is the confluence of the Klamath and Salmon Rivers. Other small towns along the way are Seiad Val-

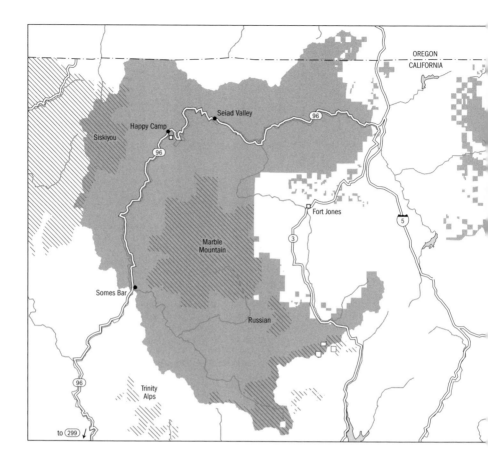

ley and Happy Camp. You can put your boat in the river at several places along the highway. Many of the gravel bars you see along the river are reminders of gold-panning days. Off of State Route 96 are other interesting things to do. One is to take a short hike off of side road 48N20 to impressive Horsetail Falls; one is to hike a circular trail to the summit of 8,990-foot Cedar Mountain; another is to drive a scenic back road for six miles from Laird's Camp over Gold Digger Pass to the Lava Beds National Monument, taking a quarter-mile side trip to lovely Panhandle Lake. For geological excitement, drive down west of Medicine Lake to see the shiny Crater Glass Flow, and a little farther on to the Callahan Lava Flow, with ice caves and lava tubes. Hole-in-the-Ground is a large volcanic depression west of Mt. Hebron.

A beautiful lake outside the wilderness areas is Juanita Lake, which is about 1.5 miles south of the Oregon border. From the campground here you can hike to Ball Mountain. Ornithologists flock to Butte Valley in October for the peak of the bird migration along the Pacific Flyway.

Marble Mountain Wilderness has the most spectacular mountain scenery

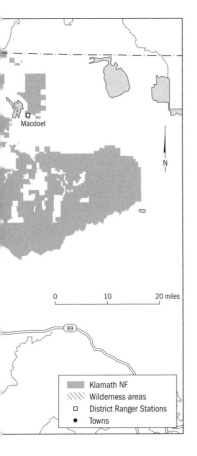

with 89 lakes and numerous wildflower meadows interrupting the forest cover. Marble Mountain with its white and gray marblelike rocks is in the northern part of the wilderness. It is composed of marine invertebrate fossils. Contrasting with the multicolored rocks are verdant meadows and deep blue lakes. Lakes for every taste are here, from half-acre Jewel Lake to 67-acre Ukonom Lake; from Gem Lake, which is only three feet deep, to Maneaten Lake, which has a depth of 112 feet; from Hooligan Lake at an elevation of 5,150 feet to Heather Lake at 7,400 feet. Impressive Chimney Rock, at the southern edge of the wilderness, towers above the adjacent Dollar, Crapo, and Morehouse Meadows. A hike from Chimney Rock to Diamond Lake permits you to see one of the loveliest trees in the forest: the silver fir (fig. 10). On Boulder Peak you will find whitebark pine and foxtail pine.

Russian Wilderness, dominated by Russian Mountain at 8,196 feet, is an area of steep slopes and U-shaped glacial cirques. This region perhaps has the greatest concentration of coniferous plants anywhere in California. Twenty species have been recorded.

The Siskiyou Wilderness is home to the legendary Bigfoot. Clear Creek National Recreation Trail crosses this wilderness for 20.5 miles.

The Trinity Alps Wilderness has the headwaters of the West and South Forks of the Trinity River and the West and South Forks of the Salmon River, all designated Wild and Scenic Rivers.

Figure 10. Silver fir.

Lake Tahoe Basin Management Area

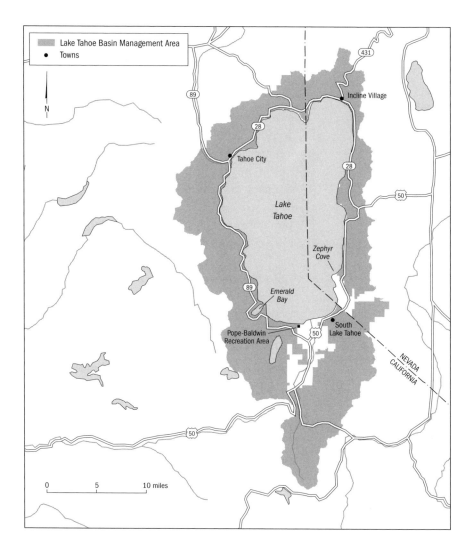

Legend:
- Lake Tahoe Basin Management Area
- • Towns

N

Incline Village

431

89

28

Tahoe City

Lake
Tahoe

28

28

50

Zephyr
Cove

89

Emerald
Bay

South
Lake Tahoe

50

Pope-Baldwin
Recreation Area

NEVADA
CALIFORNIA

50

0   5   10 miles

LOCATION: In west-central California, mostly around the southern half of Lake Tahoe. U.S. Highway 50 and State Routes 28 and 89 are the major access roads. Area Office: 35 College Drive, South Lake Tahoe, CA 96150, www.r5.fs.fed.us/ltbmu.

SPECIAL FACILITIES: Boat ramps; swimming beaches; visitor center.

SPECIAL ATTRACTIONS: Tallac Historic Site; MS *Dixie II* riverboat.

When explorer John Fremont came over a mountain ridge and saw the majestic blue water of Lake Tahoe in the vast basin below him in 1844, he was the first non-Indian to record this remarkable body of water. The Washoe Indians had known and used the lake and its surrounding forests for many decades, but Fremont's discovery was soon to open the area to masses of recreation-hungry people. It was not until the 1930s that a paved road completely encircled the lake. Now, dozens of roads, particularly along the southern shore, are crowded with people and buildings. Only about 5,500 permanent residents live here, but many people come from all over to ski in the adjacent mountains, hike in nearby forests, and just simply to enjoy the beautiful lake. Casinos on the Nevada side of the south shore have added to the congestion.

In the early 1940s, all of the land adjacent to the lake was either private property or in state ownership. Since that time, the U.S. Forest Service has acquired some lakefront property at Zephyr Cove, in the Pope-Baldwin Recreation Area at South Lake Tahoe, and along State Route 28 on the road toward the Mt. Rose junction.

It is the super quality of Lake Tahoe's water that attracts visitors. The water is said to be as pure as distilled water, about 97% pure. It is so clear that on a typical still day, a white disk the size of a 12-inch dinner plate can be seen at a depth of 60 feet in the lake. The surface water of the lake is at an average elevation of 6,225 feet, making it one of the highest large mountain lakes in the country. Stretching about 21.5 miles north to south and 12 miles east to west, the lake contains approximately 4 billion gallons of water, enough to cover a flat area the size of California to a depth of 14 inches. During the summer, the upper 12 feet of the lake forms a layer that may warm up to 68 degrees F. Below 700 feet and during the winter, the water remains a constant 39 degrees F. The deepest part of Lake Tahoe is 1,645 feet, making it the third deepest lake in North America after the Great Slave Lake in Canada (2,010 feet deep) and Crater Lake in Oregon (1,930 feet deep). Sometimes Emerald Bay (pl. 20) and some of the protected inlets develop a thin layer of ice in winter, but the entire lake has never frozen over. The outlet of the lake is the Truckee River, which flows through Reno and empties into Pyramid Lake in Nevada.

About 150 million years ago when the Sierra Nevada was being uplifted, land between two ridges sank to form a trough that filled with water where Lake Tahoe is today. During intense volcanic activity nearly 50 million years ago, lava poured into the trough and dammed up the north end, causing the water to rise several hundred feet above the present-day level. The overflow flowed out through a channel east of Mt. Pluto, one of the most active volcanoes. As recent as 25,000 years ago, glaciers on the surrounding mountains gouged out huge U-shaped valleys such as those occupied today by Emerald

Bay. The glacial deposits that were left behind blocked the original outlet of Lake Tahoe.

During the early 1900s, wealthy families built impressive vacation homes on the banks of Lake Tahoe. Three of these estates were acquired by the Forest Service between 1965 and 1971 and have now been made available to the public as part of the Tallac Historic District off of Route 89. The Pope Estate was built by the world-famous equestrian Tevis family and was later sold to lumber magnate George A. Pope. The lush gardens, complete with pools and gazebos, were the pride and joy of Mrs. Pope. The estate is used as the interpretive center for the Historic Site and houses several exhibits. It also features tours and living history programs.

The Baldwin Estate, built in 1921, now is the educational center of the Tallac Historic District and has museum exhibits of the Washoe Indians and of Lucky Baldwin. Valhalla, constructed in 1924, is now used as a community events center.

Near the Tallac Historic District is a visitor center with the short Lake of the Sky Nature Trail. In the visitor center is the Stream Profile Chamber. Here you can get an underwater cross section of Taylor Creek. A quarter-mile trail to the Taylor Creek Marsh is a good introduction to the wetlands of the area.

One of the oldest resorts around Lake Tahoe was the Glen Alpine Springs Resort, built in the early 1870s by Nathan Gilmore, a farmer from Ohio. The springs were believed to have medicinal value; Gilmore began bottling some of the spring water, and visitors came to be healed by the water. After 1920 other buildings were constructed, having been designed by Bernard Maybeck who also designed the San Francisco Palace of Fine Arts. The deteriorated "resort building may be seen by driving north on State Route 89 to Fallen Leaf Lake Road, then following that road for seven miles to the Desolation Wilderness Trailhead. From there it is 1.5 miles to the resort.

The Pope-Baldwin Recreation Area at the south end of Lake Tahoe has picnic areas, campgrounds, and swimming beaches, and includes Camp Richardson Resort. Biking trails wind throughout the Forest Service areas. Just south of Zephyr Cove at the southeast corner of Lake Tahoe is the Nevada Beach Campground operated under permit by the Lake Tahoe Basin Management Area.

Zephyr Cove is a center of activity. This has been a resort area since 1862. The current Zephyr Cove Resort is managed under special permit by the Forest Service. A rustic lodge has a full-service restaurant. On the grounds are old-style cabins for rent, and an RV park and campground are nearby. Visitors to the cove may water ski, parasail, charter a fishing boat, rent a boat, sail the Woodwind II catamaran, or enjoy the comforts of a Mississippi River–style boat, MS *Dixie II*. The first MS *Dixie II* was brought to Lake

Tahoe from Mississippi in 1948 by Jim Moss who had hoped to have a floating casino. After the Nevada Gaming Commission turned down Moss' request for a casino, the *Dixie* sank near Cave Rock and Mr. Moss reportedly disappeared. The Amundson family resurrected the *Dixie* for use as a houseboat. Cruises on the houseboat became so popular that it became a business. Over the years the original MS *Dixie* wore out and was replaced by a new cruise vessel, the MS *Dixie II*, in 1994. You can cruise Emerald Bay daily, take a champagne brunch cruise on Sundays, a dinner-dance cruise on Saturdays, and breakfast cruises on designated days. A glass-bottomed area is in the riverboat to observe the depths of Lake Tahoe.

Lassen National Forest

SIZE AND LOCATION: 1,375,000 acres in northeastern California, between Redding and Susanville, surrounding Lassen Volcanic National Park. State Routes 36, 44, and 89 are the major highways in the forest. District Ranger Stations: Chester, Fall River Mills, Susanville. Forest Supervisor's Office: 2550 Riverside Drive, Susanville, CA 96130, www.r5.fs.fed.us/lassen.

SPECIAL FACILITIES: Boat ramps; swimming beaches.

SPECIAL ATTRACTIONS: Volcanic phenomena, particularly in the Hat Creek area; Subway Cave; Lassen National Scenic Byway.

WILDERNESS AREAS: Caribou (20,546 acres); Ishi (41,099 acres); Thousand Lakes (16,335 acres).

Lassen is the land of recent as well as ancient volcanic activity. The Lassen National Forest completely surrounds Lassen Volcanic National Park where Mt. Lassen erupted as recently as 1915. From the northern end of the Sierra Nevada to the southern end of the Cascade Range, all types of volcanic phenomena can be seen and explored, particularly in the Hat Creek area.

For a great introduction to these volcanic features, start by hiking the Spatter Cone Trail that begins across from the Hat Creek Campground alongside State Route 44/89. Pick up a trail guide before starting this two-mile round trip that passes by many formations caused by volcanoes. The first part of the trail is uphill, passing through a Jeffrey pine forest, and because the weather is hot during the summer, it is advisable to take water with you. An interesting plant known as squawmat sprawls over the ground under the pines. By their very nature, volcanoes that have erupted leave a very rough and rocky terrain, so the hiker on this trail should proceed with caution. The trail passes 16 spatter cones and a myriad of volcanic domes, blowholes, and lava tubes. Steep-sided piles of lava that are shaped like small volcanoes are called spatter cones, and one of them along the trail is 33 feet across. During the eruption of volcanoes, great quantities of steam are forced out through openings known as blowholes, and some of these are in evidence along the trail. One of the most common plants seen along the trail is the green-leaved manzanita.

Two miles north of the Spatter Cone Trail is Subway Cave, which is a 1,300-foot-long tube through which lava flowed less than 2,000 years ago. You may enter the cave through the Devil's Doorway and travel down Lavacicle Lane, past such foreboding-sounding features as Stubtoe Hall and Lucifer's Cul-de-sac with its huge room called the Sanctum. You may emerge into the sunlight at Rattlesnake Collapse. The cave may be backbreaking in places where it is only four feet high, but you will be able to walk erect in other areas where the height of the cave is as much as 17 feet. Because Subway Cave is not lighted, it is necessary to carry a flashlight with you.

Lassen National Forest has other features that beckon the outdoor enthusiast. Dozens of crystal-clear lakes and wildflower meadows are dotted throughout the mountains. Three wilderness areas are great places to enjoy them.

The Caribou Wilderness is adjacent to the eastern boundary of Lassen Volcanic National Park and has many reminders of volcanic action. Red Cinder Mountain, at 8,374 feet, and Black Cinder Rock Mountain, at 7,760 feet, are two of the highest mountains in the wilderness, both of them along its western border. One popular hike is a six-mile round tripper, beginning at the Hay Meadow Trailhead and touching the edge of Evelyn Lake, Posey Lake, and Long Lake, just east of Black Cinder Rock Mountain.

The trail passes through forests of red fir, white fir, Jeffrey pine, western hemlock, and western white pine. The Cone Lake Trailhead gives access to the northern part of the wilderness area; it leads directly to the beautiful Triangle Lake with its wonderful array of water lilies after about a two-mile hike. The headwaters of the Susan River originate from the myriad of lakes on the east side of the wilderness area near the Silver Bowl Campground.

More rugged in its terrain at the southern edge of the Cascade Range is the Thousand Lakes Wilderness northwest of Lassen Volcanic National Park; some peaks rise above timberline. A 10-mile trail from Magee Trailhead to the Cypress Trailhead passes just to the east of Magee Peak and Crater Peak, the two highest peaks in the wilderness area at 8,550 and 8,677 feet, respectively. This challenging trail then goes between Gray Cliff and Red Cliff before touching the edge of Magee Lake and Everett Lake as it alternates its way between coniferous forests and mountain meadows. At the highest elevations, nice stands of red fir may be seen. Lake Eiler at the foot of Eiler Butte contains trout. In the area is the magnificent pearly-breasted goshawk.

The Ishi Wilderness is in the southwestern corner of the forest. A good place to enter is from the Black Rock Campground where a trail leads westward along Mill Creek. This is a fairly low elevation wilderness with Pinnacle

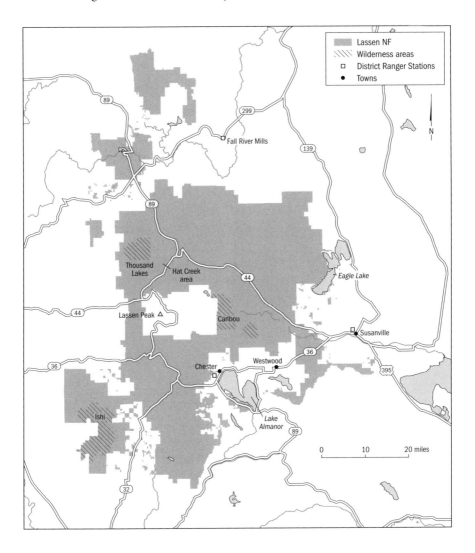

Peak at 3,293 feet the highest point. The area just northwest of Pinnacle Peak is another good place to hike into the wilderness; this trail follows Deer Creek and passes several caves and lava pillars. Both Deer Creek and Mill Creek support salmon and steelhead trout.

Heart Lake National Recreation Trail follows Martin Creek for about 3.5

miles. Excellent views of Lassen Peak may be enjoyed at several places along the trail.

Another favorite trail is the six-mile-long Spencer Meadow National Recreation Trail, which begins at a parking lot at Childs Meadow on the east side of State Route 36/89 just east of the Doe Mountain Picnic Area. The trail passes a bubbling spring before entering a forest of tall incense cedars. After passing along the west side of Patricia Lake, the trail enters the gorgeous Spencer Meadows just outside the southern boundary of Lassen Volcanic National Park.

One of the easiest trails is the Bizz Johnson Trail, which uses an old railroad bed as it makes its way for 26 miles along the Susan River. It is easily accessed from the town of Susanville.

For a beautiful scenic drive, take the 187-mile Lassen National Scenic Byway. From Susanville the byway crosses Pine Creek Valley and passes the Bogard Campground as it circles north around Lassen Volcanic National Park. Subway Cave and the Spatter Cone Trail can be accessed from this byway. Then after bisecting the national park, the byway circles south past Doe Mountain and Lost Creek Plateau before crossing the northern end of Lake Almanor.

Two large lakes, the natural Eagle Lake and artificial Lake Almanor, are mostly outside the boundaries of the forest, but the Forest Service maintains recreation areas on some of the shorelines. Both lakes are suitable for boating, swimming, waterskiing, fishing, and other water sports activities. A panoramic view that includes Eagle Lake, Mt. Lassen, and Mt. Shasta is available at the solar-operated Antelope Fire Lookout a few miles west of Eagle Lake.

For winter sports enthusiasts, there are eight winter staging areas in the forest.

Los Padres National Forest

SIZE AND LOCATION: 1,752,000 acres in west-central and southern California, between Monterey and Ventura. Other nearby cities: Ojai, Santa Barbara, Solvang, Santa Maria, San Luis Obispo. The forest lies west of Interstate 5; State Routes in the forest include 1, 33, 41, 154, 166, and 192. District Ranger Stations: King City, Frazier Park, Ojai, Santa Barbara, Santa Maria. Forest Supervisor's Office: 6755 Hollister Avenue, Suite 150, Goleta, CA 93117, www.r5.fs.fed.us/lospadres.

SPECIAL FACILITIES: Boat ramps; swimming beaches.

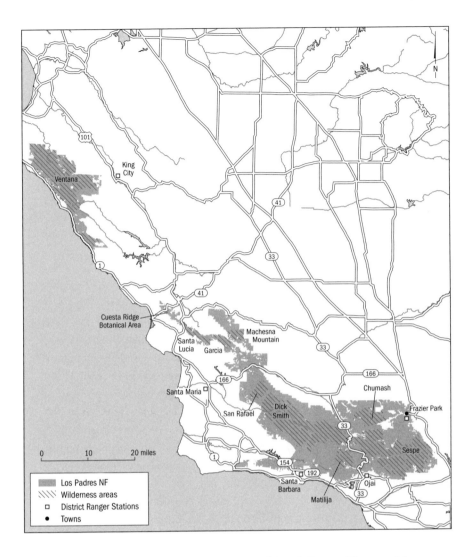

SPECIAL ATTRACTIONS: Jacinto Reyes National Scenic Byway; Figueroa Mountain Recreation Area; Santa Ynez Recreation Area; Wild and Scenic Rivers; scenic Big Sur coast.

WILDERNESS AREAS: Chumash (38,150 acres); Dick Smith (67,800 acres); Garcia (14,100 acres); Machesna Mountain (19,880 acres, partly in Bureau of Land Management land); Matilija (29,600 acres); San Rafael (197,380 acres); Santa Lucia (20,412 acres, partly in Bureau of Land Management land); Sespe (219,700 acres); Ventana (202,178 acres).

The Los Padres National Forest extends from Monterey to Ventura, paralleling the Pacific Ocean for nearly 250 miles. The forest has two noncontiguous areas. The northern section encompasses the Santa Lucia Range and extends from frontage on the Pacific Ocean to 5,844-foot Junipero Serra Peak. The more extensive southern unit includes the La Panza Range, the Santa Ynez Mountains, and the Sierra Madre with Mt. Pinos the highest in the forest at 8,831 feet. State Route 166, which follows the Cuyama River, separates the La Panza Range from the Sierra Madre.

The national forest runs the gamut for vegetation communities, from sandy beach communities in the Big Sur area along the Pacific Ocean near sea level to dense Jeffrey and ponderosa pine forests on the higher mountains. Limber pines are on the top of Mt. Pinos. Foothills vegetation is mostly dry chaparral, which gives way at a higher elevation to communities dominated by blue oak and foothill pine. Numerous rare trees and wildflowers are in the national forest.

For a wonderful scenic drive, take State Route 33, the Jacinto Reyes National Scenic Byway, from Meiners Oak to Ventucopa. Named for a rancher who was also a ranger in the Cuyama Ranger District (now the Mt. Pinos Ranger District) in the 1920s, the road passes through pleasant scenery from Ojai Valley over Pine Mountain down into Cuyama Valley. Be sure to stop and examine the historic Wheeler Gorge Campground and Guard Station that is a remnant of the Civilian Conservation Corps days of the early 1930s. Near the Ozena Picnic Ground is the historic Reyes house. Two rooms of this adobe house that was built in 1850 are original. Not far away are the 100-foot tall white rocks of Piedra Blanca. The weirdly awesome Cuyama Badlands are also nearby, as is the San Rafael Wilderness Area. Although ranging in elevation from 1,166 to 6,800 feet, most of the San Rafael Mountains are covered with chaparral, oak woodlands, and grasslands. One area of steep cliffs that stretches for 17 miles is called Hurricane Deck and is known for its unusual wind-eroded sandstone formations. The Sisquoc Condor Sanctuary where the nearly extinct California condor nests is here but not available to the public.

East of State Route 33 and south of the Lockwood Valley Road is the Sespe Wilderness where 42 miles of the Wild and Scenic Sespe Creek have carved strange sandstone formations. There are petroglyphs on many of the sandstone rocks. Another condor sanctuary is in this wilderness, as well. North of the Lockwood Valley Road and up to Quatel Canyon is the Chumash Wilderness. Mount Pinos and Cerro Noroeste are at the northernmost boundary of this wilderness.

Outside the Chumash Wilderness, a narrow road runs from the tiny community of Frazier Park to the summit of Mt. Pinos. From this high peak, or from the lookout on top of Frazier Mountain nearby, you may be lucky enough to see a California condor. If so, you will be impressed by the bird's nine-foot wing span. While you are in the Mt. Pinos area, you may wish to drive the Tecuya Ridge Road in Pleito Canyon near Antimony Peak for a look at reputedly the largest bigcone Douglas fir in the country. This mammoth tree measures 24 feet in circumference. Bigcone Douglas fir differs from the more common Douglas fir by its four- to eight-inch-long cones and its extremely sharp-tipped needles. Regular Douglas fir has cones less than four inches long and rather blunt needles.

West of State Route 33 is the Dick Smith Wilderness. Long-distance hikers will enjoy the trek to the top of Madulce Peak. The eastern part of the wilderness has sandstone formations with lots of Douglas firs growing around them.

North of Ojai is the Matilija Wilderness with a scenic canyon along Matilija Creek. The gorgeous Matilija poppy (pl. 21) is found in this region. This uncommon species, whose flowers have six white, crinkly petals up to four inches long, is unique in having the largest flowers of any native plant in California. Before leaving the southern part of the Los Padres National Forest, consider a trip to the well-preserved Manzana Schoolhouse, which dates back to 1893. Constructed of foothill pine and California sycamore, the school was built to accommodate children living along the Manzana Creek and Sisquoc River. The school ceased operation in 1902 when only one student was left. The school is just outside the San Rafael Wilderness and reached by about a six-mile hike.

A small but interesting part of the Los Padres National Forest is in an area enclosed by U.S. Highway 101 and State Routes 41 and 1. Here is Cuesta Ridge, and it is possible to drive the eight miles of dirt road along the crest. The ridge is composed of serpentine soils in which only a limited number of plant species may grow. You will see a very large stand of the rare Sargent cypress growing with the fascinating Coulter pine. The best grove of these trees is in the designated 1,300-acre Cuesta Ridge Botanical Area. If you decide to wander around beneath these trees, you may be fortunate to see other rare plants such as Hickman's checker mallow, San Luis Mariposa lily, and Brewer's spineflower. From various points along the road you will get splendid views of the Pacific Ocean and Morro Rock.

The Santa Lucia Range occupies the most northern part of the Los Padres and begins some 25 miles south of Monterey. The Ventana, Machesna Mountain, Garcia, and Santa Lucia Wildernesses provide plenty of opportunity for backcountry hiking. Instead of doing the wilderness, however, you may wish to drive the Cone Peak Road, which is situated atop one of the steepest slopes around. It is only a distance of 3.5 miles from the base of the ridge near the Pacific Ocean to the top of 5,155-foot Cone Peak. The significance of the Cone Peak area is that one of the rarest trees in the country, the bristlecone fir (fig. 11), is found only in this vicinity. This striking species has a narrow, steeple-shaped crown with drooping branches that often reach the ground. This is the only fir whose needles have a spiny tip.

Scenic State Route 1 hugs the Pacific Ocean at the western edge of the Los Padres. In a small canyon south of Redwood Gulch is the southernmost grove of the coast redwood. As you continue heading north on the highway, more

Figure 11. Bristlecone fir.

redwoods will be encountered in moist forests and on north slopes where dense fog provides ideal growing conditions for these giants. In the vicinity around Big Sur are Sand Dollar Beach and Pfeiffer Beach and Jade Cove, all on Forest Service land. At Jade Cove you may search in the pebbles along the ocean and in the adjacent cliffs for jade, which is plentiful here. You will likely find the gray-green form known as Monterey jade. For a little exercise, take the Salmon Creek Trail to beautiful Salmon Creek Falls, which is nestled among the southernmost grove of Douglas firs on the California coast.

Mendocino National Forest

SIZE AND LOCATION: 886,048 acres in the North Coast Range of northwestern California, between Redding and Ukiah. State Routes 20 and 36 are near the south and north boundaries. District Ranger Stations: Willows, Covelo, Upper Lake. Forest Supervisor's Office: 825 N. Humboldt Avenue, Willows, CA 95988, www.r5.fs.fed.us/mendocino.

SPECIAL FACILITIES: Boat ramps; swimming beaches; winter sports areas.

SPECIAL ATTRACTIONS: Thomes Gorge; Harvey Peak.

WILDERNESS AREAS: Yolla Bolly–Middle Eel (153,841 acres, partly in the Shasta-Trinity National Forest and partly in the Six Rivers National Forest); Snow Mountain (36,370 acres).

If your outdoor interests rely on the backcountry, the Mendocino National Forest is your cup of tea. It boasts of not having a single paved road that traverses from one side of the forest to the other! Although there are numerous dirt and gravel roads, many of them are rough, narrow, and suitable only during good weather.

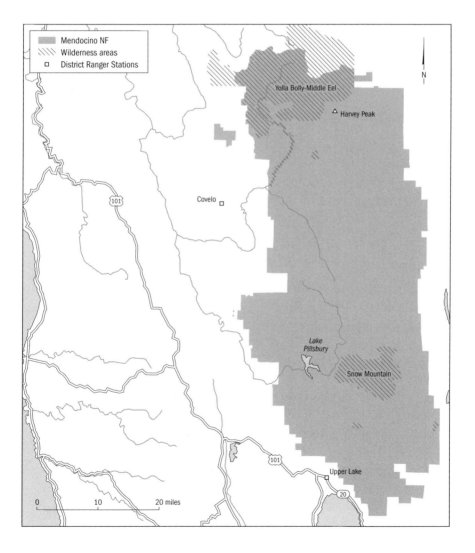

Hang gliders come from all over to glide 4.5 miles from Hull Mountain to the Gravelly Valley Airstrip or to sail two miles from Elk Mountain to the Middle Creek Campground.

The most rewarding trail in the Mendocino is in the Yolla Bolly–Middle Eel Wilderness. It is the 8.8-mile Ide's Cove Trail. Said to follow old Indian trails, this loop begins in chaparral but soon climbs up rugged Harvey Peak, at 7,361 feet, where stands of foxtail pine occur. Steep-walled glacial basins provide stunning scenery. On the way, this trail passes Square Lake and Long Lake. For shorter trails in the wilderness, try the three-mile Peterson Trail and the .7-mile, straight-up Black Butte Trail. The Peterson Trail begins at the Straight Arrow Camp and goes alternately through ponderosa pine forests, wildflower meadows, and rocky glades before ending at the clear waterholes in Thomes Creek. You pass the old Peterson cabin and site of their orchard. Black Butte Trail climbs 400 feet in one-third mile, over craggy rocks, through a red fir and pine forest, ending at an unobstructed view from the barren summit. Two-thirds of the Yolla Bolly–Middle Eel Wilderness is in the national forest, and the headwaters of the Middle Fork of the Eel River is here.

Snow Mountain is the centerpiece of the Snow Mountain Wilderness, which straddles the summit of the North Coast Range. The foothills of the mountain are mostly chaparral interspersed with oak woodlands. Ascending the mountain, you first encounter forests of Douglas fir, white fir, red fir, and ponderosa pine, as well as numerous steep slopes and narrow, rocky ravines. Expanses of bare rock spread out over the mountain's summit. The Middle Fork of Stony Creek flows through the wilderness. A trail to the top of East Peak ends with a 360° view of mountains, Clear Lake, and the Sacramento Valley.

If you do not want to hike in the wilderness area, take the Thomes Gorge Trail. For 4.2 miles this trail passes from chaparral to a digger pine forest and past some intriguing vernal pools. It ultimately drops into spectacular Thomes Gorge.

Several special areas for plants exist in the Mendocino National Forest. One such area occurs on the eastern side of the forest along Frenzel Creek near the Little Stoney Campground. This is an area where serpentine rocks are exposed, and this usually means there will be interesting plants. In the area is one of the northernmost locations for the Sargent cypress, including one tree that is about 70 feet tall with a two-foot diameter. Growing with this species is the similar-appearing and closely related McNab cypress. The McNabs are just as thick, but usually considerably shorter than the Sargent. You may also see the rather rare, prostrate Solano milkweed and an equally rare plant known as the Snow Mountain buckwheat.

For the rockhound, there are agates and jasper to search for along Big Stony Creek and jadeite and nephrite along the Eel River. The district rangers can give you directions to these fascinating sites.

Urias S. Nye, a German immigrant sheep rancher, built a log cabin in the late 1860s in the north Sacramento Valley. The cabin still stands as the oldest building in the national forest. It is about 25 miles southwest of the community of Elk Creek.

Modoc National Forest

SIZE AND LOCATION: 1,654,392 acres in northeastern California, on either side of Alturas. U.S. Highways 97 and 395 and State Routes 139 and 299 are the major access roads. District Ranger Stations: Cedarville, Adin, Alturas, Tulelake. Forest Supervisor's Office: 800 W. 12th Street, Alturas, CA 96101, www.r5.fs.fed.us/modoc.

SPECIAL FACILITIES: Boat ramps; swimming beaches.

SPECIAL ATTRACTIONS: Medicine Lakes Highlands; Devil's Garden Natural Area; Pacific Flyway; Burnt Lava Flow Virgin Area.

WILDERNESS AREA: South Warner Mountain (70,614 acres).

The Modoc National Forest lies in the remote northeast corner of California. Most of it was covered by an immense lava flow millions of years ago. Elevations in the forest range from 4,300 to 9,892 feet.

The eastern part of the forest east of Alturas contains a spur of the Cascade Range to the north known as the Warner Mountains. These mountains drop steeply on their eastern side, whereas the western side has a more gentle, rolling topography. The southern part of the Warner Mountains has been designated the South Warner Mountain Wilderness, a region 18 miles long and about eight miles wide. From Clear Lake, at 5,800 feet in the southwestern corner of the wilderness, the terrain rises to Eagle Peak, the highest peak in the forest at 9,892 feet. The western side of the wilderness boasts numerous clear streams and wildflower meadows. Brushy areas in the foothills consist mostly of bitterbrush and curl-leaf mahogany. As elevation increases, forests of ponderosa pine, white and red firs, incense cedar, and aspen give way to near-barren rocks at the summit with a sprinkling of lodgepole and western white pines. Patterson Lake is nestled at the foot of Warner Peak. Summit Trail follows the mountain crest for 27 miles through the wilderness, from Patterson Meadow to Porter Reservoir, skirting the western flank of

Eagle Peak and passing Squaw Peak and Warner Peak. Unusual geological features seen along the way are called Devil's Knob and The Slide. More than 15 miles of this trail are at the 9,000-foot level. Owl Creek Trail follows the east face of the wilderness area. From the Mill Creek Campground are short trails to Clear Lake and Mill Creek Falls. A natural stone bridge is near the Emerson Campground.

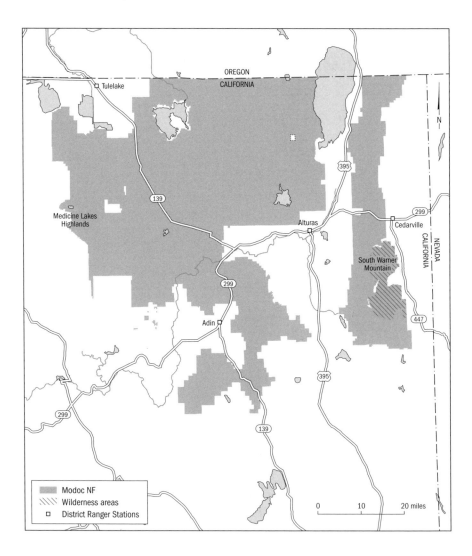

The north end of the Warner Mountains, out of the wilderness area, served as an emigrant trail to Oregon laid out by Peter Lassen in 1846. It crosses the top of the mountain at a place called Fandango Pass. A large emigrant train in the 1850s came to this point and, seeing Goose Lake in the distance, these pioneers thought they had come upon the Pacific Ocean. They made camp and, in their excitement, danced the fandango well into the night. As they danced, a band of Indians crept out of the forest and killed all but two of the party.

Several miles north of Fandango Pass near Mt. Vida are the remains of an old gold and silver mining camp that ceased operation in 1934. The Highgrade National Recreation Trail crosses the west side of Mt. Vida as it makes its way through stands of ponderosa, lodgepole, and western white pines and white firs. This 5.5-mile trail offers spectacular views of Goose Lake.

The relatively flat central part of the Modoc National Forest is a basalt-capped plateau known as the Devil's Garden. The largest continuous stand of western juniper in the world, comprising 800 acres, has been designated the Devil's Garden Natural Area. The east side of Devil's Garden has nearly pure stands of ponderosa and Jeffrey pines, while white fir and incense cedar are also found on the northern slopes. At lower elevations are black, big, and silver sagebrushes, rabbitbrush, and several kinds of bunchgrass. The natural area is on the west side of Goose Lake, seven miles from McGinty Point, which protrudes into the lake. Also at McGinty Point are Modoc Indian petroglyphs that have been there since the late 1800s.

South of the Tulelake Ranger Station, between State Route 139 and U.S. Highway 97, are the Medicine Lake Highlands, the site of numerous major volcanic eruptions during the past few thousand years. Near the center of this area a shield volcano collapsed to form a depression. Later, lava flowed from the rim of the crater, forming Glass Mountain, Pumice Stone Mountain, and Burnt Lava Flow. Clear Medicine Lake, with pumice beaches, has filled the depression. It is at least 100 feet deep and has no known natural outlet. Glass Mountain is a huge flow of rhyolitic obsidian and glossy dacite approximately 7,672 feet high. On the west slopes is a one-acre area called the Hot Spot where the pumice just beneath the surface crust is too hot to handle when first exposed. The small rocks can be molded into a putty. Burnt Lava Flow Virgin Area, a region of 900 acres, includes High Hole Crater. It is a jumble of black lava interspersed by small stands of sugar, ponderosa, and lodgepole pines. An ice cave at the northwest corner is a collapsed lava tube. If you are in this area at the right time of the year, you may see two of California's more uncommon wildflowers—the Ash penstemon and the small-leaved agastache.

Four miles west of the Canby Bridge, off of Forest Road 84, a short hike may be taken to a sharp bend in the Pit River. Here you can barely make out wagon-wheel ruts carved into the canyon rock where the Lassen Emigrant Trail party tried to negotiate the river. Fishing for rainbow, brown, and brook trout, bass, and catfish is good in the Pit River.

The Pacific Flyway migratory route that extends from Alaska to Mexico passes to the west side of the Warner Mountains where the Modoc National Forest meets the Modoc National Wildlife Refuge. Nearly 240 species of birds have been recorded here.

Plumas National Forest

SIZE AND LOCATION: 1,162,863 acres in northern California, from the foothills above Oroville to U.S. Highway 395. U.S. Highway 395 and State Routes 70 and 89 are the major highways. District Ranger Stations: Blairsden, Oroville, Quincy. Forest Supervisor's Office: 159 Lawrence Street, Quincy, CA 95971, www.r5.fs.fed.us/plumas

SPECIAL FACILITIES: Boat ramps; swimming beaches.

SPECIAL ATTRACTIONS: Feather Falls Scenic Area; Lake Basin Recreation Area; Butterfly Valley Botanical Area; Antelope Lake Recreation Area; Feather River National Scenic Byway; Middle Fork Feather National Wild and Scenic River; Bucks Lake Recreation Area.

WILDERNESS AREA: Bucks Lake (23,958 acres).

Ever since I started gathering maps and brochures about our national forests, I knew I would have to have Feather Falls in the Plumas National Forest on my list of things to see as soon as I could. The falls is the sixth highest in the United States. Park a short distance north of Feather Falls Village, a few miles northeast of the community of Oroville. After zigzagging downhill for about a mile, make your way along a refreshing mountain stream and pass a bubbling spring. Long before reaching the end of the nine-mile loop trail, Feather Falls Trail, you will hear the thundering of the water hitting the catch pool below the falls. An observation deck built by the Forest Service is atop a rocky pinnacle. Across from the deck, rushing over a granite precipice, is Feather Falls. Six hundred forty feet below, the water splashes into the catch pool with amazing force. The Forest Service has designated the area around the falls the Feather Falls Scenic Area. Besides Feather Falls, the scenic area contains a bare, round-topped granite rock on the west side of the Middle Fork of the Feather River. Bald Rock Dome, as it is called, must not be missed, nor should another beautiful waterfall along the Middle Fork just beyond Bald Rock Dome. You can hike to these two sites on a twisting trail, Bald Rock Dome Trail, that leads from Boar Creek Road. The Middle Fork Feather River seems to go on forever; all 108 miles of it have been designated a National Wild and Scenic River. The branches of the Middle Fork are great for fishing and are a challenge to rafters and kayakers. Several parts of the river that are in deep canyons contain huge boulders and nearly impassible waterfalls. Only the expert rafter should attempt these. For more gentle rafting, try the upper regions of the Middle Fork in the English Bar Zone.

If you are able to stay overnight in this area, the remote Milsap Bar Campground where the more tranquil South Branch joins the Middle Fork is rec-

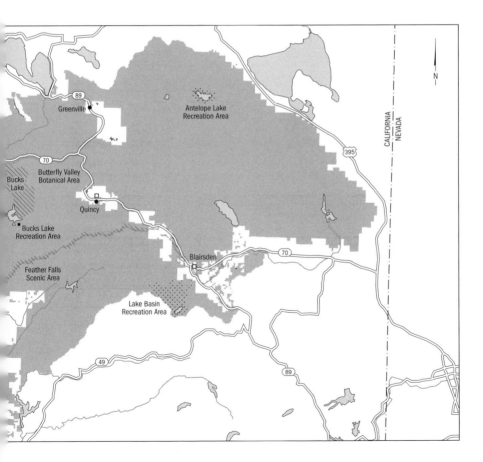

ommended. Three miles away is South Branch Falls. You will parallel a stream that is lined with willows and alders. South Branch Falls is a series of nine small and medium falls that tumble over a series of rocky domes until the last clear pool is reached. The drops of the nine falls range from 30 to 150 feet.

Bucks Lake Wilderness in the northwest corner of the Plumas is popular among hikers. The wilderness contains several small lakes and ponds and ranges in elevation from 2,000 feet at the Feather River Canyon to 7,017 feet at the towering Spanish Peak. Naturalists will marvel at the plant communities in this wilderness. At the higher elevations are dense stands of red fir. As the elevation decreases, the communities change to mixed coniferous forests to scrubby oak woodlands to brushy chaparral dominated by scraggly shrubs. Alpine meadows with many species of dwarf wildflowers are here and there above 6,000 feet, and there are even a couple of quaking bogs with sedges the prominent plants. A long cliff separates the higher elevations of the wilderness from the lower, and the Pacific Crest Trail follows the crest of the cliff for nearly 13 miles.

One of the nicest places is the Lake Basin Recreation Area whose southern boundary is five miles north of the settlement of Bassetts. Gold Lake and Long Lake are the two largest lakes in the Recreation Area. Towering above Long Lake is Mt. Elwell, at 7,818 feet. While there, you can enjoy the short Fern Falls Overlook Trail. At the Lake Basin Campground, two rocks with several petroglyphs may be viewed from observation decks.

Just east of the Lakes Basin Recreation Area are two features that should not be missed. One is a short trail to Frazier Falls. The waterfall cascades 176 feet into a catch pool. The other is the Red Fir Nature Trail, a circular route off of the dirt County Road 822 to Mills Peak. The interpretive trail passes through an ancient forest of red firs, some of them more than 150 feet tall.

Antelope Lake Recreation Area is family oriented; fishing in the lake is always worthwhile, and there are ample opportunities to camp, picnic, swim, and boat. Near the Lone Rock Campground is a nature trail, and during the summer campfire programs are conducted in the amphitheater. A log cabin that stands at the east end of the recreation area is where a rancher and his family lived many years ago. The graves of two children are on a small rise southeast of the cabin.

For those not inclined to hike, the Feather River National Scenic Byway passes through some of the most beautiful scenery in the Plumas. Although the byway extends for 130 miles from just north of Oroville to Hallelujah Junction on the California–Nevada border, much of the route that follows the North Fork of the Feather River is in the Plumas National Forest. As you pass through the narrows of the North Fork Canyon, you will see digger

Plate 1 (top). Portage Glacier, Chugach National Forest (Alaska).

Plate 2 (center). Mountain Goat, Chugach National Forest (Alaska).

Plate 3 (bottom). Mendenhall Glacier, Tongass National Forest (Alaska).

Plate 4 (top). Prickly poppy, Apache-Sitgreaves National Forest (Arizona).

Plate 5 (center). Blue flax, Phelps Cabin Research Natural Area, Apache-Sitgreaves National Forest (Arizona).

Plate 6 (bottom). Stiff gilia, Coconino National Forest (Arizona).

Plate 7. West Fork Trail,
Oak Creek Canyon,
Coconino National Forest
(Arizona).

Plate 8. South Fork, Cave Creek,
Coronado National Forest (Arizona).

Plate 9. Ocotillo, Coronado
National Forest (Arizona).

Plate 10 (top left).
Ring-tail cat, Kaibab
National Forest
(Arizona).

Plate 11 (top right).
Perry's penstemon,
Prescott National
Forest (Arizona).

Plate 12 (bottom).
Diamond Point,
Tonto National
Forest (Arizona).

Plate 13 (left). Apache plume, Tonto National Forest (Arizona).

Plate 14 (below). Superstition Mountains and saguaro, Peralta Canyon, Tonto National Forest (Arizona).

Plate 15 (top left). Bighorn sheep, Inyo National Forest (California).

Plate 16 (top right). Hikers climbing Mt. Whitney, Inyo National Forest (California).

Plate 17 (center). Hoover Wilderness, Inyo National Forest (California).

Plate 18 (bottom left). Bristlecone pine, Inyo National Forest (California).

Plate 19 (bottom right). Tufa formations at Mono Lake, Inyo National Forest (California).

Plate 20 (above).
Emerald Bay,
Lake Tahoe Basin
Management Area
(California).

Plate 21 (left).
Matilija poppy,
Los Padres
National Forest
(California).

Plate 22 (top). Cobra lilies (pitcher plants), Plumas National Forest (California).

Plate 23 (center, second from top). Pacific (mountain) dogwood, Plumas National Forest (California).

Plate 24 (center, third from top). Spotted owl, San Bernardino National Forest (California).

Plate 25 (bottom). Foxtail pine just above the Pacific Crest Trail, Trinity Alps, Shasta-Trinity National Forest (California).

Plate 26 (top). Backbone Ridge, Shasta-Trinity National Forest (California).

Plate 27 (inset). Bigelow's sneezeweed, Tahoe National Forest (California).

Plate 28 (top left). Chinese houses, Tahoe National Forest (California).

Plate 29 (top right). Middle Fork of the Salmon River, Boise National Forest (Idaho).

Plate 30 (bottom). Rafting on the Payette River, Boise National Forest (Idaho).

Plate 31. Upper Bloomington Lake at sunset, Bear River Range, Caribou-Targhee National Forest (Idaho).

Plate 32. Waterfalls along the Middle Fork of the Salmon River, Salmon-Challis National Forest (Idaho).

Plate 33 (top). Boulder Chain Lakes, White Cloud Peaks, Salmon-Challis National Forest (Idaho).

Plate 34 (bottom). View from Borah Peak, Salmon-Challis National Forest (Idaho).

Plate 35 (left). Along the Sawtooth National Scenic Byway, Sawtooth National Forest (Idaho).

Plate 36 (below). Lamoille Canyon, Humboldt-Toiyabe National Forest (Nevada).

Plate 37. Flowers in creek, Toquima Range, Humboldt-Toiyabe National Forest (Nevada).

Plate 39. Heart-leaved arnica, Malheur National Forest (Oregon).

Plate 38. South Sister from Sparks Lake, Deschutes National Forest (Oregon).

Plate 40 (top).
Wildflowers and Mt.
Hood, Mt. Hood
National Forest
(Oregon).

Plate 41 (inset).
Salal, Siskyou National
Forest (Oregon).

Plate 42 (left). Pacific tree frog, Siskyou National Forest (Oregon).

Plate 43 (center, this page and facing page). Lostine River Valley, Wallowa-Whitman National Forest (Oregon).

Plate 44 (bottom). Waldo Lake, Willamette National Forest (Oregon).

Plate 45. Pipsissiwa, Wildcat Mountain Natural Area, Willamette National Forest (Oregon).

Plate 46. Blue-eyed Mary, Wildcat Mountain Natural Area, Willamette National Forest (Oregon).

Plate 47. Woolly sunflower, Wildcat Mountain Natural Area, Willamette National Forest (Oregon).

Plate 48. Flaming Gorge National Recreation Area, Red Canyon, Ashley National Forest (Utah).

Plate 49. Vermilion Cliffs, Dixie National Forest (Utah).

Plate 50. Fish Lake, Fishlake National Forest (Utah).

Plate 51 (top). Castle Valley, LaSal Mountains, Manti-LaSal National Forest (Utah).

Plate 52 (center). Warner Lake with Haystack reflection, Manti-LaSal National Forest (Utah).

Plate 53 (bottom). Paintbrush below Mt. Timpanogos, Uinta National Forest (Utah).

Plate 54. Bridal Veil Falls, Uinta National Forest (Utah).

Plate 55. Big Cottonwood Canyon, Wasatch-Cache National Forest (Utah).

Plate 56. Jardine juniper, Wasatch-Cache National Forest (Utah).

Plate 57. Black bear, Colville National Forest (Washington).

Plate 58. Paintbrush and lupine, Colville National Forest (Washington).

Plate 59. Fireweed, Colville National Forest (Washington).

Plate 60 (top). Bunchberry, Lake Twenty-two Natural Area, Mt. Baker–Snoqualmie National Forest (Washington).

Plate 61 (center). Snowshoe hare, Lake Twenty-two Natural Area, Mt. Baker–Snoqualmie National Forest (Washington).

Plate 62 (bottom). One busy pika, Lake Twenty-two Natural Area, Mt. Baker–Snoqualmie National Forest (Washington).

pines on the upper slopes. They are recognized by their long, gray-green needles and gigantic cones that may reach a length of 10 inches and a diameter of eight inches.

Many years ago I read in *Fremontia,* the Journal of the California Native Plant Society, about a marvelous area for plants in the Plumas called Butterfly Valley. The featured plant in this designated Botanical Area is the incomparable cobra lily (pl. 22), an insectivorous species with a flaring hoodlike leaf and forked tongue, features that resemble its namesake. My visit to the Plumas National Forest was not complete until I found this small area tucked in a corner of the forest about seven miles northwest of Quincy. In all, more than 360 species of plants are known from Butterfly Valley, including several considered endangered or threatened in California.

Butterfly Valley

As soon as I first read about Butterfly Valley in *Fremontia,* the journal of the California Native Plant Society, I knew I would have to go there. This area of bogs and seeps in the middle of a ponderosa pine forest is in the Plumas National Forest in the northern part of California's Sierra Nevada.

Butterfly Valley is so botanically unique that is has been designated a Botanical Area by the U.S. Forest Service and is under complete protection from logging of trees and picking of wildflowers. Although the major attraction of Butterfly Valley is the large concentration of the insectivorous cobra lily, more than 360 kinds of plants have been found in this area. Among these are five insectivorous plants in addition to the cobra lily, 12 different wild orchids, and several species considered by the California Native Plant Society to be endangered or threatened in California.

On a pleasant August morning, my wife and I drove northwest on State Route 70 about seven miles from the charming mountain community of Quincy. Just past the marked Roundhouse Road, we turned left on a narrow asphalt road and followed it a couple of miles until it became gravel at a fork in the road. Avoiding the left fork, which was marked Private Property, we veered to the right. A Forest Service sign indicated "Botanical Gardens 1 Mile, Twain [a small village] 6 Miles."

Our narrow forest road passed through a woods dominated by ponderosa and sugar pines, with a good number of white firs, Douglas firs, incense cedars, and California black oak scattered throughout. Ponderosa pine, often called yellow pine by local Californians, shares the dominance of the forest at Butterfly Valley with other tall-growing conifers of sugar pine, white fir, Douglas fir, and incense cedar. The deciduous trees, which are much fewer in number, are mostly California black oaks, Pacific dogwoods (pl. 23), and bigleaf maples. The only shrubs, and they are both plentiful, are the green-leaved manzanita, with its thick, leathery leaves and exquisite red trunks, and the white-leaved manzanita. Wildflowers, while not densely populating the understory, are well represented by purple fritillary, wild iris, crimson columbine, dwarf blue larkspur, two kinds of lupines, western shooting-star, and the very common mule's-ears, whose golden yellow flower heads provide vivid color to the woods in early July. Two species that are striking because of their white-striped leaves are rattlesnake plantain orchid and white-veined wintergreen.

At the next junction we made a sharp left and soon saw our first cobra lilies in the Darlingtonia Bog. Standing water was beneath the cobra lilies, and myriad other wetland plants, notably sedges and rushes, were growing with the cobra lilies. This boggy habitat, formed from hillside seeps, paralleled the forest road for a hundred feet or so, at times extending a few feet up small slopes west of the road. The boggy terrain that provides ideal habitat for the cobra lily is caused by numerous hillside seeps. Standing water as much as 12 inches deep occurs here and there. The cobra lily, with its hood-like leaf complete with forked tongue, has a fanciful resemblance to the reptile. Although many of the plants associated with the cobra lily are sedges and rushes, there are colorful wildflowers in the bog. Among these are four true members of the lily family—bog asphodel, sticky tofieldia, beavertail grass, and *Hastingia alba*. Other wildflowers include California grass-of-Parnassus, Parnassus-leaved violet, Sierra gentian, the rather rare Plumas marsh aster with its lovely blue heads, western sneezeweed that resembles a small sunflower, and Douglas' water hemlock, a white-flowered member of the carrot family. Although most of the plants in the Darlingtonia Bog are nonwoody, the shrubby Labrador tea and western blueberry occur sporadically. Two sundews are also here. Their gland-tipped hairs that glisten in the sunlight ensnare minute insects whose organic matter is used by the sundews.

Less than a quarter-mile farther on, an open area of about 10 acres, called Sweetwater Marsh, appeared in a small valley off the left side of the road. After walking through a narrow border of alders, I was in the wet marsh with water creeping over the tops of my boots. There were some cobra lilies in the marsh, but mostly the vegetation was composed of grasses, sedges, and rushes, punctuated by many wildflowers as well as two kinds of sundews, also insectivorous. The large, open expanse of Sweetwater Marsh is inviting, but the continuous cover of vegetation obscures the very wet substrate that the plants grow in. It is best not to walk into the marsh because the disturbance will cause damage to the plant life. From the periphery we observed cobra lily, white brodiaea, bog saxifrage, a St. John's wort, yellow monkey-flower, broad-leaved yampah, and sheep parsnip. The latter two species are in the carrot family.

A short distance past Sweetwater Marsh, in a deeper one-acre depression filled with water, was Pond Reservoir, whose muddy borders supported still another community of wetland plants. In the standing water when we were there was an arrowhead, water plantain, scarlet smartweed, a large-fruited *Carex*, water cinquefoil, two buttercups, water shield, a pondweed, watercress, and the tiny water starwort. Like so many aquatic species that seem to have broad geographical ranges, almost every one of these listed above occurs in the midwestern and eastern United States. Also found in the water are two bladderworts, bringing to five the number of plants in Butterfly Valley that trap and consume insects.

Common in the mudflats surrounding the open water was a variety of sedges and rushes, northern bog violet, marsh purslane, and primrose monkey-flower.

Continuing along the forest road, we came upon a moist, heavily shaded area in the forest that contains the gorgeous Sierra woodfern, lady fern, and California grape fern. This delightful area is known as Fern Glen. The ferns grow among a wonderful assemblage of woodland wildflowers. The overstory is mostly ponderosa pine, but the soil is much more moist from hillside seeps than in other parts of the ponderosa pine forest. A plethora of wildflowers throughout the growing season provides abundant color. Included among these are wild ginger, two woodland starflowers, Lassen daisy, broad-leaved aster, red larkspur, false Solomon's-seal, fawn lily, twisted stalk, and Sierra lily. One attractive shrub at Fern Glen is the Sierra Nevada pink currant.

Near the southern end of the botanical area, a small community dominated by beargrass (actually in the lily family and not the grass family) adds still another habitat type in Butterfly Valley. This rather open area, known as Beargrass Glade, is dominated by beargrass, a huge plant with long, narrow, grasslike leaves from which a large spike of white flowers appears during August. Growing with the beargrass are Washington lily; Oregon white-topped aster; pearly everlasting, whose white, paperlike flower heads persist for weeks; and a shrub known as Sierra laurel.

Butterfly Creek flows along the northern edge of the botanical area, and Blackburn Creek flows along the southern end. Both are tributaries of the Spanish River, which eventually empties into the Feather River. These two streams are lined with dense thickets of willows, alders, and red osier dogwood.

San Bernardino National Forest

SIZE AND LOCATION: 630,971 acres in southern California, between the eastern suburbs of Los Angeles and Palm Springs. Major access roads are Interstates 10 and 15, U.S. Highways 60 and 395, and State Routes 18, 30, 74, 138, 173, and 243. District Ranger Stations: Sky Forest, Fontana, Idyllwild. Forest Supervisor's Office: 1824 S. Commercenter Circle, San Bernardino CA 92408, www.r5.fs.fed.us/sanbernardino.

SPECIAL FACILITIES: Winter sports areas; boat ramps; swimming beaches.

SPECIAL ATTRACTIONS: Rim-of-the-World National Scenic Byway; Palms to Pines National Scenic Byway; Arrowhead Geological Site; Black Mountain Scenic Area; Gold Fever Auto Tour.

WILDERNESS AREAS: Sheep Mountain (41,883 acres, partly in the Angeles National Forest); San Jacinto (32,248 acres); San Gorgonio (94,702 acres, more than one-third on Bureau of Land Management land); Cucamonga (12,781 acres, partly in the Angeles National Forest); Bighorn Mountain (38,485 acres, mostly on Bureau of Land Management land).

Interstate 10 from Palm Springs to the suburbs of Los Angeles stays in a broad valley with high mountains towering above both sides of the freeway. To the north is the east-west trending Transverse Range that includes the San Bernardino and San Gabriel Mountains. To the south, the San Jacinto Mountains are part of the north-south trending Peninsular Range that eventually extends into Mexico. The San Bernardino National Forest encompasses almost all of these mountains and is one of the most heavily visited in the country.

The forest has steep, rugged topography with several lakes interspersed and clear mountain streams everywhere. Because the dry Santa Ana winds blow across the area in the fall and winter, much of the lower elevations in the forest are inhabited by shrubby chaparral species or, on the far eastern edge of the forest, desert species. Part of the desert area is in the Bighorn Mountain Wilderness on Bureau of Land Management land, but the wilderness extends into the San Bernardino National Forest up to Granite Mountain. As elevation increases, forested communities dominate. These include oak woodlands, where the most often observed species are scrub canyon live oak, California scrub oak, interior live oak, coast live oak (fig. 12), and evergreen live oak, and piñon–juniper communities that are dominated by piñon pine and western juniper. Majestic coniferous forests clothe the higher elevations, with many species of trees present. Common are sugar pine, ponderosa pine, Jeffrey pine, limber pine, Coulter pine, white fir, bigcone Douglas fir and, on the highest peaks, gnarly specimens of lodgepole and knobcone pines.

Figure 12.
Coast live oak.

The eastern side of the San Gabriel Mountains is in the San Bernardino National Forest and includes the very popular Cucamonga Wilderness and the Sheep Mountain Wilderness. The trail to Cucamonga Peak from the Joe Elliott Campground will test the more rugged hiker. Mount Baldy, the highest peak in the San Gabriel Mountains at 10,064 feet, is in the Sheep Mountain Wilderness. Here you may also find the Nelson bighorn sheep and even a California spotted owl (pl. 24).

Just north of the wilderness area and easily reached by a very short hike from State Route 138 are the fascinating Mormon Rocks. Here are sandstone walls pockmarked with hundreds of small holes and caves that provide a habitat for wildlife activity. Owls and pack rats use this strange area, and the rocks are teeming with lizards.

The most well-known landmark in the San Bernardino Mountains is the well-defined shape of an arrowhead on the side of the mountain. It can be seen for miles. The shape is the result of a dense concentration of a white sage species and lupines surrounded by the shrubby chamise plant. This phenomenon is located just above the old Arrowhead Springs Resort.

For a comprehensive tour of the San Bernardino Mountains, take the fabulous Rim-of-the-World National Scenic Byway, following State Route 18 up switchbacks to the settlement of Crestline and then heading east. The Arrowhead Geological Area is off to the right, with Arrowhead Lake to the left. After driving past ski resorts, you come to Lakeview Point for one of the best views in the forest. Mountains loom up everywhere, and Big Bear Lake can be seen in the distance. If you have time and need to move about, take the scenic Camp Creek National Recreation Trail from the Lakeview parking area to Clark's Ranch on Deer Creek. It is about a five-mile hike.

A little farther along, the scenic byway skirts the northern side of immense Big Bear Lake. Stop at the Big Bear Work Station in Fawnskin, and pick up a folder for the Gold Fever Auto Tour. This 11-mile route over dirt roads goes from the ranger station into Holcomb Valley where gold was mined well over a century ago. The brochure points out such historical sites as the Metzger

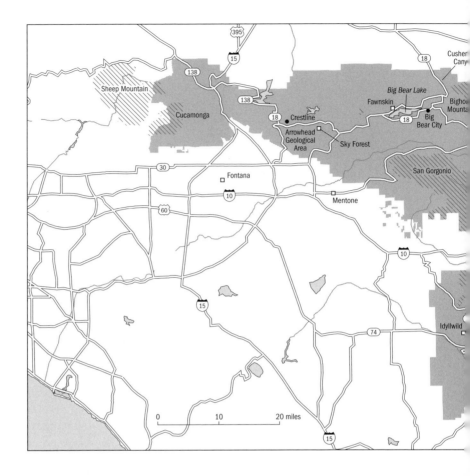

Mine, Two Gun Bill's Saloon, and a Hangman Tree. In the Holcomb Valley area you may be puzzled to see treeless areas covered with thousands of tiny quartzite particles. These small areas, which make up the Pebble Plains, have had an unusual geological history, and the soil type present is home to several plants that can live nowhere else.

At Big Bear City, branch off of State Route 18 and continue the Rim-of-the-World National Scenic Byway on State Route 38, climbing to Onyx Summit for a fabulous view of the desert in the distance. At Barton Flats there are campgrounds and picnic areas and the attractive but small Jenks Lake. Two easy half-mile trails lead from here. The Ponderosa Vista Trail leads to a scenic overlook, and the Journey Through the Whispering Pines Trail wanders through forests of pine and oak.

After going around the northern boundary of the San Gorgonio Wilderness, the scenic byway descends and leaves the forest at the Mill Creek Work Station at Mentone. If you are interested in climbing to the summit of Mt.

N

n Jacinto

10

74

Santa Rosa
Wilderness

Gorgonio, take Forest Trail 1E08 from Camp Creek Falls east of the Mill Creek Work Station.

Two other things of botanical interest in the Big Bear Lake area are worth doing. Just south of Boulder Bay, hike for about a mile to see the largest lodgepole pine tree in the world. Then drive State Route 18 east of Big Bear Lake to Cushenbury Canyon where the limestone ravines are home to five very rare plants. A very rough dirt road leads to the largest Joshua tree in the world. It is an enormous old thing, but many of the branches are lifeless and the plant does not look healthy.

From Palm Desert, the Palms to Pines National Scenic Byway travels from desert habitats on the Bureau of Land Management land at the outskirts of Palm Desert and begins a long but scenic climb into the San Jacinto Mountains of the San Bernardino National Forest. The extreme southeast corner of the national forest has been designated the Santa Rosa Wilderness Area, and most of it is desert. In the area are ocotillos, agaves, and several kinds of cacti that you can see if you take the Cactus Springs Trail from the trailhead on the byway. From the resort community of Idyllwild, there are many options. You may hike into the San Jacinto Wilderness, particularly to the summit of a granite dome known as Tahquitz Peak, or you can venture into the Black Mountain Scenic Area and fish in Lake Fulmor or drive a very rough mountain road from Metate Pass to Boulder Bay Campground and then onto another marvelous vista point known as Fairview Point. On the way you pass Cinco Poses Springs.

Before taking the Palms to Pines National Scenic Byway, you may wish to drive south out of Palm Desert to the Agua Caliente Indian Reservation and get permission to hike the Palm Canyon Trail. After four miles of hiking in the canyon, you enter Forest Service land. Along the trail are specimens of the magnificent California fan palm. As the lower leaves of this tall palm die, they remain on the tree in the form of a dense skirt.

Pebble Plains

The pebble plains in California's San Bernardino National Forest—scattered openings where clay soil is intermixed with particles and small rocks of quartzite—confront plants with a desertlike habitat. Totaling about 550 acres, the plains consist of two dozen patches, ranging in size from 1.5 to 75 acres. Additional patches once existed but have been obliterated by human activity, particularly the creation of Big Bear Lake early in the twentieth century.

Geologist Peter Sadler suspects that the red clay and quartzite pebbles were laid down millions of years ago, during a warmer, wetter period, when the region was relatively flat. Lava flows that overlie some of the clay and pebbles date back six or seven million years—before the San Bernardino Mountains began to uplift. As the mountains formed, the clay and pebbles in some places eroded away or were buried by other materials. Sadler believes that some of the pebble plains are surviving examples of the original deposits, but that others may consist of redeposited sediment. Erosion continues to carry some quartzite pebbles down the slopes of Gold Mountain.

Because of the high elevation, summer temperatures range from 80 degrees F during the day down to 40 degrees F at night (temperatures drop below freezing at least 250 nights a year). The alternate formation and melting of ice crystals pushes soil upward, bringing pebbles to the surface and creating peculiar, finger-shaped ridges in the clay. Annual precipitation on the pebble plains is 10 to 20 inches, mostly falling in winter as rain or snow but with a few heavy summer thunderstorms. Water drains poorly through the fine-grained clay soil, tending to run off or stand in shallow pools. In summer such pools evaporate rapidly, and the land is extremely dry. The reason these areas are not invaded by the adjacent pine forest, botanist Tim Krantz suggests, is that the clay dries out so quickly that pine seedlings cannot survive.

District botanist Maile Neel has recorded 45 kinds of plants from the pebble plains, one-fifth of them considered rare by the California Native Plant Society. Some of the rarities are known only from the pebble plains, whereas others are also found some distance away, farther south in California or in Mexico.

Nearly half the plant species on the plains are annuals that flower and form seeds early in spring and then disappear before the onslaught of summer. Among them is Baldwin Lake linanthus, a two-inch-tall member of the phlox family that has creamy yellow flowers. An annual that blooms later is eye-strain monkey-flower, so named because of its small stature and tiny flowers. The monkey-flower grows at the edges of seeps, where water from springs crosses the clay and pebbly soil. In the spring, these moist areas are bright with flowers, including yellow violets, collinsias, and purple rock cresses.

The perennials have developed various tactics to withstand the interrelated stresses of heat and aridity. Gray, hairy leaves reflect the sun's intense rays, while conserving moisture. Species with hairy leaves include the Bear Valley sandwort, whose small, white, five-petaled flowers appear in June and July, usually following a summer shower. A dwarf member of the carnation family, this sandwort is found on every one of the pebble plains and only there. Among the species whose leaves are gray as well as hairy are Parish's rock cress, which is a purple-flowering mustard, and the ashy-gray paintbrush, which has greenish yellow flower clusters in its northern range and dull-crimson flower clusters in its southern range.

About one-fourth of the species are succulents, which store water in their leaves or stems. An example in Munz's hedgehog cactus, found in the San Bernardino Mountains and in Baja California, Mexico. The leaves of some perennials are matted together in a little cluster that hugs the ground. One is the Pebble Plains buckwheat, found on all the Pebble Plains (often it is the most common plant) and nowhere else. It blooms in July, August, and September by sending up a wiry stalk that bears a head of tiny white flowers. Fuzzy rattail—which grows in the San Bernardino Mountains and in Baja California—is gray, hairy, *and* matted.

The Forest Service is striving to protect the surviving Pebble Plains. But because they are treeless, they are attractive to people riding off-the-road vehicles, which grind down the plants and leave deep ruts during rainy weather. Wild burros also roam the area, and although they pose a lesser threat, they, too, trample the delicate vegetation and rough up the soil, inviting invasion by nonnative species.

Cushenbury Canyon

Only 50 miles east of Los Angeles, the San Bernardino National Forest provides a wide range of vegetation, from the creosote bush and burrobrush of the Mojave Desert to the ground-hugging alpine plants that grow above timberline on 11,499-foot Mt. San Gorgonio, the highest of the San Bernardino Mountains. Between these extremes are extensive woodlands of piñon pine and juniper; mixed conifer forests of white fir, sugar pine, and ponderosa pine; and subalpine forests with limber and lodgepole pines. The terrain consists mostly of granite and quartzite rock, punctuated by outcrops of carbonate rock, including limestone and dolomite. These outcroppings provide the only home of five very rare plants: Parish's daisy, Cushenbury buckwheat, Cushenbury milk vetch, the San Bernardino Mountains bladderpod, and Cushenbury oxytheca.

The carbonate outcrops lie in a 35-mile band running east-west along the northern slopes of the San Bernardino Mountains, at elevations between 3,500 and 8,000 feet. The five rare plants grow at scattered sites within this zone, most often in the understory beneath piñon pine and California juniper, alongside the more common mountain mahogany, Mormon tea, and Mojave yucca. All but the San Bernardino Mountains bladderpod can be found in a deeply incised ravine known as Cushenbury Canyon (the bladderpod is confined to adjacent Big Bear Valley).

Parish's daisy is a 10-inch-tall perennial with rose-colored flower heads and narrow leaves covered by soft, silvery hairs. It is named for the nineteenth-century California explorer-botanist S. B. Parish, who first described it and some of the other five rarities.

Cushenbury buckwheat (Federally Endangered) has tiny, white-woolly leaves, which grow in dense mats up to 20 inches wide. In May and June, clusters of cream-colored flowers rise on four-inch-tall stalks above the cushion of leaves; the flowers turn purplish as they begin to wither. The sprawling Cushenbury milk vetch (Federally Endangered) radiates stems about 12 inches long. Small clusters of purple, sweetpea-shaped flowers appear near the ends of the stems. The San Bernardino Mountains bladderpod (Federally Engandered) is an eight-inch-tall member of the mustard family that bears small yellow flowers, silvery-hairy leaves, and an inflated seedpod. Rarest of all, and the only annual, is the Cushenbury oxytheca (Federally Endangered), a four-inch-tall, white-flowerered plant related to the buckwheat. Also present is the Federally Threatened Parish's daisy.

The carbonate rock that supports these plants also poses the greatest threat to their survival, for it is a desirable commodity. It is so pure that pharmaceutical companies use it as an ingredient in antacids, and other industries use it in sugar refining and rubber manufacturing, as flux for steel, as a whitener for paper, and for fixing dyes in fabrics. The carbonate is also mined for conversion to cement, including the smooth, final coat applied to swimming pools.

Forest plants are unable to cover the unsightly vertical walls left by mining operations. The mining also generates 13 tons of waste material for every ton of ore produced. The gigantic piles of overburden that accumulate are inhospitable to vegetation, and "fugitive" dust is readily blown onto surrounding vegetation, soil, and roadways. Adjacent to cement-making operations, a quarter-inch layer of cement often covers the ground and low-growing vegetation.

The five rare plants that grow on the carbonate rock, already barely clinging to existence because of their restricted habitat, have little chance of surviving continued mining operations. Before 1975, there were no federal regulations on what miners could do to the terrain, and even now the laws on reclamation are weak and poorly enforced.

The mining companies mounted a major effort to convince the U.S. Fish and Wildlife Service not to list the five rare plants as either Federally Endangered or Federally Threatened. One argument used was that these plants are not restricted to carbonate rocks and therefore could be found elsewhere, but botanists have no evidence for this.

In the meantime, the mining industry continues to take advantage of the very outdated 1872 Federal Mining Act, which was designed to encourage private individuals and companies to use public lands more fully. Those who have a plan to carry out mining can apply for a patent on public land. If the patent is approved by the federal government, the land becomes converted to private land at the price of only two dollars an acre.

Sequoia National Forest

SIZE AND LOCATION: 1,136,095 acres in central California at the southern end of the Sierra Nevada, east of Fresno, Visalia, and Porterville and northeast of Bakersfield. State Routes 155, 178, 180, 190, and 198 are the major roads in or near the forest. District Ranger Stations: Dunlap, Springville, Lake Isabella, Kernville. Forest Supervisor's Office: 900 West Grand Avenue, Porterville, CA 93257, www.r5.fs.fed.us/sequoia.

SPECIAL FACILITIES: Boat ramps; marinas; motorcycle and off-highway vehicle areas; winter sports areas.

SPECIAL ATTRACTIONS: Giant Sequoia groves; four National Wild and Scenic Rivers; Boyden Cavern.

WILDERNESS AREAS: Golden Trout (303,511 acres, partly in the Inyo National Forest); Dome Land (130,081 acres, partly on Bureau of Land Management land); South Sierra (60,084 acres, partly in the Inyo National Forest); Jennie Lakes (10,289 acres); Monarch (44,896 acres, partly in the Sierra National Forest); Kiavah (43,803 acres).

The Sequoia National Forest extends from the North Fork of the Kings River south to the Kern River and Piute Mountains, from the foothills of the San Joaquin Valley to the crest of the Sierra Nevada. Enough is here to satisfy almost any aspect of outdoor activities.

Giant sequoia trees, the world's most massive living organisms, are confined to a 15-mile-wide strip on the western slopes of the Sierra Nevada for a distance of about 250 miles. These behemoths live in groves, and there are more of them in the Sequoia National Forest than anywhere else. All of the

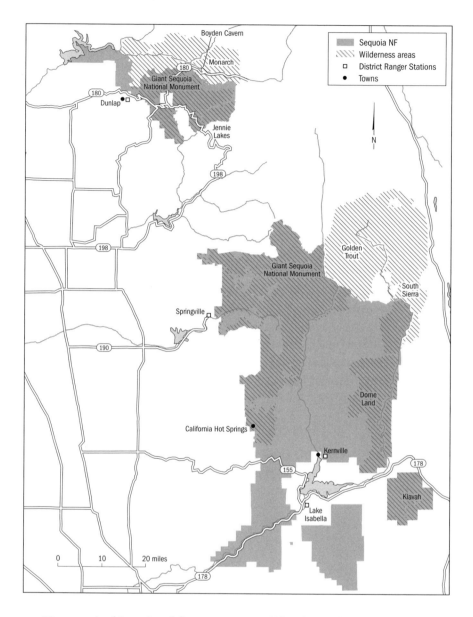

38 groves in this national forest were consolidated into the Giant Sequoia National Monument in 2000. This national monument of 327,000 acres is managed by the U.S. Forest Service. Perhaps the best way to enjoy a great concentration of these trees is to walk the half-mile-long Trail of a Hundred Giants in the Long Meadow Sequoia Grove. Here are 268 sequoias, nearly half of them more than 10 feet in diameter at 4.5 feet above the ground. The largest tree in the grove is 20 feet across and towers 220 feet above the forest floor. The Redwood Campground (a misnomer because these trees are not redwoods), which is adjacent to the grove, is a few miles west of Johnsondale.

Another pleasant walk may be taken through the Freeman Creek Grove, and the trail here parallels and crosses this sparkling mountain stream.

Two sequoias are, or were, of special interest in the forest. The largest living sequoia in any national forest, called the Boole Tree, is north of Converse Mountain off the Hume Lake Road. If you are up to hiking several miles, you can reach this giant that stands 269 feet tall and has a basal circumference of, if you can believe it, 112 feet!! After seeing this tree, you may wish to swim in nearby 887-acre Lake Hume.

The other sequoia you need to see is the stump of the General Noble Tree—a mighty stump it is. It is called the Chicago Stump because the tree was cut down 50 feet above the ground after several days' efforts, hollowed out, and transported to Chicago in 1893 for the World Columbian Exposition. The impressive stump is reached by a short and easy trail off the Verplank Road near State Route 180.

Sequoia National Forest has hiking trails for everyone. The more rugged hiker will head to the six wilderness areas where trails ascend mighty mountains, descend into rocky gorges, pass by tranquil wildflower meadows and blue lakes, and make their way through endless forests of pines and firs.

We personally like the small Jennie Lake Wilderness in the Lake Hume area, perhaps because it is less intimidating. There is fairly easy hiking to Jennie and Weaver Lakes and difficult hiking to the top of 10,365-foot Mitchell Peak. In this wilderness you find lots of soft-looking western white pines, lodgepole pines, and spire-topped red firs.

Monarch Wilderness is also relatively small, but we do not recommend it for tender feet. If you can make it to Spanish Mountain, you will be rewarded with the most magnificent scenery to be found anywhere. The deep, rocky canyons are particularly brightly colored when the sun reflects off the rocky surfaces.

Golden Trout Wilderness is named for California's state fish. In addition, the Federally Threatened Little Kern Golden Trout is here. Fishermen should note that there are special restrictions when fishing for these fish. When viewed from the air, the wilderness area looks like a large basin totally surrounded by jagged mountain peaks. As you hike toward timberline, the bushy-looking foxtail pine will be seen. Lower down from the foxtails are extensive forests of Jeffrey pines, and below them, open woods of scattered piñon pines. The North Fork and South Fork of the Kern River, both designated National Wild and Scenic Rivers, flow through this wilderness. The North Fork has Class IV and V rapids and waterfalls.

South Sierra Wilderness is on the southeastern side of the national forest. The terrain on the Sequoia National Forest side of the wilderness does not have the rugged canyons that are found on the Inyo side. Eleven miles of the Pacific Crest Trail are in this wilderness.

Near the southeastern corner of the Sequoia National Forest is Dome Land Wilderness, named for its uncountable number of bare granite domes. This wilderness is drier than the others and supports much less vegetation. In fact, one-fourth of the wilderness is in the Bureau of Land Management's California Desert District. Nonetheless, where the South Fork Kern River crosses the Dome Land, it has carved unbelievably deep, rocky gorges.

For those of us who prefer less strenuous trails outside the wilderness areas, we suggest a part of the Cannell National Recreation Trail that begins two miles north of Kernville on Mountain Highway 99. The initial part of this trail involves a gradual climb with views into the Kern River Valley. If you want to work up more of a sweat, continue on the steeper part of the trail to Pine Flat. From here you can continue on along and across Cannell Creek, or you can retrace your steps back to Mountain Highway 99.

Fourteen miles beyond the Cannell Trailhead on Mountain Highway 99 is the Flynn Creek drainage. Two trails from here are of special interest. The Packsaddle Cave Trail climbs steeply through a digger pine forest to a cave that has a few stalactites and stalagmites, although vandals have removed many of them. The Flynn Trail has a steep ascent to Speas Ridge but, if you can get there, you will be rewarded by the presence of the rare Brewer's oak. You will know it by its round-lobed leaves that feel like velvet.

If you are into unusual plants, you can follow an easy trail up the north slope of Bald Eagle to see a fine grove of the gray-leaved Piute cypress. Foxtail and digger pines are on this slope as well.

For a rewarding easy trail the whole family will enjoy, climb to the summit of Sunday Peak from Forest Highway 90.

If you want to explore a "civilized" cave with a guide, you must drive through the Sequoia National Park and back into the National Forest to the depths of the 8,000-foot deep King River Canyon. There you find Boyden Cave, situated beneath the 2,000-foot-high marble walls of Kings Gates. On the 45-minute tour through the cave, which has a constant temperature of 55 degrees F, you will marvel at such sights as The Pancake Room, the Upside Down City, Baby Elephant and Mom, and the Christmas Tree. A fee is charged for going through the cave. The cave is open to the public from April through October.

If you are interested in water-related activities, including boating, the 11,000-acre Isabella Lake south of Dome Land Wilderness may be for you. But in winter, Sherman Pass, atop a mountain that seems to rise up out of desertlike terrain, has the best snowmobiling areas in the forest. A good ski area is found at Greenhorn Summit. The Forest Service permits motorcycle and off-highway vehicle riding in special areas, and there are equestrian trails galore.

Shasta-Trinity National Forest

SIZE AND LOCATION: 2,153,544 acres in north-central California. Interstate 5 bisects the forest, while State Routes 3, 89, and 299 provide good access. District Ranger Stations: Mountain Gate, Hayfork, Weaverville, Mt. Shasta. Forest Supervisor's Office: 3644 Avtech Parkway, Redding, CA 96002, www.r5/fs/fed/us/shastatrinity.

SPECIAL FACILITIES: Boat ramps; swimming beaches; winter sports areas.

SPECIAL ATTRACTIONS: Whiskeytown-Shasta-Trinity National Recreation Area.

WILDERNESS AREAS: Castle Crags (8,627 acres); Mt. Shasta (38,200 acres); Yolla Bolly–Middle Eel (153,841 acres, partly in the Mendocino and Six Rivers National Forests and 8,500 acres on Bureau of Land Management land); Trinity Alps (502,764 acres, partly in the Klamath National Forest); Chanchelulla (8,200 acres).

In the first decade of the twentieth century, President Theodore Roosevelt proclaimed the Shasta and the Trinity as two new national forests in California. In 1954, these two forests were combined to form the massive Shasta-Trinity National Forest. The Trinity Mountains dominate the western Trinity half of the forest, and mighty Mt. Shasta is the centerpiece of the eastern half. Wherever you wander in the forest, you usually are not far away from a view of snow-capped Mt. Shasta. At 14,162 feet, it is topped in the lower 48 states only by the 14,496-foot Mt. Whitney. Mt. Shasta Wilderness surrounds the mountain and also includes seven glaciers, waterfalls, lava flows, hot springs, and other reminders of volcanic activity. More rugged hikers may climb to the summit of Mt. Shasta. If you are able to make it to the top, you will smell sulfur that is belching from fumaroles.

Castle Crags Wilderness is a neat but rugged region of sheer granite cliffs and tantalizing spires that reach to 7,200 feet. Although the spires may look inviting to the experienced climber, you must not try to ascend them because of their very crumbly nature. Just be content to follow the 19 miles of the Pacific Crest Trail (pl. 25) that is in the wilderness.

An exciting 17-mile back road runs from the nearly extinct town of Helena to the Hobo Gulch Trailhead into the Trinity Alps region. The road follows Backbone Ridge (pl. 26) above the East Fork for most of the way, passing the ruins of several abandoned mines. After about six miles on the road, take the two-mile Digger Pine Flat Trail westward to a large stand of digger pines, an attractive tree with its large, heavy cones. After twisting over Stoveleg Gap, you will eventually arrive at Hobo Gulch. Many trails radiate

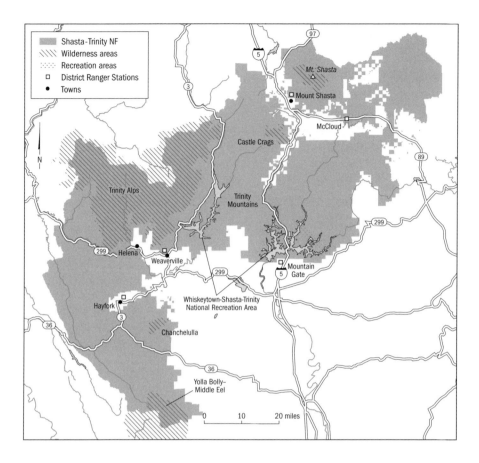

from Hobo Gulch. One of the most interesting heads due north along the scenic Trinity River, because eight miles from the trailhead, in the middle of Pfeiffer Flat, you come upon a one-room log cabin. The Jorstad Cabin was constructed in 1937 and occupied until 1985 by George Jorstad who worked his mining claim during the summer and then returned to San Francisco in the winter where he was employed as a journalist.

Eight miles southeast of the community of Hayfork, just off the road that parallels Hayfork Creek, is a splendid natural bridge. A picnic area has been provided for visitors to this fascinating geological attraction. Southeast of the natural bridge is tiny Beegum Campground, which is beautifully situated near a wonderful rocky canyon along Beegum Creek.

Another historic cabin, this one reconstructed, is at the end of a quarter-mile trail from the Ah-Di-Na Campground. Originally on the site of an Indian settlement, this cabin was a fishing retreat for the William Randolph Hearst family. It is about 18 miles south of McCloud.

Samwell Cave, at one time an Indian holy place, may be of interest to spelunkers. To visit this cave, which is off the Gilman exit of Interstate 5, you need to obtain a key from the local ranger station.

Much of the Shasta-Trinity National Forest that is not in wilderness is part of the Whiskeytown-Shasta-Trinity National Recreation area. This vast region is centered around three large impoundments—Lake Shasta, Lake Clare Engle, and Whiskeytown Lake. All have facilities for every kind of water sport you may wish. Whiskeytown Lake is managed by the National Park Service.

The small Chanchelulla Wilderness, centered around 6,399-foot Chanchelulla Peak, has steep slopes with chaparral near the base and coniferous forests above. No established trails, no lakes, and no streams are in this rather inaccessible area.

In a lot of brushy foothills in the forest, chamise and manzanitas are the dominant species. As you ascend the mountains, you find massive forests of hemlock, sugar pine, incense cedar, Jeffrey pine, and Douglas fir above an understory of tanoak, manzanitas, chinquapin, and snowbrush. At 8,000 feet and higher, gnarly whitebark pine and white fir are common.

Sierra National Forest

SIZE AND LOCATION: 1,300,000 acres in central California on the western slope of the Sierra Nevada, east of Merced and Fresno and south of Yosemite National Park. State Routes 41 and 168 are the main accesses. District Ranger Stations: Prather, North Fork. Forest Supervisor's Office: 1600 Tollhouse Road, Clovis, CA 93611, www.r5.fs.fed.us/sierra.

SPECIAL FACILITIES: Boat ramps; swimming beaches; motorcycle and off-highway vehicle areas; winter sports areas.

SPECIAL ATTRACTIONS: Nelder and McKinley Groves of sequoias; Yosemite Mountain–Sugar Pine Railroad; Sierra Vista National Scenic Byway; Sierra Heritage Scenic Byway; Merced National Wild and Scenic River.

WILDERNESS AREAS: Ansel Adams (230,258 acres, partly in the Inyo National Forest); John Muir (580,293 acres, partly in the Inyo National Forest); Dinkey Lakes (30,000 acres); Kaiser (22,700 acres); Monarch (44,896 acres, partly in the Sequoia National Forest).

The Sierra National Forest seems to have everything. It boasts of 63 campgrounds, 11 reservoirs, 5 wilderness areas (comprising 43% of the forest),

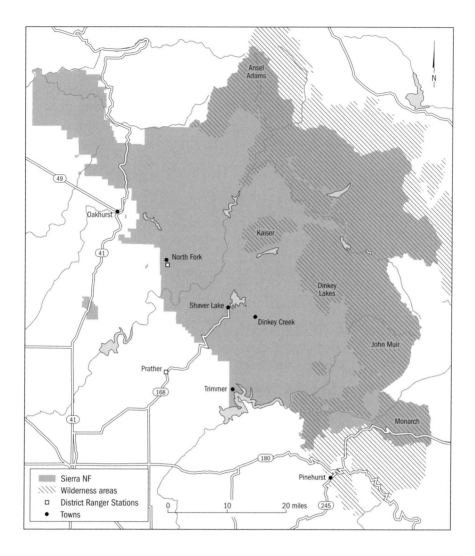

480 lakes, and 1,800 miles of rivers and streams. Outstanding in the Sierra National Forest are the sequoia groves. A few miles south of Yosemite National Park is a dirt road, Nelder Grove Road, that leads to a secluded grove of sequoias known as Nelder Grove. Here, in the solitude of nature, is this forest of giants that will create in you a new sense of reverence and inner peace. In this grove are 101 of these mature mammoths. Bull Buck, the most massive one, stands 247 feet tall and has a circumference around its base of 99 feet. It is estimated to be about 2,700 years old.

A second sequoia grove, the McKinley, lies southeast of the Dinkey Lakes area. Hikers will find this grove easy to maneuver with its paved trail that traverses through the magnificent trees.

A marvelous trail outside of the sequoia groves is the Lewis Creek National Recreation Trail, which for 3.7 miles follows the route of the Madera–Sugar Pine Lumber Company flume. Halfway along the trail is the 80-foot Corlieu Falls, and farther on is the 30-foot Red Rock Falls. Another nice trail is the two-mile White Bark Vista Trail that leads off from Kaiser Pass. It ends with a spectacular view nestled among whitebark pines.

Fishermen will have trouble trying to decide in which of the myriad lakes and streams to cast their lines, but one of the favorite spots is Bass Lake just east of Oakhurst. Campgrounds are available, and the lake is great for swimming and waterskiing. The South and Main Forks of the Merced River are also ideal for various water sports. Seventy-nine miles of the Main Fork have been designated as a Wild and Scenic River, as has the entire 43-mile length of the South Fork. As you wander along the shores, you may see red-legged frogs, water shrews, and willow flycatchers.

For a change of pace, you may want to take your family on the Yosemite Mountain–Sugar Pine Railroad for a four-mile ride along a narrow gauge railroad with cars pulled by a Shays steam engine. The route follows the Madera–Sugar Pine Lumber Company route. Board the train at the station about four miles south of the Yosemite National Park's South Entrance. There is a charge for riding the train.

If you are into wilderness hiking, the Sierra National Forest is for you. Massive John Muir Wilderness follows the crest of the Sierra Nevada for 100 miles, from Mammoth Lakes to Kings Canyon National Park; only a part of this wilderness is in the national forest. You can see alpine areas, mountain meadows, crystal-clear lakes, and bubbling streams. The 14,486-foot Mt. Whitney, the highest mountain in the lower 48 states, is here. The wilderness may be accessed off of the Kaiser Pass Road or the Dinkey Creek Road.

Half as large but equally scenic is the Ansel Adams Wilderness. Much of the wilderness is composed of the Minaret Range. The higher elevation areas have virgin stands of red fir, various pines, and quaking aspens. Granite outcrops and rocky gorges add to the beauty. Small glaciers are in evidence on the colder north-facing slopes of the mountains. Some of the lakes in this wilderness form the headwaters of the Middle and North Forks of the San Joaquin River.

The Monarch Wilderness is in the Kings River drainage and is the most rugged of them all. In fact, that part of the wilderness in the Sierra National Forest has no trails. Steep cliffs lead up to 11,077-foot Hogback Mountain.

Kaiser Wilderness climbs gradually from the north edge of Huntington Lake to 10,320-foot Kaiser Peak. You begin the trail in a nice forest of Jeffrey pine; at the top is a magnificent stand of red fir. Access is from Huntington Lake and also along the Kaiser Pass Road and the Stump Springs Road. While you are in the Huntington Lake area, you may wish to hike the one-mile trail to Rancheria Falls, which is 50 feet wide and 150 feet high.

Dinkey Lakes Wilderness is in more rolling terrain but still at high elevation. The entire wilderness is above 8,000 feet. Mountain meadows alternate with granite outcrops throughout the area. You may access this wilderness from areas along the Dinkey Lakes Road and along State Route 168.

For a scintillating trip through the Sierra National Forest via conventional vehicle, you must take the newly designated Sierra Vista National Scenic Byway. Start at the North Fork Ranger Station with a quarter-mile nature hike that is suitable for the handicapped. Then drive to the Redinger Overlook where you get a bird's-eye view of Redinger Lake, the San Joaquin River, and the Sierra front country. At the Clearwater Fire Station, take time to marvel at the Jessie Ross Cabin. This 1.5-story log structure was built in the late 1860s by Jessie Ross, who made his living selling packing supplies to miners. When he built this cabin at a site one-half mile south of where it has been restored today, it could be reached only by a primitive trail in the mountain. At Mile High Vista is a spectacular view of the Minarets and 13,157-foot Mt. Ritter. A short side trip takes you to a nifty granite arch. As you cross Portuguese Creek, you are in the midst of a dense forest of lodgepole pine. Take a short hike to some glacier-carved granite domes called The Balls. A little farther along the road is Globe Rock, a balanced rock worthy of a close look. The scenic byway climbs to Cold Springs Summit where you may then take the road that goes past Fresno Dome and Nelder Grove.

The Sierra Heritage Scenic Byway begins in the town of Clovis in the San Joaquin Valley and follows State Route 168 to Kaiser Pass Road and White Bark Vista. Along the way are rare plants, high Sierra lakes, and scenic vistas. One of these rare plants is found along State Route 168 southeast of Shaver Lake, in an area known as the Big Sandy Bluffs. This is a shrub known as carpenteria, found only in California; it grows in considerable numbers in what has been designated a Botanical Area. Carpenteria grows to a height of 15 feet and, during May and June, produces masses of large white flowers with a yellow center. John C. Fremont discovered this plant in 1845, and it is confined to chaparral in an area 17 miles by 12 miles. If you get to the Sierra National Forest in May or June, be sure to look for this rarity.

If you travel by motorcycle, there are 60 miles of motorcycle trails at the Miami Motorcycle area. If off-highway driving is your thing, there are 12 areas in the Sierra National Forest set aside, including the 30 miles of the Dusy-Ershim Route that is a narrow corridor between the John Muir and Dinkey Lakes wilderness areas.

Nelder Grove

When the naturalist and philosopher John Muir was locating and inventorying the stands of California's giant sequoias during the early 1870s, he came upon a handsome grove a few miles south of the now famous Mariposa Grove of Yosemite National Park. As he wandered through the pristine forest, which was bisected by a placid, picturesque creek, he was startled to come upon a small, well-constructed log cabin. In front of it, perched on a bark stool, was an old and heavily weathered pioneer. Several years later, Muir recalled his encounter with this former, mostly unsuccessful gold prospector:

> The name of my hermit friend is John A. Nelder, a fine, kind man, who in going into the wood has at last gone home; for he loves nature truly, and realizes that these last shadowy days with scarce a glint of gold in them are the best of all. He bade me welcome, made me bring my mule down to his door and camp with him, promising to show me his pet trees.

Today, this sequoia grove and the creek that runs through it bear Nelder's name.

Giant sequoias, whose diameter may reach 30 feet and whose height approaches 300 feet, are the largest living things in the world. The redwoods of coastal California and southern Oregon are taller, but they lack the great girth of the sequoias. Both trees belong to the family Taxodiaceae, which also includes the swamp-inhabiting bald cypress of the southeastern United States. Giant sequoias have thicker, redder bark and more sharply pointed leaves than the redwoods. Another difference is that redwoods are able to sprout from their stumps, whereas sequoias cannot.

The gigantic size of sequoias can be attributed to their longevity and the ample supply of moisture and nutrients where they live. Unlike animals, most plants continue to grow throughout their lifetime. If moisture and nutrients are plentiful, they may grow rapidly. Sequoias survive fires because their soft, spongy bark is up to two feet thick. The wood itself is decay resistant, making the trees less susceptible to damage by insects and fungi (even fallen sequoia logs may remain intact for hundreds of years before beginning to show signs of deterioration). Thus the trees can live to ripe old ages, growing taller and thicker each year. The General Sherman Tree in Sequoia National Park, apparently in its prime of life, is about 3,500 years old and still going strong. Speculation that some sequoias alive today may live to be 10,000 years old provokes heated debate among biologists.

A sequoia usually does not begin to form cones until it is at least 125 years old. On average, a mature tree produces some 1,500 cones per year. Each cone takes about 18 months for its 150 to 300 seeds to mature, and the seeds may not be shed for as long as 20 years. The living part of the seed is only about a quarter-inch long, yet it has the potential of developing into a tree that may contain 50,000 cubic feet of vegetable matter. Approximately two-thirds of the seeds are abnormal, and one-fourth of the rest are likely to be damaged by the Douglas chickaree (reddish squirrel) and other animals. The seeds that do germinate successfully remain in a precarious position, as desiccation of the soil during a seedling's first few years will probably cause it to die.

The sequoia seed germinates with the formation of a tiny taproot, sometimes while snow still covers the ground. After a few years, the taproot is replaced by an extensive network of feeder roots that eventually may extend outward as much as 300 feet and cover three acres. The bark remains gray until the tree is about 20 years old; then it takes on its characteristic fibrous texture and rich red color.

For the first 150 to 300 years of its life, the giant sequoia is shaped like a Christmas tree, tapering to a graceful, pointed top, but as the years progress, the branching pattern results in a broadly rounded top. Because of its great height, the giant sequoia is vulnerable to lightning strikes, which may knock out part of the crown. Many old trees also exhibit "snag-tops," which are dead branches in the crown. They result when ground fires damage part of the water-transport system within the trunk, cutting off the supply of water to the affected branches high up the tree.

Nelder Grove is one of 75 distinct stands of giant sequoias that dot the western slopes of the Sierra Nevada for nearly 250 miles, from about 20 miles west of Lake Tahoe south to within 40 miles northeast of Bakersfield. My first hike through Nelder Grove's tranquil Shadow of the Giants Trail came less than an hour after I had been jostled in a crowd of tourists as we were bused in and out of the Mariposa Grove in Yosemite. Mariposa Grove is parklike, with trees that have never felt the woodsman's ax and an understory kept clean through the removal of brush and the trampling by thousands of visitors each year. Beneath the tall trees in Nelder Grove, in contrast, is a dense growth of small trees, shrubs, and wildflowers that would be difficult to traverse were it not for the hiking trail.

Within two decades after John Muir had chanced upon his hermit friend, about half of the giant sequoias in the grove were cut off as much as 20 feet above the ground, and the great tops of the trees crashed to the forest floor, often splintering into uselessness. Giant limbs and huge stumps were left in the forest by the lumbermen who destroyed one-third of the giant sequoias in the nation between 1880 and 1930. Fewer than 50 mature sequoias were spared in Nelder Grove. Today there are 101 mature sequoias.

The trees that fell, either through natural causes or people's destructive force, and the dense brushlike vegetation that ensued, provided fuel for fires, which under natural conditions rage through sequoia groves as often as five times each century. Trees that grow with the sequoias, such as white firs, sugar pines, and incense cedars, are able to survive only a few of these fires, but sequoias, with their thick, asbestos-like bark, show only burn scars after a fire and are seemingly only a little worse for wear. A sequoia more than 2,000 years old may have felt the heat of 100 fires in its lifetime, but it continues to grow and reproduce.

As long as sequoias, firs, pines, and cedars—all capable of growing taller than 200 feet—live unmolested in a grove, they create a thick canopy that little sunlight can penetrate. In addition, their masses of fallen leaves accumulate on the forest floor. Under these conditions of dense shade and heavy ground litter, sequoia seeds will not germinate. This is true of Mariposa Grove, where lumbering has never occurred and where fire has been suppressed since the establishment of Yosemite National Park. Ironically, the effort to protect such mature stands of sequoias has threatened their regeneration.

But where sunlight pours in and the soil is scarified by fire or human disturbance, sequoia seeds may germinate and develop into saplings. Nelder Grove is such a place. At the grove, hikers can see trees that are less than 100 years old, a new generation of future giants.

Six Rivers National Forest

SIZE AND LOCATION: 980,000 acres in northwestern California, east of Crescent City and Eureka, occupying a rather narrow strip for 150 miles south from the Oregon border. U.S. Highway 199 and State Routes 36, 96, and 299 are the major roads. District Ranger Stations: Gasquet, Willow Creek, Bridgeville, Orleans. Forest Supervisor's Office: 1330 Bayshore Way, Eureka, CA 95501, www.r5.fs.fed.us/sixrivers.

SPECIAL FACILITIES: Boat ramps; winter sports areas; visitor center.

SPECIAL ATTRACTIONS: Several National Wild and Scenic Rivers; six designated Botanical Areas; Smith River National Recreation Area; Smith River National Scenic Byway; Trinity River National Scenic Byway; Bigfoot National Scenic Byway.

WILDERNESS AREAS: North Fork (7,999 acres); Siskiyou (152,860 acres, partly in the Klamath and Siskiyou National Forests); Trinity Alps (502,764 acres, partly in the Klamath and Shasta-Trinity National Forests); Yolla Bolly–Middle Eel (mostly in the Shasta-Trinity and Mendocino National Forests).

Like the three tenors, the Six Rivers National Forest is incomparable. The erosional processes of the Smith, Klamath, Mad, VanDuzen, Trinity, and Eel rivers have created a rough and rugged terrain, much of which is accessible only by hiking or horseback riding.

Although this forest is home to tall coastal redwoods, which seem to be trying to reach the sky, Douglas fir is the predominant tree in the mountains. Almost all Douglas fir forests have an understory of tanoak and madrone. In the high country of the Six Rivers National Forest, white fir becomes significant. Mixed evergreen forests, oak woodlands, and grasslands occur at lower elevations, and the rivers and streams provide rich riparian habitats including wet meadows.

Because of the six rivers and their tributaries, water-related activities top the list of things to do. Fishing for steelhead, salmon, and trout is the best in the country. In the wilder parts of the rivers, whitewater rafting will challenge even the best.

The forest has designated six botanical areas to protect plants and their communities. Within these areas you find distinctive plant communities that support a high number of rare species. For example, the North Fork Botanical Area protects 21,370 acres of the North Fork of the Smith River watershed. Typical forest communities dominated by Jeffrey pine are on the ridgetops, with smaller stands of lodgepole pine and knobcone pine interspersed. The graceful and regional Port Orford cedar is the primary conifer in riparian areas; beneath them the western azalea adds a riot of color in season. Most of the rarer species occupy very open, rocky habitats where they grow from serpentine soils, which are anathematic to most plant species. Common wildflowers you may see include the frosted paintbrush, serpentine phacelia, cliff fleabane, and coast flat-stemmed onion. If you are lucky enough to be in the right place at the right time and have a good plant guide with you, you may run across uncommon serpentine plants such as Howell's sandwort, nodding arnica, Bolander's lily, and Vollmer's lily. Nodding eriogonum is even rarer, and Mcdonald's rock cress is listed as Federally Endangered.

One of the nicest is the Myrtle Creek Botanical Area, with an easy half-mile foot trail that runs from a redwood forest through a tanoak community to a sparsely populated knobcone pine forest. The beauty of the trail is enhanced along Myrtle Creek by the presence of Port Orford cedar and red alder. Seepage areas are home to the insectivorous Cobra lily and the pretty five-finger maidenhair fern. The ground-hugging shrub called modesty, a plant related to mock orange, is here, as are the beautiful California rose-bay and western azalea shrubs. Redwood ivy, one of the so-called inside-out flowers, grows beneath the redwoods. For an even easier way to observe the Cobra

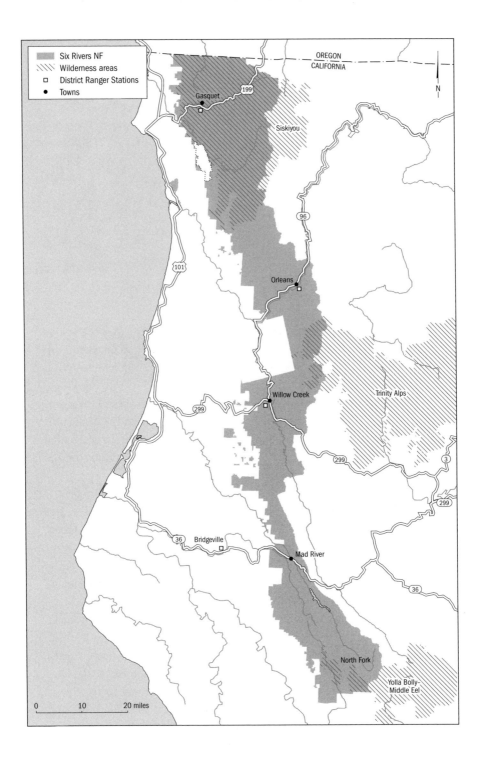

Six Rivers NF
Wilderness areas
□ District Ranger Stations
● Towns

OREGON
CALIFORNIA

N

Gasquet
199
Siskiyou

96

101

Orleans

Willow Creek
299

299

Trinity Alps

3

299

Bridgeville
36

Mad River

36

North Fork

Yolla Bolly-
Middle Eel

0   10   20 miles

lily, take the Darlingtonia Trail just east of the Smith River National Recreation Area Visitor's Center. Two platforms along the trail allow you to look into this boggy habitat without damaging the fragile vegetation or getting your feet wet.

The significance of Bear Basin Butte Botanical Area lies in the presence of 14 different kinds of coniferous trees. In addition to the usual pines, firs, and incense cedar, this 7,500-acre area in the western Siskiyou Mountains has the locally endemic and graceful Brewer's weeping spruce, the equally graceful Port Orford cedar, and one of the rarest of all the conifers, the Alaska yellow cedar (fig. 13).

Figure 13.
Alaska yellow cedar.

The Horse Mountain Botanical Area (1,080 acres) provides marvelous views from the mountain summit. To the west, extending as far as you can see, is the Pacific Ocean, while to the east are the towering snow-covered peaks of the Trinity Alps.

The geological highlight of the Lassics Botanical Area is Lassic Peak, composed of mudstone and sandstone. The 3,640 acres around the peak compose the botanical area; so unique are the rocks that it is also a designated geological area! The 1,060-acre Broken Rib Botanical Area rounds out the areas of significant botanical interest.

For the hiker, the most popular hiking trail, although steep and rugged, is the Devil's Punchbowl Trail. Climbing 1,500 feet in about three miles, this trail offers the best views of the surrounding mountains of any in the forest. Two very picturesque lakes are also visible from the trail. Other trails, most of them rated moderate to difficult, occur in the four forest wilderness areas.

If you are not into strenuous hiking or just want to have a more relaxing day, there are three scenic byways that allow the forest visitor a chance to experience Six Rivers' magnificent scenery from the comfort of a vehicle. Twenty-seven miles of the Smith River National Scenic Byway stay near the river for scintillating views of wildly splashing rapids and deep, rugged canyons. This scenic route follows U.S. Highway 199 all the way to the Oregon border from Crescent City. The Trinity River National Scenic Byway parallels the Trinity River for equally beautiful scenery, including cliffs with gray-green gray pines growing from them. Another nice route is the Bigfoot National Scenic Byway that follows State Route 96 from Willow Creek.

Stanislaus National Forest

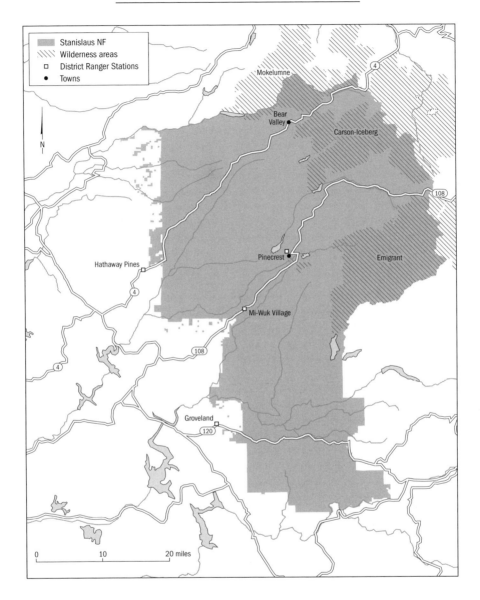

SIZE AND LOCATION: 899,000 acres in central California, east of Sonora, on the western slope of the Sierra Nevada. State Routes 4, 108, and 120 serve the area. District Ranger Stations: Groveland, Hathaway Pines, Mi-Wuk Village, Pinecrest. Forest Supervisor's Office: 19777 Greenley Road, Sonora, CA 95370, www.r5.fs.fed.us/stanislaus.

SPECIAL FACILITIES: Boat ramps; swimming beaches; winter sports areas.

SPECIAL ATTRACTIONS: Reconstructed Mi-Wuk Indian Village; Merced National Wild and Scenic River; Tuolumne National Wild and Scenic River; Trail of the Gargoyles; Columns of the Giants.

WILDERNESS AREAS: Carson-Iceberg (161,181 acres, partly in the Toiyabe National Forest); Emigrant (112,277 acres); Mokelumne (99,161 acres, partly in the Eldorado National Forest).

The wild and rugged Stanislaus National Forest is wedged between the Eldorado and Toiyabe National Forests to the north. Yosemite National Park borders nearly the entire eastern edge of the forest to the south. Nearly one-third of this large forest is designated wilderness area.

The Emigrant Wilderness is located entirely within the Stanislaus. It lies against the northwest corner of Yosemite National Park on the western slope of the Sierra Nevada. Covering an area 25 miles long and 15 miles wide, this is a wild region of alpine lakes, deep canyons, rocky domes, volcanic peaks, and wildflower meadows. This is one of the areas through which the gold seekers made their way across the Sierra Nevada. Four hundred eighty-five people, 135 wagons, and more than 3,000 head of livestock crossed the old Emigrant Pass at the eastern edge of the wilderness between 1851 and 1854. Hiking over the pass today makes you realize how rugged these pioneers had to be. At Saucer Meadow is the grave of one of these pioneers. Other areas in the wilderness that conjure up memories of the past are Emigrant Basin, Emigrant Meadow, and Starvation Lake. One of the best places to enter the wilderness is from the Deadman Campground at the northern end of the region. For nearly two miles the trail follows Kennedy Creek through Kennedy Meadow. Then the left fork continues along Kennedy Creek for about six miles to lovely Kennedy Lake, beautifully situated at the foot of 10,718-foot Kennedy Peak. The right fork skirts the east side of Relief Reservoir before coming to Saucer Meadow. The trail then follows Summit Creek for more than 10 miles before reaching Emigrant Pass. Along the way the trail passes several gorgeous wildflower meadows and several lakes, including the elegant Emigrant Meadow Lake. From Sonora Pass on State Route 108, just outside the wilderness, you can access the Pacific Crest Trail and, after four miles, the trail passes east of Leavitt Peak, the highest in the wilderness at 11,570 feet.

Mokelumne Wilderness is shared with the Eldorado National Forest. The Stanislaus part of this wilderness lies north of and adjacent to State Route 4, which passes through Hermit Valley. From near the Mosquito Lakes Campground along this road is an interesting trail that goes northwest through Sandy Meadow, in full bloom during the summer, and then into Jackson Canyon before coming to the secluded Wheeler Lake. The entire round trip is about 10 miles. The North Fork of the Mokelumne River traverses the wilderness.

Just east of where the river becomes a part of the Salt Springs Reservoir is a glistening pool of blue water known as Blue Hole.

Also from near the Mosquito Creek Campground is a trail that leads south into the Carson-Iceberg Wilderness which is shared with the Toiyabe National Forest. This trail parallels Pacific Creek before entering the wilderness as it swings east of Bull Run Peak. From here the trail eventually makes its way to the northeast end of the narrow Spicer Reservoir. Just before reaching the reservoir is a branch trail that goes through rugged Jenkins Canyon before finally emerging at the Clark Fork Campground on Clark Fork Road. Part of the name of the Carson-Iceberg Wilderness is derived from a huge rock formation, The Iceberg (fig. 14), at the southern end of the wilderness. A one-mile trail along Disaster Creek from the end of Clark Fork Road takes the hiker to the west-

Figure 14. The Iceberg, view from Iceberg Meadow.

ern side of The Iceberg, but it is prominent at the end of the Clark Fork Road.

Before the gold seekers came through this area, Mi-Wuk Indians lived in these forests. The Indians lived in villages as they hunted and fished and processed plants for food, fiber, and medicine. The Mi-Wuk lived here until 150 years ago before being displaced by the gold seekers, but some never left. Others have returned to the area to live. The Forest Service has reconstructed a Mi-Wuk Village across the street from the Summit Ranger Station in Pinecrest. A guide booklet interprets the 14 stops along the easy quarter-mile trail. You will see such things as a sweat house, a conical one-family bark house, a granary, and a bedrock mortar used for pulverizing acorns. The village is beneath typical midmontane vegetation of sugar pine, ponderosa pine, lodgepole pine, white fir, and incense cedar.

Many different rock formations are in the Stanislaus. An interesting array of rocks called the Dardanelles can be viewed from the Dardanelle Overlook along State Route 108. To see some strange formations, take the Trail of the Gargoyles that takes you past hardened mud flows, glacial boulders, solidified ash, and iron-filled columns. The trail begins six miles up the Herring Creek Road from State Route 108. A guide leaflet explains these phenomena, which have such fanciful names as Devil's Horns, Gargoyle Ridge, Strip of Fire, River of Stone, Satan's Slipper, Pages of the Earth, and the Wall of Noses.

An outstanding feature in the Stanislaus is the Columns of the Giants just west of the Pigeon Flat Campground off of State Route 108, about 23 miles east of Strawberry. Similar to the more famous Devil's Postpile National Monument, this area of 105 acres contains outstanding examples of basaltic columns, some approaching 60 feet in height.

The Merced, Tuolumne, Mokelumne, Clavey, and Stanislaus Rivers have carved several deep and scenic canyons into the western slope of the Sierra Nevada. Eleven miles of the Merced River and 29 miles of the Tuolumne River in the Stanislaus National Forest have been designated Wild and Scenic Rivers. Whitewater rafting, with Class IV and V rapids, may be enjoyed in the Tuolumne and the North Fork of the Stanislaus Rivers.

Two favorite hikes for scenery are the rather easy one-mile trail from Lake Alpine to Duck Lake and the 1.5-mile trail from Lake Alpine to Inspiration Point for a super view. Part of the curved Bourland wooden train trestle dating back to 1923 but abandoned in the 1950s may be seen a few miles southeast of Strawberry. Built for the West Side Lumber Company, it is the largest remaining bridge in the Sierra Nevada. The bridge is 315 feet long and 75 feet tall. It is usually accessed either via the Cottonwood Road out of Tuolumne or Forest Road 31 out of Long Barn.

About eight miles southeast of the Groveland Ranger Station off of Moore Creek Road is an open limestone grotto known as Bower Cave. In its heydey from 1861 to 1874, the cave was run by entrepreneurs who sold food to the gold miners. When a dance floor was laid in the cave, the area became a center of activity. Eventually a hotel and service station were added nearby. Today, only foundations of these early structures remain, and trees grow from the cave's floor. Those wishing to explore this area will need a permit from the Groveland Ranger Station.

Tahoe National Forest

SIZE AND LOCATION: 795,205 acres in north-central California, north and west of Lake Tahoe. Interstate 80 and State Routes 20, 49, and 89 are major roads. District Ranger Stations: Comptonville, Foresthill, Sierraville, Truckee. Forest Supervisor's Office: 631 Coyote Street, Nevada City, CA 95959, www.r5 .fs.fed.us/tahoe.

SPECIAL FACILITIES: Boat ramps; swimming beaches; winter sports areas; off-highway vehicle staging areas; visitor center.

SPECIAL ATTRACTIONS: Northernmost stand of giant sequoias; Yuba-Donner National Scenic Byway.

WILDERNESS AREA: Granite Chief (19,048 acres).

Located in the central Sierra Nevada, the Tahoe National Forest is a land of contrasts. The eastern side of the forest has gently rolling topography. The western side is more dramatic with deep, dissected river canyons penetrating the western slopes of the Sierra Nevada. The Middle Fork of the American River forms much of the southern boundary of the forest, and the Yuba River bisects the forest at about its middle. The North Yuba cuts across the northern edge of the national forest. The lowest elevation along the Middle Fork is about 1,300 feet, whereas Mt. Lola tops out at 9,143 feet.

Gold seekers passed this way 150 years ago, seeking a route over the Sierra Nevada. The George and Jacob Donner parties were part of this group and only 45 of the 89 members survived the blizzard in 1846. Their final camps are properly commemorated in the forest.

To get a grand orientation to the Tahoe National Forest, take the 160-mile-loop Yuba-Donner National Scenic Byway. Leaving Truckee and heading west on Interstate 80, the byway climbs past Donner Lake to the rugged Donner Pass. Skiing opportunities in the winter abound at Donner Summit. Near the Indian Springs Campground is an off-highway vehicle staging area. The byway leaves Interstate 80 and follows State Route 20 to Nevada City. On the way is a marvelous overlook that provides a vast view to the north. You may want to hike the Rock Creek Nature Trail, which is about five miles east of Nevada City.

The quarter-mile Rock Creek Nature Trail crosses Rock Creek on a small bridge and then parallels the creek for a short distance. Along the way you will enjoy the variety of vegetation—bigleaf maples, white alders, California yews, California hazelnuts, the gorgeous reddish orange-barked madrones (fig. 15), green-leaved manzanitas, bear clovers, wild roses, and bracken ferns

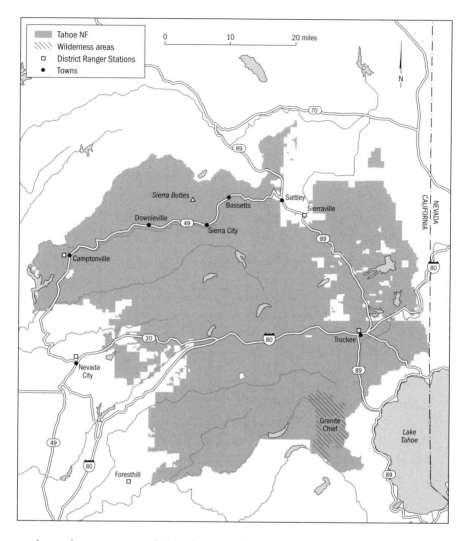

beneath an overstory of white firs, Douglas firs, sugar pines, yellow pines, and incense cedars. The smells of the coniferous trees are intoxicating.

From Nevada City, the Yuba-Donner National Scenic Byway turns north on Highway 49 toward Downieville. A side road (Forest Highway 8) leads to the large Bullard's Bar Reservoir where picnic areas and campgrounds are plentiful. Some of the campgrounds are accessible only by boat. Superb views of the lake may be enjoyed from Sunset Vista and from the top of the dam. Fishermen are drawn to Bullard's Bar by the prospect of Kokanee salmon, rainbow and brown trout, largemouth and smallmouth bass, and catfish.

Near Comptonville is a short drive to the Oregon Creek Covered Bridge. This must not be missed. The old wooden bridge over Oregon Creek was constructed by hand in 1871 and is more than 100 feet long. It was a major link between the California mining communities and the Comstock Mines of Nevada via the Henness Pass Road.

After enjoying the covered bridge, continue on the scenic byway as it now parallels the North Yuba River. As this scenic part of the route climbs over Indian Hill and passes Indian Rock and Grizzly Peak, there are several beautifully situated picnic areas and campgrounds. Between Downieville and Sierra City are the first views of the castellated Sierra Buttes to the north. At Sierra City, a two-mile road leads to Wild Plum Campground and Picnic Area where a short trail winds to Wild Plum Falls.

Figure 15. Orange-barked madrone (aka Pacific madrone).

Back on State Route 49, the byway passes very close to Loves Falls between Sierra City and Bassetts. At Bassetts, take County Road 620 north, following breathtaking Salmon Creek for a while. You will end up at Sand Pond where you may take a nature trail around a high-mountain wet meadow.

After hiking the Sand Pond Nature Trail, ogling the serrated summit of the Sierra Buttes and enjoying the placid Sardine Lakes, drive back to Bassetts and continue on State Route 49, climbing to Yuba Pass where there is a nicely placed campground. As the scenic byway descends from Yuba Pass, there is an overlook where the vast Sierra Valley stretches out below for miles. Near the community of Sattley, the scenic byway joins State Route 89 for its return to Truckee. At Cottonwood Campground, there is a short trail that interprets the trees of the forest. Just before reaching Truckee, stop at the Donner Camp Picnic Area where a part of the ill-fated Donner party made their last camp in 1846.

Granite Chief is the only wilderness area in the Tahoe, but it is among the best. This area of granite cliffs, glacier-carved valleys, wildflower meadows, and dense coniferous forests begins six miles west of Lake Tahoe. It can be accessed from the Deer Park Picnic Area on State Route 89 or from the end of the Mosquito Ridge Road just past the French Meadow Reservoir.

The Mosquito Ridge Road, although paved, is a hair-raising, white-knuckle, narrow road that parallels the Middle Fork of the American River for 50 miles, connecting the community of Foresthill with the French Meadow Reservoir. After driving 36 miles on this road, you will come to one of the most exciting parts of the Tahoe National Forest. Called the Placer Grove of Big Trees, the area harbors the northernmost stand of giant sequoias in the world. Six of these trees are along the trail, and they are 50 miles north of the next known population of these giants. The largest in volume of the six trees is the Pershing Tree, which is 225 feet tall with a diameter of 12 feet. The Joffre Tree is 25 feet taller, but much more slender. As you walk this peaceful trail, you will see other behemoths of Douglas fir, white fir, yellow pine, sugar pine, and incense cedar. The trail crosses a creek lined with western azalea, western blueberry, and red osier dogwood. Wildflowers on the forest slopes include false Solomon's-seal, twisted stalk, western dogbane, wintergreen, and rattlesnake plantain orchid.

Floaters will find pleasure in the serene water of the South Fork of the American River, and whitewater enthusiasts can challenge their skills on the North Fork where there are Class IV drops such as Zigzag, Achilles Heel, and Bogus Thunder.

For a touch of history, take the Boca Exit off of Interstate 80 and drive a short distance to the Boca Town Site and Cemetery. Between 1868 and 1927, Boca was a thriving town famous for lumber, ice, and brewing. A short loop trail runs through the old town.

About 13 miles north of Downieville is the Devil's Postpile. From the parking area, an uphill trail of 150 yards comes to some remarkable basalt columns formed when the upward movement of lava hardened and became shaped into posts. A short distance east of the Devil's Postpile is Chimney Rock, a volcanic rock that stands 25 feet tall on a base that is 12 feet in diameter.

Sand Pond

In 1884 prospectors entered a tranquil valley in the northern Sierra Nevada of California. A forest of lodgepole pines, set off by a cluster of picturesque lakes, known as the Sardine Lakes, covered the moist, 20-acre basin. From various vantage points in the valley the men could see the serrated crests of the Sierra Buttes rising in the distance. But the peaceful setting was about to be disturbed: the prospectors found gold in the quartz of the surrounding hills, and soon the Young America Mine was in full production.

To get at the gold, the miners cut down much of the lodgepole pine forest. Without the trees to absorb the moisture, most of the basin began to develop into a wetland. And once the mine was worked out and the prospectors moved on, beavers added to the wetland by felling more trees. Then, in the early twentieth century, cleanup crews hauled away the tailings from the gold mine, leaving a depression that created a lake. The result was far more inviting than this history would suggest. The lake, known as Sand Pond, is a popular swimming area in the Tahoe National Forest. A marsh has developed just east of the pond, and only a few lodgepole pines remain as a relic of the earlier forest.

I found the drive to Sand Point breathtaking. From Interstate 80 near Lake Tahoe, I proceeded north on State Route 89 to Sierraville. From there, for the next five miles, I enjoyed the view of a broad valley where cattle roam and graze. Near the small community of Sattley, I turned west on State Route 49 and followed the circuitous ascent to Yuba Pass, climbing from 4,940 feet in the valley to 6,700 feet at the pass. A U.S. Forest Service campground is beautifully situated at the top of the pass. Descending westward, I entered the charming village of Bassetts, where I caught my first glimpse of the scintillating Sierra Buttes. From Bassetts, I took the Gold Lake Highway north for nearly two miles, then followed the Forest Service Sardine Lakes Road to Sand Pond and the Sardine Lakes.

The Forest Service has constructed a mile-long looped nature trail that begins about 200 feet east of the Sand Pond parking lot. The trail traverses the marsh from one end to the other along an extensive boardwalk, after which it passes through a coniferous upland forest on its south side. After twisting around large boulders, dropping down out of the upland woods, and crossing a stream, the trail ends back at the Sand Pond parking lot.

Just north of Sand Pond, but in Plumas National Forest, are two other areas worth visiting. One is a short trail to Frazier Falls, a 176-foot cascade. The other is the Red Fir Nature Trail, off a dirt County Road 501, Frazier Falls Road. The trail loops through an ancient forest of red firs, some of them more than 150 feet tall.

Except for a few lodgepole pines, the marsh at Sand Pond is a dense covering of grasses, sedges, and rushes among small pools of standing water. Sedges of the genus *Carex* are the most diverse group of plants. Bluejoint grass is common, its silver green spikelets borne on slender, four-foot-tall stems. Soft rush, with its hollow stems, grows in tufts as large as two feet in diameter; other, lower-growing rushes appear singly or in small colonies. Near the edge of the marsh, and also scattered within it, is rose spiraea, a lovely pink-flowering shrub. Wildflowers include yellow monkey-flower, willow herb, and Bigelow's sneezeweed (pl. 27).

The shallow pools contain leafy pondweed, whose two-inch-long leaves float on the water. Bladderworts, whose intricately branched, bladder-bearing stems are completely submerged, send up small aerial stems with inch-long yellow flowers. Bordering the open water and sometimes growing in it is threeway sedge. If you look straight down on this foot-tall sedge, only three of the plant's 12 or 15 leaves will be apparent, so perfectly are the others aligned beneath them.

Just before reaching the boardwalk that goes around the marsh, you pass through a small forest of lodgepole pine, quaking aspen, and white fir. The understory is a diverse mix of ferns and wildflowers. The most conspicuous plant is corn lily, a perennial that can grow four feet tall and whose 15-inch-long, 10-inch-wide leaves appear corrugated because of their thick veins. Other wildflowers to look for are western mountain aster, meadow rue, pink wintergreen, and Chinese houses (pl. 28) (a plant in the snapdragon family).

A stream that borders one edge of the marsh and eventually empties into Sand Pond is lined with a dense thicket of shrubs. The major species are alder, rose spiraea, red osier (the same species found in the eastern United States), and at least three species of willows (Scouler, arroyo, and *Salix lasiandra*). Several sedges and rushes from the marsh grow entangled beneath the shrubs.

The higher and drier terrain east and south of the marsh supports Jeffrey pine, white fir, incense cedar, and Douglas fir. In the shrub layer are green-leaf manzanita, bog bilberry, western azalea, and leather oak. Wildflowers include largeleaf avens, a pink honeysuckle, California goldenrod, false Solomon's-seal, and Washington lily.

NATIONAL FORESTS IN IDAHO

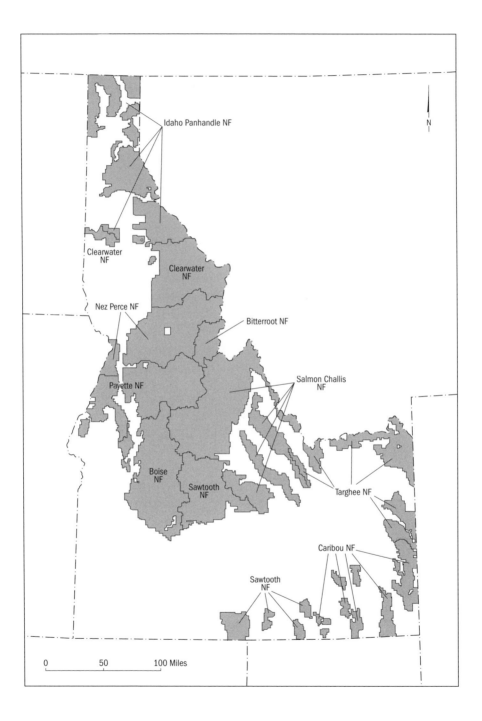

Idaho Panhandle NF

Clearwater NF

Clearwater NF

Nez Perce NF

Bitterroot NF

Payette NF

Salmon Challis NF

Boise NF

Sawtooth NF

Targhee NF

Caribou NF

Sawtooth NF

N

0   50   100 Miles

The national forests in Idaho are divided among U.S. Forest Service Intermountain Region 4 and Northern Region 1. Those national forests in the Intermountain Region are Boise, Caribou-Targhee, Clearwater, Payette, Salmon-Challis, and Sawtooth. Those in the Northern Region are Idaho Panhandle, consisting of the Coeur d'Alene, Kaniksu, and St. Joe. In total, the national forests in Idaho contain 21.3 million acres and 9 wildernesses. The Intermountain Regional office is in the Federal Building, 324 5th Street, Ogden, UT 84401.

Boise National Forest

SIZE AND LOCATION: Approximately 2.6 million acres in south-central Idaho. Major access routes are Interstate 84, U.S. Highways 20 and 95, and State Routes 2, 21, 55, and 68. District Ranger Stations: Cascade, Emmett, Idaho City, Lowman, Mountain Home. Forest Supervisor's Office: 1249 S. Vinnell Way, Suite 200, Boise, ID 83709, www.fs.fed.us/r4/boise.

SPECIAL FACILITIES: Winter sports areas; boat ramps; swimming beaches.

SPECIAL ATTRACTIONS: Sawtooth National Recreation Area; Ponderosa Pines Scenic Route; Wild and Scenic East Fork of the Salmon River; Wild and Scenic Middle Fork of the Salmon River.

WILDERNESS AREAS: Sawtooth (217,088 acres, partly in the Challis and Sawtooth National Forests); Frank Church–River of No Return (2,866,698 acres, partly in six national forests).

The jagged peaks of the Sawtooth Range form the eastern boundary of the Boise National Forest. Some of the more rugged areas are in the Sawtooth National Recreation Area and the Sawtooth Wilderness. Sparkling Ardeth and Spangle Lakes are in the Sawtooth Range, as are Blizzard and Greylock Mountains. Hiking trails lead to all of these attractions.

At the northeastern corner of the Boise is a part of the Frank Church–River of No Return Wilderness. The Wild and Scenic East Fork of the Salmon River is in the northern edge of the national forest, and the Wild and Scenic Middle Fork of the Salmon River (pl. 29) runs across the eastern edge. Beautiful Marble and Indian creeks cross the wilderness area.

The more accessible parts of the Boise National Forest begin less than five miles from the capital city of Boise. The Boise has numerous inviting streams, rolling high meadows, and forested canyons. Desert valleys in the

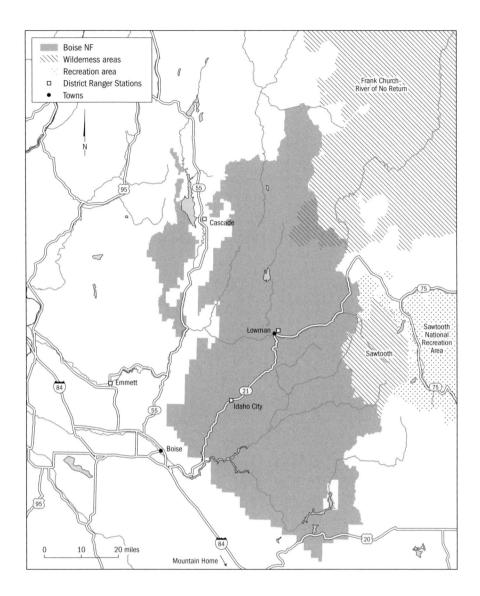

Boise NF
Wilderness areas
Recreation area
□ District Ranger Stations
● Towns

Frank Church-
River of No Return

Cascade

Lowman

Sawtooth
National
Recreation
Area

Sawtooth

Emmett

Idaho City

Boise

Mountain Home

0   10   20 miles

lower elevations provide an interesting contrast. Gold was discovered in 1866 in the Boise Basin, and evidence of the mines may still be seen in areas. You can visit the sites of several boom towns, including Brummer, Thunder, Banner, Pioneerville, Graham, and Quartzburg. Black bears, mule deer, and elk are common in the national forest, whereas in the higher mountains, bighorn sheep and mountain goats roam. The most prevalent forest trees are ponderosa pine, Engelmann spruce, lodgepole pine, and subalpine fir, and grasses and sagebrush dominate the desert valleys.

Lowman is in the very center of the national forest, and State Route 21 from Boise to Lowman and beyond has been designated the Ponderosa Pines Scenic Route. Where the highway from Boise enters the national forest, it follows the course of Moores Creek to Idaho Falls in the Boise Basin. The continuation to Lowman is very scenic as it passes between Pilot Peak to the west and Sunset Peak to the east. After a series of switchbacks, the highway reaches Lowman. From Lowman are three roads that venture into much of the rest of the national forest. The right fork, which is still State Route 21 and paved for a few miles, follows the sparkling South Fork of the Payette River. Eventually, another right fork continues along the Payette River (pl. 30) and goes to a superb campground west of Sawtooth Lake at the edge of the Sawtooth Recreation Area.

Another road leaves Lowman and heads northeast along Clear Creek, then follows Bear Valley Creek where the two creeks come together. This road passes through Bruce Meadows and then comes to impressive Dagger Falls near the upper end of the Middle Fork of the Salmon River. The Dagger and Boundary Creek campgrounds will allow you ample time to explore this exciting area just outside the Frank Church–River of No Return Wilderness. Although Dagger Falls drops only 15 feet, it is in a spectacular setting. If you decide to hike 5.5 miles north to Velvet Falls, you will not be disappointed.

A third road from Lowman heads due west. Several hot springs are next to the South Fork of the Payette River. Two waterfalls are along the river as well. Big Falls drops 40 feet into a catch pool, and Little Falls, with a drop of only 10 feet, is disproportionately broad. Where Big Pine Creek enters the South Fork is a 65-mile-long Forest Road 555 that goes to the northern boundary of the Boise where Yellow Pine Campground is located. This road comes to large Deadwood Reservoir where many water activities can be enjoyed. The road then parallels the Deadwood River and Johnson Creek before reaching the campground.

Another drive to take starts at the Lucky Peak Reservoir east of Boise. The road along the north side of the reservoir branches off of State Route 21 and takes a circuitous route as it stays close to the Middle Fork of the Boise River. Two options exist. One is to stay near the river until it reaches the ghost town of Atlanta where you can explore for signs of past habitation. The other is to take Forest Road 156 south past Steel Mountain until the road reaches Featherville and then continue south along the banks of the South Fork of the Boise River to the Anderson Ranch and Little Camas Reservoirs. The Danskin Range is just west of these reservoirs.

A smaller unit of the Boise lies west of State Route 55 about 30 miles north of Boise. Sage Hen Reservoir is the major attraction here.

Caribou-Targhee National Forest

SIZE AND LOCATION: Approximately 2.7 million acres in southern and eastern Idaho, with extensions into Utah and Wyoming. Major access routes are Interstate 15, U.S. Highways 20, 26, 30, 89, and 191, and State Routes 28, 31, 32, 33, 34, 36, 38, 47, 84, and 87. District Ranger Stations: Ashton, Dubois, Island Park, Montpelier, Idaho Falls, Soda Springs, Driggs, Pocatello, Malad. Forest Supervisor's Office: 1405 Hollipark Drive, Idaho Falls, ID 83401, www.fs.fed.us/r4/caribou.

SPECIAL FACILITIES: Winter sports areas; boat ramps; swimming beaches.

SPECIAL ATTRACTIONS: Minnetonka Cave; Mesa Falls National Scenic Byway.

WILDERNESS AREAS: Winegar Hole (10,715 acres); Jedediah Smith (123,451 acres).

For most of the twentieth century, the Caribou and Targhee were separate national forests in eastern Idaho. Both have been placed under one administrative unit, and the Idaho portion of the old Cache National Forest was added to the Caribou.

Caribou National Forest

The Caribou National Forest consists of several disjunct units in southeastern Idaho. The Caribou Range is the major mountainous area of the national forest, although several smaller ranges are present.

West of St. Charles and huge Bear Lake is a prime forest attraction. Minnetonka Cave is a limestone cavern about a half-mile long with nine rooms, one of them 300 feet in diameter and 90 feet high. The cave is filled with stalactites and stalagmites, some of them old and some still being formed. Fishhook- and corkscrew-shaped helictites are also present. The cave is located in St. Charles Canyon with Cloverleaf Campground nearby. A forest ranger conducts guided tours during the summer. The cave is well illuminated.

From Bloomington, a few miles north of St. Charles, a scenic road along Bloomington Creek goes near the very scenic glacial-formed, 10-acre Bloomington Lake (pl. 31). A quarter-mile hike will take you to the lake. A road from Paris that winds up Paris Canyon to an ice cave is worth checking out. The cave has ice the year round. Trails lead to Paris Peak and Midnight Mountain.

Portions of the Malad Range west of the Bear River are in the Caribou National Forest. Several roads enter this section of the national forest from

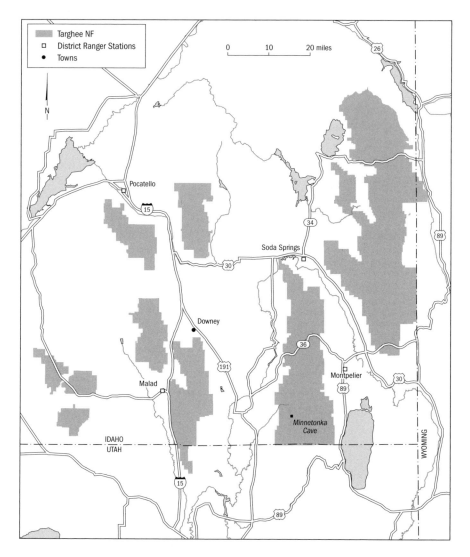

Targhee NF
□ District Ranger Stations
● Towns

0   10   20 miles

N

Pocatello

Soda Springs

Downey

Malad

Montpelier

Minnetonka
Cave

IDAHO
UTAH

WYOMING

either Malad City or Downey. A paved road crosses the Malad Range from Malad City, passing Henderson and Deep Creek Reservoirs. Old Baldy Peak may be seen to the east just before the road enters Weston Canyon and leaves the forest. At the southern end of the national forest, a few miles into northern Utah, is Gunsight Peak with its notched summit.

Some interesting places are in the Bannock Range a few miles south of Pocatello. Our forest service guide took us on a fascinating trip that started on State Route 38. We took a paved side road along the East Fork of Mink Creek to the Scout Mountain Campground situated beneath huge Douglas firs. The 8,710-foot Scout Mountain is reached by a two-mile trail from the campground. Prominent Tom Mountain can be seen farther to the south.

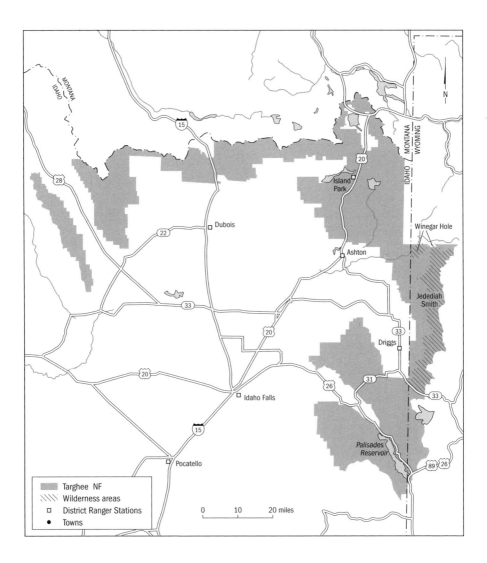

Along the West Fork of Mink Creek is a fine natural area that includes a riparian community along the creek and on the slopes of Slate Mountain. The east side of the natural area faces south and is drier than the north-facing slopes on the western side. The drier areas contain great amounts of black sagebrush and Sandberg's bluegrass with scattered specimens of Utah juniper. Also present are bitterbrush, big sagebrush, arrowleaf balsamroot, and Wyeth's false buckwheat. On rockier areas are chokecherry, creeping barberry, cut-leaf balsamroot, and mule's-ears. The western edge of the natural area supports forests of Douglas fir and quaking aspen, with wax currant, pinegrass, blue wild rye, western coneflower, and butterweed groundsel in the understory.

The largest part of the Caribou National Forest is in the Caribou Range at the eastern edge of Idaho, extending into Wyoming. Huge Palisades Reservoir is on the northeastern corner of the national forest. A boat ramp is at the campground. Also from this campground is a scenic paved road through the northern end of the Caribou Range. The road follows several different picturesque creeks. From the road are trails to Red Peak, Big Elk Mountain, Little Elk Mountain, and Poker Peak. South from the highway is a primitive road through Anderson Gulch to Caribou City, now a placer gold ghost town but once a thriving community of 1,500 people. In a few miles along the highway is a trail to Keenan City, another abandoned gold-mining community.

State Route 34 crosses the Caribou Range from Freedom to Wayan following Tincup Creek. The trail up to Tincup Mountain is about three miles long, but the scenery along the way is top-notch.

Along the original Oregon Trail, there was a 50-mile stretch of alkaline desert on the western side of the Continental Divide, from South Pass to Green River. To avoid this desolate area where there was little vegetation for the animals to eat, the Lander Trail was constructed between 1858 to 1860. A small section of the old Lander Trail crosses the Caribou Range from the Stump Creek Guard Station near Auburn to State Route 34 near Wayan. You can see parts of the old trail in front of the guard station. Just north along the trail is the site of the Oneida Salt Works, which was established in 1866. From the salt works, the Lander Trail goes through meadows and sagebrush foothills of Stump Creek to steep and narrow Terrace Canyon where the pioneers had to build rock ledges to form the roadbed. They also had to scrape away some of the rock face. You may see evidence of this as you hike through Terrace Canyon. Here and there along the trail are aspen carvings that are blazes made on the tree trunks. The oldest carving dates back to 1887. The grave of emigrant J. W. Lane who died in 1859 is near the trail where it leaves the Caribou National Forest.

At the south end of the Caribou Range, U.S. Highway 89 is very scenic in Montpelier Canyon.

Targhee National Forest

The 1.8 million-acre Targhee National Forest is mostly in eastern Idaho, with a small extension into extreme western Wyoming adjacent to Yellowstone and Grand Teton National Parks. The forest is situated along the Continental Divide and includes all or parts of several mountain ranges including the Lemhi, Beaverhead, Bitterroot, Centennial, Henry's Lake, Teton, Big Hole, Caribou, and Snake River. Most of the forests in the Targhee are dominated by lodgepole pine (fig. 16).

A good place to start your exploration of the Targhee is at the western edge on State Route 28. This highway passes through Birch Creek Valley south of Gilmore. Below Gilmore Summit is a side road to the abandoned townsite of Hahn. This primitive road then goes through Spring Mountain Canyon, climbing rapidly to Big Windy Peak where there is a nice view. A little farther on State Route 28 is a road to the Birch Creek Charcoal Kilns. In 1881, lead and silver ore was discovered in this area and, in a few years, 16 charcoal kilns were built out of bricks to supply fuel for a nearby smelter.

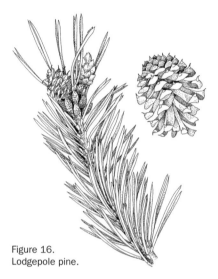

Four of these kilns remain and are worthy of a visit. About five miles farther southeast on State Route 28 to the east are the Birch Creek pictographs, which are located in rock shelters at the beginning of Indian Head Canyon in the Beaverhead Mountains just outside the forest boundary. The pictographs are drawn in red hematite at the back of the rock shelters. On the west side of State Route 28 is a large, roadless area dominated by 12,197-foot Diamond Peak. This region has sharp cliffs that rise steeply above Birch Creek Valley.

Figure 16.
Lodgepole pine.

The floor of the canyons has a cover of various kinds of grasses, sagebrush, and mountain mahogany, and the lower slopes have forests of Douglas fir and lodgepole pine. Higher elevations support subalpine fir and limber pine. Mostly solid rock areas exist above 10,000 feet. Little vegetation occurs at these elevations.

From the town of Dubois on Interstate 15, take State Route 22 through the Snake River Plains, entering the national forest in the Beaverhead Mountains. There are beautiful sites all the way to Medicine Lodge Pass, a former stagecoach stopping area. Two side roads are worth taking before reaching the pass. One follows Webber Creek and enters the national forest at the northern edge of the Italian Peaks area. The forests in this wild area are dominated by Douglas fir and Engelmann spruce on the middle elevation slopes and subalpine fir and limber pine on higher sites. The other side road is up Irving Creek to Red Canyon and the Red Conglomerate Peaks. Douglas fir and mountain mahogany line the canyons, whereas on the higher mountains, alpine vegetation is mixed with limber and lodgepole pines.

Interstate 15 crosses the Targhee National Forest between Humphrey and

Spencer as it parallels Beaver Creek. Signal Peak on the east side of the highway is the center of a beautiful mountainous area traversed by Camas Creek. Douglas fir, subalpine fir, and quaking aspen are the important forest species.

U.S. Highway 20 passes through the eastern side of the Targhee. Henry's Lake, not in the national forest, is at the northern end of the crest. State Route 87 on the northeast side of Henry's Lake goes up over Raynolds Pass. About half of the plant communities near the pass are forests of Douglas fir, lodgepole pine, and quaking aspen, and the other half is mostly sagebrush–grassland community.

As you travel from Yellowstone National Park west on U.S. Highway 20 and cross Targhee Pass, the area to the north is known as Lionshead. The terrain is steep and rugged with mountains rising sharply above Henry's Lake. Continuing south on U.S. Highway 20, you see Mt. Jefferson's 10,196-foot summit to the west. A back road over Red Rock Pass north of Mt. Jefferson leads to this area. Lodgepole pine covers the lower and midslopes, whereas alpine fir, Engelmann spruce, and limber pine are at higher elevations. Across the highway to the east is Two Top Mountain. The Twin Creek Road accesses this area. The National Recreation Snowmobile Trail is on the mountain slopes.

Forest Road 059 is a short loop route that passes a major attraction called Big Springs. Near here is the Johnny Sack Cabin, built by a master craftsman German immigrant between 1932 and 1934. Its significance is its architectural qualities. Much of the original furniture was made by him. The cabin is open to the public during the summer and is operated by the Fremont County Historical Society.

The east end of Island Park Reservoir is in the national forest. A 2,500-acre flat, open area is southwest of the reservoir with a continuous cover of pinegrass and elk sedge. Engelmann spruce surrounds this very pristine marsh.

At Henry's Fork Reservoir, take State Route 47, the Mesa Falls National Scenic Byway, heading southeast to the community of Warm River. After 13 miles you will come to the Grandview Vista Point overlooking the Snake River Butte area and Henry's Fork River to the west. Lower Mesa Falls, approximately 400 feet below the vista point, drops 65 feet in a thunderous cascade. Upper Mesa Falls, about one mile up-river, plunges 115 feet over a sheer rock wall. Sheep Falls is approximately six miles up-river from Lower Mesa Falls. Access to Upper Mesa Falls is via a paved road. At Upper Mesa Falls, Big Falls Inn was reconstructed in 1999 and is now a popular visitor attraction and information center.

East of Warm River is marvelous scenery on the way toward Yellowstone National Park, and there are hiking trails to Snow Creek Butte, Moose Creek

Butte, and Huckleberry Ridge. At the junction of State Routes 47 and 36, State Route 36 heads east, eventually entering Yellowstone National Park. Just before entering the park, while still in the Targhee, is Cave Falls Campground at the edge of the Winegar Hole Wilderness. Secluded Junco Lake, Fish Lake, and Lake-of-the-Woods are great places to look for trumpeter swans, loons, and sandhill cranes. Lake-of-the-Woods is a 200-acre lake only three miles from Grand Teton National Park.

Jackass Meadows are south of the Winegar Hole Wilderness Area. This is a plateau dissected by a canyon formed by Squirrel Creek. Huckleberries are common here. Forests surrounding the meadows consist of Douglas fir, lodgepole pine, and subalpine fir.

Between State Routes 32 and 33 and Grand Teton National Park is a long part of the Targhee National Forest on the western slopes of the Teton Mountains. Much of this area is in the Jedediah Smith Wilderness. One forest road from Driggs follows Teton Creek to the Teton Creek Campground. There are nonmotorized trails along both the North Fork and South Fork of the creek. The trail along South Fork goes to Alaska Basin, an alpine area with colorful arcticlike wildflowers in a very pretty setting. The 11,923-foot Buck Mountain is on the border between the national forest and Grand Teton National Park. Three miles south and east of Driggs is a forest road up Darby Canyon. The road ends near the wilderness boundary, which is about two miles from the Wind Caves. You need to be an experienced spelunker to explore the caves.

State Route 33 at Victor turns into State Route 22 at the Wyoming-Idaho border and follows Trail Creek as the road climbs to Teton Pass. At Victor, State Route 33 turns to the west toward Swan Valley. The road goes up over Pine Creek Pass. The Big Hole Mountains with its highest peak, Garns Mountain, is to the north of the highway.

Once in Swan Valley, State Route 26 follows near the South Fork of the Snake River southeast to the dam for Palisades Reservoir. The Snake River forms the southern boundary of this part of the Targhee. The natural Upper and Lower Palisade lakes are eight miles north of Palisades Reservoir but can only be reached by trails. There is a lovely waterfall in Waterfall Canyon south of the upper end of Upper Palisade Lake.

Clearwater National Forest

SIZE AND LOCATION: 1.8 million acres in north-central Idaho, from the Montana border to the Palouse Prairie north of Moscow. Major access routes are U.S. Highways 12 and 95 and State Routes 6, 8, 11, and 13. District Ranger Stations: Potlatch, Orofino, Kamiah, Kooskia, Lolo. Forest Supervisor's Office: 12730, Highway 12, Orofino, ID 83544, www.fs.fed.us/r4/clearwater.

SPECIAL FACILITIES: Lolo Pass Visitor Center; Lochsa Historical Ranger Station.

SPECIAL ATTRACTIONS: Lolo Trail.

WILDERNESS AREAS: Selway-Bitterroot (1,340,460 acres, partly in the Bitterroot and Nez Perce national forests).

History is everywhere in the Clearwater National Forest. The Nez Perce (Nee-Me-Poo) Indians, through use, created many of the trails that cross the forest. Lewis and Clark camped along and followed the old Indian trails in 1805 and 1806. Gold miners rushed to the forest in the 1860s to seek their fortunes and founded raucous towns such as Pierce and Moose City.

U.S. Highway 12, the Lewis and Clark Highway, enters the Clearwater National Forest at Lolo Pass and is a scenic route through the national forest. It parallels the high ridgeline route Lewis and Clark followed. The Lewis and Clark and Nez Perce National Historic Trail are parts of the route known as the Lolo Trail. The route is free of snow generally from July through September. Passenger cars and RVs are not recommended, but vehicles with high clearance can make the trip.

At the Lolo Pass Visitor Center you can become familiar with the area. Nearby Packer Meadow is where Lewis and Clark had one of their campsites. Access to the Lolo Trail is a few miles southwest of the pass. If you choose to try and follow the trail that Lewis and Clark hiked all the way across the forest, you will end up at the Lolo Campground nearly 100 miles to the west several days later. Along this primitive trail you climb over Papoose Saddle, take the south fork at Cayuse Junction, and come to the Indian Post Office where you see rock cairns perhaps piled up by Indians to mark where the Lolo Trail turns off of the divide. Farther on, the trail passes a formation known as Devil's Chair before reaching Indian Grave Peak where there is a grave of a 14-year-old boy nearby. The trail then circles around Bald Mountain and through Noseeum Meadows to Chimney Butte. After climbing over Sherman Saddle, Deep Saddle, and Green Saddle, the trail comes to beautiful Weitas Meadows. It then takes a southward turn to Mex Mountain and follows Dollar Creek and Eldorado Creek before coming to the Lolo Campground.

Most of the visitors to the national forest will stay on U.S. Highway 12. All along the highway are trails south into the Selway-Bitterroot Wilderness. After the visitor center, your next stop should be the DeVoto Memorial Grove and its many magnificent old-growth western red cedars. U.S. Highway 12 next follows the Lochsa River. A back road south from Star Meadows leads to Sneakfoot Meadows, a lovely wildflower area surrounded by subalpine firs. The gorgeous beargrass is in the understory.

At Warm Springs Pack Bridge is a trail south to Jerry Johnson Hot Springs, a popular day-use destination. If you continue two miles past the hot springs, you see the Jerry Johnson Falls just inside the wilderness area. Next on U.S. Highway 12 is the half-mile Colgate Licks National Recreation Trail.

The Lochsa Historical Ranger Station has been in operation since the 1920s. The building was restored in 1976 to its original appearance. The highway then enters Black Canyon and has a pull-out to view the 175-foot Shoestring Falls to the east. If you walk a tenth of a mile from the pull-out, you see the double Wild Horse Creek Falls from the road. Another pull-out allows you to view the 80-foot Horsetail Falls. Where U.S. Highway 12 crosses Tumble Creek is a trail to Bimerick Falls and Bimerick Meadow.

A few miles farther along near the Lochsa River are plant communities more like those along the Pacific Coast but rare east of the Cascade Range. Plants in this region include Pacific dogwood, red alder (fig. 17), western sword fern, alder-leaved serviceberry, Pacific yew, Douglas hawthorn, kinnikinnick, beargrass, Oregon grape, black cottonwood, red osier dogwood, snowberry, ocean spray, syringa (the state flower of Idaho that resembles a mock orange), and paper birch.

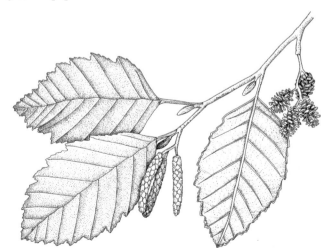

Figure 17.
Red alder.

From the town of Superior, Montana, on Interstate 90, is a forest road that heads south through the Lolo National Forest and into the Clearwater and that has several points of interest. Moose City is the site of an abandoned gold-mining town. At the Weitas Creek Campground is a hiking trail to the remote Weitas Guard Station cabin. At the north end of Isabella Creek is the Heritage Grove with more big western red cedars.

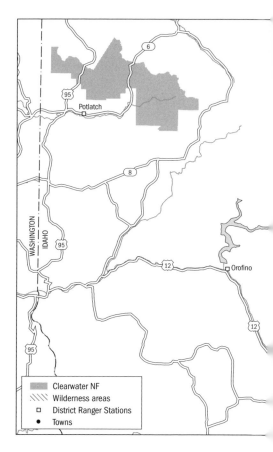

One of the finest natural areas in any national forest is Aquarius, a 3,900-acre area adjacent to the Aquarius Campground. This is the best example of coastal vegetation in the northern Rocky Mountains. The forests consist of western red cedar, western larch, western white pine, grand fir, Douglas fir, and red alder. Many of the plants on the forest floor are generally uncommon in this part of Idaho. Some of them are mountain boykinia, broad-leaved penstemon, western starflower, Idaho barren strawberry, cluster lady's-slipper, and coastal shield fern. The terrain is steep and rugged, with some of the slopes extending directly to the edge of the Clearwater River.

Lookout towers not in use by the Forest Service may be rented overnight at Weitas Butte, Austin Ridge, Cold Springs Cabin, Wallow Mountain, and Castle Butte for a unique overnight experience. Check with the District Ranger Stations for availability.

Aquarius

In the March 1986 This Land column in *Natural History*, I wrote about the lush temperate rainforests in northwestern Washington state, particularly those found in the Olympic National Forest and in rich ravines near the national forest. Twenty-five to 30 million years ago, temperate rainforests similar to those in Washington occurred in what is now north-central Idaho.

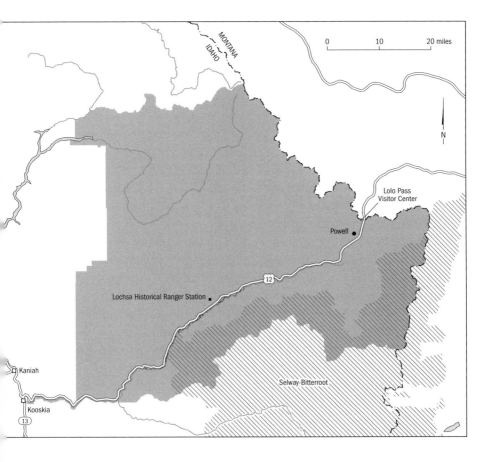

These ancient rainforests had developed before the Cascade Range had up-lifted through tumultuous volcanic eruptions. At that time, areas to the west of this region consisted of shallow seas and tidal flats.

The fossil record indicates that these ancient Idaho rainforests consisted of dawn redwoods, California redwoods, ginkgoes, bald cypresses, and rela-tives of present-day sweet gum and magnolias. Also extant were tall ferns, and dense masses of mosses grew over a rocky terrain and fallen logs. Heavy fogs, ample rainfall, and mild temperatures from the Pacific Ocean provided ideal conditions for the Idaho rainforests.

After the Cascade Range uplifted, the mountains interrupted much of the Pacific moisture, and the reduced moisture began to take its toll on some of the plants in these rainforests. Trees such as the dawn redwood and ginkgo disappeared completely from North America, whereas others that could mi-grate did so. Some, such as the California redwood, migrated to the west, whereas bald cypress and magnolia-like plants migrated to the southeast, where they remain today.

During the Ice Age thousands of years ago, some plants that had been able to survive in Idaho after the Cascade Range uplifted were driven down into canyon bottoms where the sheets of ice did not penetrate. One of these refuges was in a canyon of the North Fork Clearwater River. As the Ice Age came to an end about 12,000 years ago, temperatures rose and the climate became drier, conditions not favorable for rainforest plants.

Fortunately, some rainforest plants in the Clearwater River canyon were able to survive because of Black Mountain, which rises 7,000 feet above the river basin. Black Mountain, a part of the Bitterroot Range, interrupted moisture from Pacific storms that got past the Cascades, and this moisture found its way into the bottom of the canyon.

The rainforest is in an area of Idaho known as Aquarius, and the 3,900 acres it occupies has been designated the Aquarius Natural Area. It is a part of the Clearwater National Forest a few miles north of the community of Ahsahka and 40 miles east of Orofino, Idaho, in an area so remote that botanists had never explored here until 1968. When Idaho botanists Robert Steele and Frederic Johnson did get to Aquarius, they found a region of warm, south-facing slopes, cool north-facing slopes, perennial springs, and moist benchlands harboring old growth forests of western red cedar and many plants previously unknown from Idaho. Some plants in Aquarius are known from no other place in the world; some are found only in the Idaho rainforest and then again in the Pacific Coast temperate rainforests 300 miles away; a few are plants known from Aquarius and again from the eastern deciduous forests.

Endemic species in the Clearwater basin are Case's corydalis, a plant that is related to bleeding-hearts, Constance's bittercress, a small-flowered plant in the mustard family, and Idaho barren strawberry. Species disjunct between the western Cascades and Aquarius include deer fern, clustered lady's-slipper orchid, white shooting-star, Henderson's shooting-star, phantom orchid, bank monkey-flower, Sierra woodfern, western starflower, redwoods violet, Henderson's sedge, and licorice fern. Among plants of the eastern deciduous forest found at Aquarius are maidenhair fern, maidenhair spleenwort, and northern oak fern.

Botanists have recorded 24 species of ferns and their relatives at Aquarius, more than half the number of these plants known from the entire state of Idaho. In a checklist of plants of Aquarius given to me by Steele and Johnson, 47 of the plants also occur in the eastern United States.

Other significant features of Aquarius include the largest stand of red alder east of the Cascade Range–Sierra Nevada and the occurrence of the Pacific Coast Van Dyke salamander and several previously undescribed species of beetles and earthworms.

When I walked the trails in the forest along either side of the North Fork Clearwater River, I experienced the same feeling of being in a tropical forest that I had on my first trip to the Olympic rainforest in Washington. Dense old-growth forests dominated by western red cedar were draped with pendulous lichens and mosses. The forest floor was covered with rocks and fallen logs, many of them obscured by the growth of mosses and five-foot-tall ferns. Aquarius is, indeed, a temperate rainforest isolated in north-central Idaho.

As luck would have it, three years after Johnson and Steele discovered the botanical significance of Aquarius, two-thirds of the area was destroyed when the Dworshak Reservoir was completed in 1971. The rainforest that is left is a designated natural area in the Clearwater National Forest.

Nearly all the forested areas in Aquarius are dominated by western red cedars. On gentle slopes and on terraces above the river, huge shield ferns are the dominant plants in the understory. On moist slopes above the terraces, maidenhair fern replaces the shield ferns as the dominant plant. Where the forests are drier on the more south-facing slopes, Douglas fir also is found, with ocean spray in the shrub layer.

Within the natural area are several streams that empty into the North Fork Clearwater River. The riparian habitat along the streams usually contains red alder, black elderberry, Scouler willow, black cottonwood, red osier dogwood, and yellow monkey-flower.

The phantom orchid is unique in that it is pure white, totally devoid of chlorophyll. It derives its nutrients from the leaf litter, the same as one-flowered Indian pipe, which is also in the area.

Idaho Panhandle National Forest

SIZE AND LOCATION: Approximately 3.2 million acres in northern Idaho, with small units in Montana and Washington. Major access routes are Interstate 90, U.S. Highways 293 and 95, and State Routes 1, 3, 5, 6, 8, 31, 41, 57, and 202. District Ranger Stations: Avery, Coeur d'Alene, St. Maries, Sandpoint, Silverton, Bonners Ferry, Priest River. Forest Supervisor's Office: 3815 Schreiber Way, Coeur d'Alene, ID 83818, www.fs.fed.us/r1/pgr/idaho_pan.

SPECIAL FACILITIES: Winter sports areas; boat ramps; swimming beaches.

SPECIAL ATTRACTIONS: St. Joe National Wild and Scenic River; Route of the Hiawatha Bicycle Trail.

WILDERNESS AREAS: Cabinet Mountains (94,272 acres, partly in the Kootenai National Forest); Salmo Priest (11,949 acres).

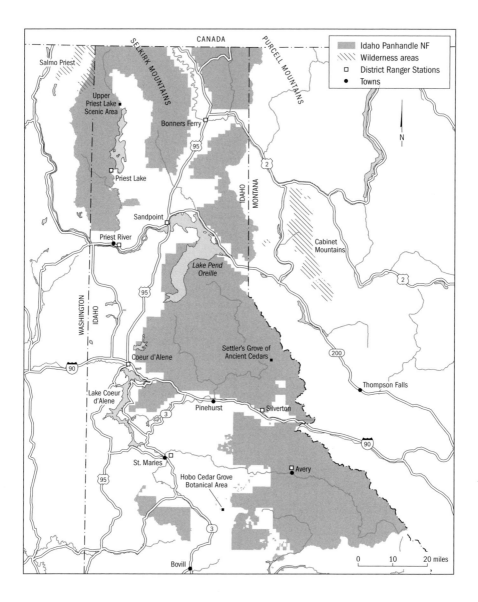

The narrow neck of Idaho that extends from the community of Potlatch to the Canadian border is known as the panhandle, and the three national forests in that area make up the Idaho Panhandle National Forest. The Kaniksu, Coeur d'Alene, and St. Joe National Forests are in one administrative unit, although at one time they were considered separate.

Kaniksu National Forest

The northernmost national forest in the Idaho Panhandle extends into Washington on the west and Montana on the east. Included within the 1.8-million-acre Kaniksu National Forest are the Selkirk and Purcell Mountains and a part of the Cabinet and Bitterroot Mountains. These mountains are extremely rugged with sharp relief and many lakes in glacial cirques. The 74 lakes have a water surface of 122,000 acres, including 85,000 acres in the large Lake Pend Oreille. Three major rivers in the national forest are Clark Fork, Kootenai, and Pend Oreille.

In most areas of the forest, four vegetation zones may be recognized. The lower foothill community includes forests of ponderosa pine and lodgepole pine with snowberry and ninebark the dominant shrubs. As you climb to the midslopes where there is more moisture, the prevalent trees are Douglas fir, grand fir, western red cedar, and western hemlock. Beneath the canopy are ninebark, snowberry, pachistema, devil's-club, and kinnikinnick. Among the wildflowers are sego lily, fireweed, trilliums, Indian paintbrushes, fawn lilies, shooting star, and lady's-slipper orchid. The subalpine zone below the mountaintops is dominated by subalpine fir and mountain hemlock with an understory of rusty menziesia, pachistema, beargrass, and whortleberry. The alpine vegetation on the mountain peaks consists of trees that are dwarfed and gnarly or has no trees at all. This bleak habitat supports alpine wildflowers that have a very short growing season in which to flower and fruit.

The Selkirk Mountains are the only place in the lower United States where mountain caribou live. Other large animals include white-tail and mule deer, grizzly bears, black bears, elk, moose, mountain goats, bighorn sheep, and mountain lions.

The Purcell Mountains occupy the northeastern corner of the Kaniksu and extend into Montana. U.S. Highway 95 cuts across the northern end of the mountains to Eastport on the Canadian border. Less than one mile south of Eastport is a bumpy gravel road that in two miles comes to the Copper Falls Trail. After a quarter-mile hike, you marvel at the 160-foot falls of Copper Creek. About 10 miles east of Eastport, in Montana, is the Northeast Peak Scenic Area. The 6,900-acre alpine area contains three rocky peaks and seven lakes.

Three miles northeast of Bonners Ferry, U.S. Highway 2 follows the Kootenai River and, after six miles, comes to the turnoff to spectacular Moyie Falls. This thunderous falls drops nearly 140 feet in a series of tiers, the largest segment with a drop of 100 feet. Two and one-half miles south of Bonners Ferry, the forest road from Moravia comes to the impressive double Snow Creek Falls. The Upper Falls plunges 125 feet, the Lower Falls about 75 feet.

A marvelous natural area exists about 12 miles southeast of Bonners Ferry. Hunt Girl Creek Natural Area is a rugged subalpine region with elevations ranging from 3,900 feet to 6,220 feet. The midslopes are dominated by western red cedar, western hemlock, western white pine, western larch, and grand fir. Above these forests are fine stands of subalpine fir, whitebark pine, and Engelmann spruce. This 1,505-acre tract also has a small lake, a narrow gorge, and several wet meadows. Boulder Mountain is at the western edge of the natural area.

Continuing south on U.S. Highway 95, a road east from Samuels leads to a small series of cascades known as Grouse Creek Falls. A very scenic forest highway leads northwest from Samuels along Pack River. In nine miles are the Jeru Creek Falls. It requires a one-mile hike down a four-wheel-drive road to see this 150-foot falls. Along the forest highway are hiking trails to Gunsight Peak, Mt. Roothaan, and Chimney Rock. Sandpoint is at the northern end of Lake Pend Oreille, and many campgrounds are around this 1,200-foot-deep lake. Two major parts of the Kaniksu National Forest are on either side of Sandpoint. State Route 200 east of Sandpoint goes around the northeast corner of Lake Pend Oreille and then follows Clark Fork to the community of Thompson Falls. North of Clark Fork are three waterfalls worth seeking. Rapid Lightning Falls is a series of cascades that drops a total of about 30 feet. Char Falls near the end of Trestle Creek Road is a more impressive horsetail falls that plunges for 75 feet. Wellington Creek Falls is a magnificent 75-footer that splashes into a scenic catch basin.

State Route 202 north along Bull River runs up the western edge of the Cabinet Mountains Wilderness Area. In the Kaniksu portion of the wilderness in Montana are nice trails to Rock Lake, St. Paul Lake, St. Paul Peak, and Snowshoe Peak.

The town of Priest River is a few miles west of Sandpoint. State Route 57 north is one of the most scenic in Idaho and provides access to wonderful attractions all the way to the Canadian border. Torelle Falls is about 8.5 miles north of Priest River. Many campgrounds are along the west side of Priest Lake, as well as boat ramps, swimming beaches, and resorts. Two significant natural areas are west of Priest Lake. Bottle Lake is a 15-acre wetland surrounded by old-growth forests of western red cedar, western hemlock, and western white pine. Lady fern, sword fern, heart-leaved arnica, bead lily, bunchberry, wild ginger, and red baneberry are just a few of the ferns and wildflowers on the forest floor. The wetlands have open water, sphagnum bogs, and a deep marsh. The abundance of water is due to a series of underground springs. At the northern edge of Bottle Lake is Tepee Creek Natural Area. The significance of this natural area is a mature stand of western white pine, with some of the trees more than 300 years old. Six miles north is the smaller and secluded Upper Priest Lake, the centerpiece of the Upper Priest Lake Scenic Area.

Lovely Hahn Lake is along the east side of State Route 57 and has fine wetland vegetation around the shore and aquatic vegetation in the water. Nearby is the Hanna Flats Nature Trail, which is a pleasant hike through wetlands and forest.

Canyon Creek Natural Area is east of the Priest River. Although this old-growth forest of western white pine, western red cedar, and western hemlock is important, the grassy bald on the upper south slope above the canyon impressed us. The bald has extensive grasslands of green fescue with beargrass mixed in. The area has a number of rock slides.

The Kaniksu Marsh Natural Area, 22 miles north of Priest River, is a 90-acre undisturbed marsh with submerged plants surrounding a stand of emergent plants. At the edge of the deep marsh are bog birch and gray alder. Keep your eyes open for the painted turtle, rare in Idaho.

Coeur d'Alene National Forest

This national forest contains approximately 750,000 acres of forested mountains and clear mountain streams, from the rugged Bitterroot Mountains along the Idaho-Montana border to the Rathdrum Prairie on the west. The national forest includes the Coeur d'Alene and St. Joe Mountains. Most of the trees in this forest are western white pine, Douglas fir, grand fir, western hemlock, subalpine fir, western larch, Engelmann spruce, western red cedar, and lodgepole pine.

Interstate 90 enters the east side of the Coeur d'Alene National Forest at Lookout Pass, which is a popular winter sports area. The highway follows along the southern border of the national forest. Just before reaching the town of Mullan, the primitive Willow Creek Road goes south to the Stevens Lakes area. A two-mile trail leads to Willow Creek Falls and another half-mile trail leads to Stevens Lake Falls. Past the falls are several scenic alpine lakes below Stevens Peak.

At the old mining community of Wallace is a good highway north into the national forest. After crossing the Coeur d'Alene River west of Kings Point, the highway continues north to Prichard. From here is a back road east along the West Fork of Eagle Creek to the Settler's Grove of Ancient Cedars. This is 173 acres of some of the oldest and largest western red cedars in Idaho. Some of the larger ones are about seven feet in diameter.

Much of Interstate 90 follows the old Mullan Road laid out by Capt. John Mullan in 1861. Mullan and his party camped at a summit east of Lake Coeur d'Alene. One of Mullan's men blazed the words "July 4, 1861, M. R." on a western white pine. This historic tree became known as the Mullan Tree, and the summit was called the Fourth of July Summit. The old tree died on November 11, 1962, and wind broke off the dead tree, leaving a 20-foot stump, still with the blaze on it.

From Pinehurst on Interstate 90, there is a back road that follows the west side of the Coeur d'Alene River. In several miles the road comes to an interesting rock formation known as Steamboat Rock. A side road from Steamboat Rock goes to Steamboat Peak and finally to the Grassy Mountain Lookout.

St. Joe National Forest

The St. Joe is the southernmost of the three national forests in the Idaho Panhandle. Its northern boundary is near Lake Coeur d'Alene. The St. Joe River crosses the entire forest from east to west. It has the highest elevation above sea level of any navigable river in the world. The upper reaches of the St. Joe River spawned a small gold rush in the 1870s.

The eastern half of the St. Joe National Forest near the Bitterroot Mountains has steep cliffs that line narrow canyons; the western half has more gently rolling terrain characteristic of the Palouse flatlands. A number of glacial lakes are in the higher elevations. Western white pine is the most prevalent tree in the national forest, but lodgepole pine, western larch, Douglas fir, and subalpine fir are common. A devastating forest fire in 1910 altered the face of the forest, and the forest is still recovering. Sixty-one men lost their lives fighting this disastrous fire.

St. Maries is a good place to begin your exploration of the St. Joe. Forest Highway 50 closely follows the St. Joe River. Fifteen miles from St. Maries are the Falls Creek Falls where the creek plunges about 30 feet. The main road ends at the Spruce Tree Campground, about 60 miles from St. Maries. This is an extremely scenic route that crosses many sparkling streams, some of which display evidence of placer gold mining in the past. From the old Red Ives Ranger Station near the end of the road, there is a rough back road along Red Ives Creek. After five miles is a trail to the Needle Peak Viewpoint for a grand view of this spire-topped peak.

Two miles west of Avery on Forest Highway 50, where the highway crosses Setzer Creek, is Forest Highway 301 south along Fishhook Creek. The Upper Fishhook Natural Area is an undisturbed stand of old-growth red cedars. It is a majestic site.

With a good forest map, you can follow Forest Road 201 from the natural area around Roundtop Mountain and Moses Butte, ending at Stubtoe Butte. From here you may hike to the Mallard Larkin Pioneer Area, a region of 46,000 roadless acres in a subalpine setting. More than half of the area is in the Clearwater National Forest. The pioneer area lies on a divide between the North and Little North Fork of the Clearwater River and contains 12 high mountain lakes.

From the Fishhook Natural Area you can take another road past Marble Mountain and then to Hobo Creek where there is a steep but short trail along the creek to a steam donkey—a steam-driven machine that was used to drag logs out of the forest. It was abandoned many years ago. Three miles south of the steam donkey is the Hobo Cedar Grove Botanical Area with its 240 acres of old-growth western red cedars.

After the cedar grove, the forest road follows Merry Creek to the old Clarkia Ranger Station. Just east of Clarkia is the Emerald Creek Garnet Area in a beautiful forest of grand fir, Douglas fir, western red cedar, and western white pine. Garnet deposits are in the East Fork of Emerald Creek. The garnets, which range in size from a grain of sand to two inches in diameter, are in the gravel and sand just above the bedrock. Star garnets with four and six rays are fairly common. You will need digging equipment and a permit from the Forest Service to find your gems. The star garnet is Idaho's state gemstone.

South of Clarkia at the small crossroads community of Bovill, take State Route 8 southeast to Elk River. South of town is the Elk Creek Recreation Area, which features eight noteworthy waterfalls. Elk Falls, with a drop of 150 feet, is the highest, followed by the more turbulent and raging 100-foot drop of Lower Elk Falls.

The Route of the Hiawatha is a bicyclist's dream. Beginning at the Roland Trailhead near Lookout Pass on Interstate 90, the bicycle route follows the former Milwaukee Railroad line for 13 miles, dropping 1,000 feet to the Pearson Trailhead northeast of Avery, Idaho. The route features train trestles 200 feet above the river and a tunnel.

Nez Perce National Forest

SIZE AND LOCATION: 2,218,040 acres in north-central Idaho. Major access routes are U.S. Highways 12 and 95 and State Route 14. District Ranger Stations: Grangeville, Kooskia, Elk City, White Bird. Forest Supervisor's Office: Route 2, Grangeville, ID 83530, www.fs.fed.us/r1/pgr/nez_perce.

SPECIAL FACILITIES: Boat ramps.

SPECIAL ATTRACTIONS: Hells Canyon National Recreation Area; Salmon National Wild and Scenic River; Snake National Wild and Scenic River; Selway National Wild and Scenic River.

WILDERNESS AREAS: Frank Church–River of No Return (2,366,698 acres, in six national forests, 105,736 acres in the Nez Perce); Gospel-Hump (200,464 acres); Selway-Bitterroot (1,340,460 acres, in four national forests, 560,088 acres in the Nez Perce); Hells Canyon (214,944 acres, in three national forests, 59,900 acres in the Nez Perce).

Nearly half of the Nez Perce National Forest is in wilderness areas so that if you are handicapped or aged, most of the magnificent mountains, lakes, and streams are inaccessible to you. Nonetheless, a few roads penetrate into the forest.

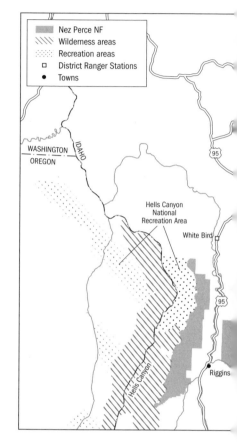

At Riggins is Forest Highway 517, Seven Devils Road west into the Seven Devils Mountains that tower 7,000 feet above the Snake River Gorge, the nation's deepest canyon. From Heaven's Gate at an elevation of 8,000 feet is a spectacular view of the area. Devil's Tooth is an extraordinary formation nearby as is He Devil Mountain, one of the Seven Devils. A three-mile trail leads to Mirror Lake, which reflects the pointed Tower of Babel in the clear water. These are all musts when you visit the Nez Perce. Forest Highway 1614 from Riggins heads east along the Salmon River, which parallels the southern boundary of the national forest. At the Allison Creek Campground is a long back road (Forest Road 221) to the gold-mining ghost town of Florence. The historic cemetery is on the site. Florence was reputedly one of the most lawless areas in the wild west during gold-mining fever days. The old Masonic Hall in Florence later became the courthouse for Idaho County. A few remnants of cabins are still here. Pilings near the townsite are the remains of dredging for gold. However, the best access to Florence is on Forest Road 221, about 41 miles from Grangeville. If you take this route, about seven miles from Grangeville is Fish Creek Meadows, a developed site with campground, five loop hiking trails, and a ski area.

The Magruder Road Corridor (Forest Road 468) is a lengthy route across wild and woolly mountain country that weaves between two huge wilderness areas. It begins at the old Red River Ranger Station and ends in Darby, Montana, having passed through parts of the Nez Perce and Bitterroot National Forests. Some of the road is unimproved and very steep and winding. After entering the forest, the first eight miles are a steady climb with grand views of Oregon Butte and Buffalo Hump in the Gospel-Hump Wilderness.

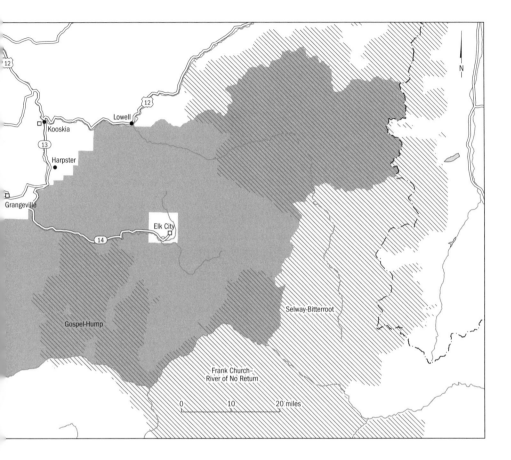

After passing through Mountain Meadows, the Magruder Road Corridor comes to Granite Spring Campground. Just beyond is the Elk Mountain Road (Forest Road 285) to the Green Mountain Lookout and its spectacular views of the Red River drainage. After a sharp descent to Bargamin Creek and Poet Creek Campground, the road again climbs to the Burnt Knob Lookout. It is only a short distance to Sabe Vista, Horse Heaven, and the Bitterroot National Forest in Montana. Great views are to be had from the 7,490-foot Sabe Vista, including damage done by the Ladder Creek Fire of 1988. Horse Heaven has a superb growth of grasses. The 1939 Horse Heaven Cabin is along hiking trail 028.

South from Elk City is the Gold Rush Loop tour that passes Orogrande and eventually comes to a fabulous high-mountain sphagnum bog and wet meadow complex known as the Big Creek Meadows. Forests surrounding these wet habitats are composed of lodgepole pine, subalpine fir, and Engelmann spruce. In the wetter portions of the natural area are shrubs such as Sitka alder, bog birch, alder-leaved serviceberry, swamp gooseberry, alpine wintergreen, shiny-leaved spiraea, and smooth Labrador tea, many of these uncommon for this part of the northwest. Wildflowers include Columbian monkshood, twin-flowered marsh marigold, Jeffrey's shooting-star, false green hellebore, and the rare mountain bog gentian, slender bog orchid, and Washington monkey-flower. Before reaching Big Creek Meadows, the loop tour has several numbered stops to witness past gold-mining activities such as dredged gravel piles, old mill sites, and the Walker Cabin, which can be rented from the Forest Service. After leaving Big Creek Meadows, the loop road continues to the Dixie Townsite and the old Dixie Cemetery where there are seven graves. The odd-shaped headstones of the graves were originally dragstones from a nearby arrastra.

The town of Kooskia is at the northwest corner of the Nez Perce National Forest. From Kooskia, U.S. Highway 12 follows the Selway River eastward. The Clearwater National Forest is to the north, the Nez Perce to the south. At the community of Lowell, take the Selway River Corridor that stays along the Selway River. The Fenn Ranger Station is a historic two-story building constructed in the late 1930s and still in use. It is on the National Register of Historic Places.

At the O'Hara Kiosk is a three-mile-long back trail south to the O'Hara Natural Area. This 7,000-acre area preserves a wide variety of wetland and forested habitats. The major trees are Douglas fir, grand fir, western red cedar, and subalpine fir. Rare plants found here that normally occur along the Pacific Coast include giant horsetail, fir clubmoss, and evergreen synthyris, a member of the snapdragon family.

Farther along the Selway River Corridor is the Selway Falls Campground near Selway River Falls. Selway Falls is most spectacular during the spring melt from April through June. Meadow Creek Trail goes south into the Selway-Bitterroot Wilderness. After about 15 miles is the Meadow Creek Guard Station, which was built in the 1920s. The main cabin consists of the office, a sleeping area, and a kitchen. The bunkhouse contains a main room and loft. The cabin may be rented from the Forest Service.

For an interesting backroad experience, take the Elk City Wagon Road that long ago led miners and freighters to the gold fields of Elk City. Beginning at Harpster on State Route 13, the wagon road ends 53 miles later at Elk City. Harpster was established in 1861 as the first stop for people going from Grangeville to Elk City. Between miles 1.0 and 2.6 on your odometer you may see the original old wagon road to the left. After crossing the Wall Creek Bridge, the road comes to the townsite of Clearwater. The old road eventually passes a four-way junction at mile 9.5, a limestone rock outcrop at 10.5, and the Initial Tree at 12.8. The 5.5-foot-diameter ponderosa pine has carved initials and dates. In all, there are 29 numbered sites on the way to Elk City.

The Salmon, Snake, and Selway, all Wild and Scenic Rivers, have stretches great for every level of whitewater rafters, including several Class V rapids.

Payette National Forest

SIZE AND LOCATION: Approximately 2.3 million acres in west-central Idaho. Major access routes are U.S. Highway 95 and State Routes 55 and 71. District Ranger Stations: Council, McCall, New Meadows, Weiser. Forest Supervisor's Office: 800 W. Lakeside Avenue, McCall, ID 83638, www.fs.fed.us/r4/payette.

SPECIAL FACILITIES: Winter sports areas; boat ramps; swimming beaches.

SPECIAL ATTRACTIONS: Salmon National Wild and Scenic River; Snake National Wild and Scenic River; Hells Canyon National Recreation Area.

WILDERNESS AREAS: Frank Church–River of No Return (2,366,698 acres, partly in six national forests).

Unsurpassed scenery is everywhere in the Payette National Forest from the Snake River and the incomparable Hells Canyon that forms the western boundary of the national forest and the National Wild and Scenic Salmon River that forms the northern boundary. The eastern one-fourth of the national forest is in the Frank Church–River of No Return Wilderness, and there are long, rugged hiking trails throughout the wilderness. Whitewater rafting will challenge the most experienced on the Salmon River. The runs of salmon and steelhead trout in the Salmon River and lower portion of the Snake River draw fishermen from all over the country. More than half of the national forest is forested, with whitebark pine and subalpine fir found above the 8,000-foot level, whereas grasslands and sagebrush occur in the dry habitats in Snake River Canyon. Gold was discovered in the Thunder Mountain area around 1900, and the sites of old mining towns can be explored.

Good paved roads are few in the Payette, but there are numerous back roads, several of them hair-raising, into the depths of the national forest. One of the paved roads is State Route 71 from Cambridge to the Oregon border where it crosses the Snake River. South of the highway are the Hitt Moun-

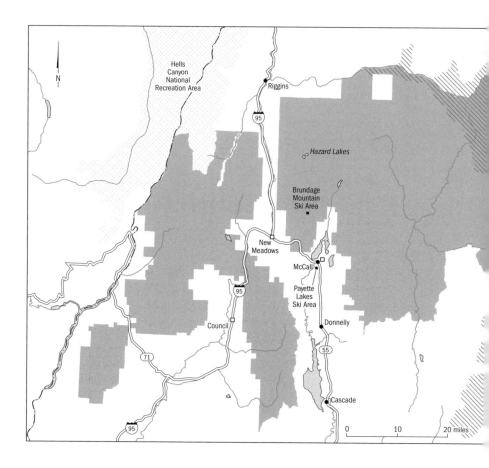

tains with 7,589-foot Sturgill Peak. North of the highway is a road to Brown-
lee Campground. One trail from the campground goes to Dukes Creek Natural Area northwest of Cuddy Peak. Volcanic cliffs rise steeply above Dukes
Creek. The natural area has undisturbed stands of grand fir, with ninebark,
mountain maple, and pinegrass prevalent in the understory.

After the highway crosses into Oregon, State Route 86 goes north to Ox-
bow. Another paved highway crosses the Snake River back into Idaho and
heads north into the Seven Devils Scenic Area in the Hells Canyon National
Recreation Area. The Seven Devils Area overlooks the deep Hells Canyon, a
canyon deeper than the Grand Canyon. Kinney Point and Sheep Rock along
the highway are great places to look down into the canyon. At Sheep Rock is
a self-guided half-mile nature trail. Plants identified along the trail are snowberry, wild currant, quaking aspen, mountain maple, ninebark, Indian paintbrush, pearly everlasting, subalpine fir, Douglas fir, curl-leaf mountain
mahogany, whitebark pine, rock beardstongue, snowbank ceanothus, and
bluebunch wheatgrass. About one-fourth of the way along the trail you can

Frank Church–River of No Return

Payette NF
Wilderness areas
Recreation areas
□ District Ranger Stations
● Towns

look north into Deep Creek Canyon where vivid reddish orange rocks make up the Red Ledge. The color is due to iron oxides.

If you want a thrill-a-minute drive that wanders through various forest types, take Forest Road 074 west off of U.S. Highway 95 about 11 miles north of New Meadows. Start early in the day. It would also help to have a current map and nerves of steel. Road junctions are many. To the north is Pollock Mountain in the Pony Creek Research Natural Area. Elevations in this natural area range from 4,000 to 8,080 feet on the mountain. Fine forests of old-growth ponderosa pine, Douglas fir, grand fir, subalpine fir, and whitebark pine are here. The rock cliffs and talus slopes on Pollock Mountain have some rare species of plants. Continue on the gravel road south and circle around Lost Valley Reservoir, then follow Grouse Creek north and west, eventually coming to Bear Creek Falls Natural Area. The important elements of this natural area are the picturesque falls, undisturbed Douglas fir forests, excellent bluebunch wheatgrass grasslands, and nearly pure stands of stiff sagebrush. Follow the road through Huntley Gulch to the Kleinschmidt Grade, which wriggles precipitously down to the Snake River. You will think you will never reach bottom.

McCall, at the outlet of Payette Lake, is the starting point for two other very scenic drives into the Payette National Forest. One is a 27-mile drive to the Lava Lakes and Lava Ridge; the other is an extremely long loop route, mostly on gravel or dirt roads. You will want to allow more than one day for this latter trip.

The road to the Lava Lakes first comes to the Payette Lakes Ski Area along State Route 55, a short distance west of McCall, then heads due north on the Goose Lake Road to the Brundage Mountain (fig. 18) Ski Area at the end of the pavement. If you brought your fishing gear, you may want to stop at Brundage Reservoir, Goose Lake (fig. 19), Duck Lake, Hidden Lake, or one of the three Hazard Lakes near the road. The Lava Butte Lakes are near the end of Forest Road 308, as is a trail to the Lava Butte Research Natural Area. Lava Butte is a subalpine peak in which the eastern side of the peak is granitic

Figure 18. Original fire lookout (ca. 1917) atop Brundage Mountain.

and the western side volcanic. A comparison of the vegetation on the two sides is fascinating. Upper forests on the peak consist mostly of whitebark pine and subalpine fir, with lodgepole pine on the northern face.

The long back road loop, Forest Road 21, begins as the paved Warren Wagon Road as far north as Upper Payette Lake. From there northward, the road is unpaved and even primitive in places. It follows the North Fork of the Payette River to the Burgdorf Guard Station. If you do not want to negotiate the rest of the long loop, you may turn north at the junction for a very scenic drive along Lake Creek to the Salmon River where you can take the Salmon River Road west to Riggins. The loop road curves along the Secesh River and through beautiful country to the old Warren Ranger Station, picturesquely nestled along Warren Creek. After climbing over Warren Summit, the road drops down to the South Fork of the Salmon River and then follows the river for two or three miles, eventually crossing it. Pilot Peak is to the east. The road now follows Elk Creek and ascends to Elk Summit and the old Big Creek Ranger Station. Three miles south is the Belvedere Creek Natural Area with a number of features: several lakes and streams; Coin Mountain and Pinnacles, both with alpine vegetation above 8,990 feet; and snowslide chutes. The road then makes a big loop through Profit Gap and stays alongside Profit Creek to the South Fork of the Salmon River where it comes to a "T." A wild side trip to the east climbs over Monument Summit and then descends to Monument Creek where there was a huge avalanche years ago that

Figure 19. Four-hundred-acre Goose Lake, popular for fishing and recreation.

destroyed a town. Roosevelt Lake and Thunder Mountain are a few miles farther east. The loop road at the "T" turns west and continues along the South
Fork for about seven miles. Three miles north is the Circle Creek Research
Natural Area along the east fork of Tailholt Creek. This natural area is a
prime example of a ponderosa pine–Douglas fir–grand fir forest with bitterbrush, snowberry, ninebark, and blue huckleberry in the shrub layer. The
back road then climbs over Lick Creek Summit and makes its way back to
McCall.

South of McCall is a paved forest highway from Donnelly on State Route
55 to Cascade. About one mile north of Cascade is the Warm Lake Road. To
the west is the Council Mountain Research Natural Area. This volcanic
mountain, at 8,126 feet, preserves a pristine forest of subalpine fir and whitebark pine.

From Warm Lake in the Boise National Forest, about 20 miles east of Cascade, is a forest road north that follows along the South Fork Salmon River
and enters the Payette National Forest near White Rock Peak and continues
to the old Krassel Ranger Station. Phoebe Meadows Research Natural Area is
accessible by a trail southeast from the station. The wet meadows have a diverse wetland flora of sedges, grasses, rushes, and willows. Forested parts of
the natural area contain Douglas fir, subalpine fir, and grand fir, with a rich
understory of spiraea, dwarf huckleberry, blue huckleberry, western ledum,
grouse whortleberry, and mountain maple.

Salmon-Challis National Forest

SIZE AND LOCATION: Approximately 4.3 million acres in east-central and central Idaho. Major access routes are U.S. Highways 20, 26, 93, and 93A and State Routes 21, 22, 23, 28, and 88. District Ranger Stations: Challis, Leadore, Mackay, Salmon, Clayton, North Fork. Forest Supervisor's Office: 50 Highway 93 South, Salmon, ID 83467, www.fs.fed.us/r4/sc.

SPECIAL FACILITIES: Winter sports areas; boat ramps.

SPECIAL ATTRACTIONS: Salmon National Wild and Scenic River; Sawtooth National Recreation Area; Ponderosa Pine Scenic Route; Salmon River Scenic Route; Custer Adventure Motorway.

WILDERNESS AREAS: Frank Church–River of No Return (2,366,698 acres, partly in six national forests); Sawtooth (217,088, partly in three national forests).

Until they were combined into one administrative unit, the Salmon and Challis were two separate forests in the central and east-central parts of Idaho. The Salmon is the more northern of the two forests.

Salmon National Forest

The Salmon National Forest extends from Lost Trail Pass on the Montana border in the Bitterroot Mountains to Gilmore Summit on the south. The Bitterroot Range on the Continental Divide forms the eastern boundary. After the Shoshone Indians lived here, Lewis and Clark came through the region in 1805, followed by trappers, traders, and miners, including Kit Carson and Jedediah Smith. With the discovery of gold in 1866 along Napias Creek, permanent residents began to establish towns such as Shoup, Gibbonsville, and Leesburg.

Much of the Salmon National Forest is in remote areas, including the western portion that is in the Frank Church–River of No Return Wilderness. Mt. Maguire, at 10,082 feet, towers over the wilderness area. In the vicinity are 105 glacial lakes. The Middle Fork of the Salmon River (pl. 32) that flows through the wilderness has challenging rapids for whitewater rafters. Hancock and Ouzel rapids are two of the major ones.

U.S. Highway 93 crosses the Continental Divide at Lost Trail Pass at the northern tip of the Salmon National Forest. This highway is the Salmon River Scenic Route, also known as the Lewis and Clark Trail. It works its way south along the Salmon River to North Fork. From North Fork, the scenic

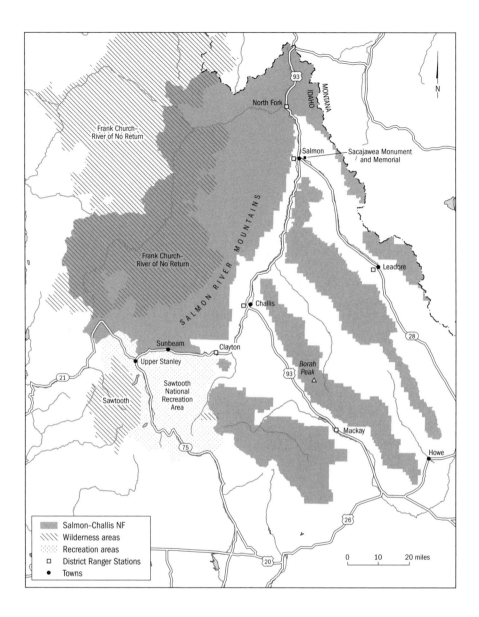

route continues to Salmon and beyond, separating the largest part of the national forest to the west with four smaller units to the east. Three monuments to Lewis and Clark are between Lost Trail Pass and North Fork. From the Twin Creek Campground is a trail along Twin Creek to the north of Allan Mountain and nearby Allan Mountain Natural Area. This area has nearly undisturbed forests. The old gold-mining town of Gibbonsville is where Dahlonega Creek empties into the North Fork.

From North Fork is a road to the west along the Salmon River known as

Clarks Reconnaissance Trail. At Indianola is a road north along Indian Creek to Ulysses Historical Site where gold was discovered in 1895. When the stamping plant at Ulysses burned in 1904, it essentially brought an end to the town. Ulysses had Idaho's largest active gold mine. In a few miles of following every curve of the Salmon River, the road comes to Shoup, which had greater success as a gold town than Ulysses. Prospectors found gold in some Salmon River gravel bars in 1868, but it was not until 1881 that a major lode was discovered. By 1882, Shoup had a post office and a large population. Remains of the old town can still be explored.

The road along the Salmon River continues past Shoup to Panther Creek at the northern edge of the Frank Church–River of No Return Wilderness. Well-preserved Indian pictographs are at several places along the Salmon and accessible from the road. The Forest Service has erected panels near the mouth of Panther Creek that describe the Indian writings. There is also a panel concerning Indian writings near the Ebenezer Bar Campground. The long Copper Mountain hiking trail extends from Panther Creek to Haystack Mountain by way of Copper Mountain.

From Panther Creek to the Corn Creek Campground, the forest road continues alongside the Salmon River and is a most scenic drive that should not be missed. A back road from the Colson Creek Guard Station along Colson Creek passes through huge, pristine coniferous forests. Eighteen miles from Shoup along Fountain Creek is a stairstep series of waterfalls.

A back road along the east side of Panther Creek leads entirely through the national forest to Morgan Creek Summit and then into the Challis National Forest before ending at U.S. Highway 93. This extremely scenic back road crosses many creeks and gulches. Where Porphyry Creek enters Panther Creek is a side road to the west. If you stay on this side road, you will end up at Middle Fork Creek Campground and a trail to Middle Fork Peak, at 9,127 feet, on the boundary of the wilderness area. On this road to Middle Fork Creek is a side road north past Quartzite Mountain to Yellowjacket Lake and Campground. Both of these back roads go into narrow passages between arms of the wilderness area.

The historic site of Leesburg is 10 miles west of Salmon via a primitive road. Located in a high-mountain basin, Leesburg developed and prospered after gold was discovered along Napias Creek in 1866. Within a year, the town had a population of about 2,000. The town has had many ups and downs since that time. Except for some renewed mining activity, all that remains are a few old cabins. Devlin Falls is one-half mile from Leesburg.

South of Salmon about six miles, the Williams Creek Road is paved as far as Williams Creek Summit. If you decide to continue on an unpaved road, you will end up at Deep Creek Campground.

State Route 28 branches off of U.S. Highway 93 at Salmon on its way to

Leadore and Gilmore Summit. A small section of the Lemhi Range is on either side of the highway. To the east is a winding unpaved road that eventually crosses Lemhi Pass on the boundary between Idaho and Montana. The Sacajawea Monument and Memorial are here, because Lewis and Clark and his band of explorers, along with their Indian guide, crossed the pass in 1805. West of State Route 28 is the western part of the Lemhi Range. Hiking trails are along Twelvemile Creek, Bear Valley Creek, Mill Lake, and Emerson Lake, and to Lem Peak, Sheephorn Peak, and Portland Mountain, all above 10,000 feet. From the gold-mining townsite of Gilmore at the south end of the Salmon National Forest is a paved road to Mesdow Lake, at 9,200 feet the highest campground in the national forest.

From Leadore, State Route 29 passes through Railroad Canyon and crosses Bannack Pass into Montana in the Beaverhead Mountains.

Challis National Forest

The Challis National Forest, with nearly 2.5 million acres, is a land of towering mountains, deep canyons, crystal-clear lakes, and turbulent rivers. Remains of old gold-mining activity are in several places, and visiting these sites conjures up visions of the past. The mountains are densely forested with stands of ponderosa pine, Engelmann spruce, lodgepole pine, and quaking aspen. On the upper slopes of the mountains are limber pine, whitebark pine, and subalpine fir.

The Challis has a number of good short hiking trails and roads to most of the points of interest. However, the Frank Church–River of No Return Wilderness occupies a part of the western edge of the national forest, including part of the wild Middle Fork of the Salmon River, a whitewater rafting river for those who want more strenuous experiences. The Salmon River Mountains make up most of the large western part of the forest, with the incomparable White Cloud Peaks (pl. 33) along their southern end. East of Challis, the national forest has two districts, one in the Lost River Range and one in the Lemhi Range. The White Knob Mountains are at the southern end of the national forest.

Drive north of Challis and take the Morgan Creek Road to scout for geodes, agates, and opals that are in the creek gravels. A mile road from Morgan Creek follows the course of West Fork. You may want to hike into the beautiful Camas Meadows where there are wildflowers galore during June, July, and August.

To explore the northern part of the River Mountains, take a road from Challis along Challis Creek northwest to Sleeping Deer Mountain on the border of the wilderness area. The small alpine lakes that surround the base of the mountain are exquisite. Where Challis Creek curves away from the road, a side road follows the creek to the Challis Creek Lakes at the foot of White Mountain. It is a reverent area.

You may get a great idea of the southern part of the Salmon River Range with its gold-bearing creeks by driving a loop road, called the Custer Adventure Motorway, from Challis to Sunbeam and back to Challis. Follow Garden Creek to the site of an old toll gate on the Challis-Bonanza Road. The gravel road then comes to turbulent but beautiful Yankee Fork and follows the river past Mt. Greylock to Custer and Bonanza, two once-thriving gold-mining towns. Custer, an important early-day mining center, was laid out in 1878 and, by 1881, had 3,500 residents. The mill and most of the buildings are gone, although the schoolhouse that was given to the Forest Service remains and is a museum worth visiting. A gravesite behind the school holds the graves of three girls killed in a snowslide in 1891. Down the road a couple of miles is Bonanza, the site of another prosperous gold-mining town. The old cemetery is fenced and protected by the Forest Service.

Where Yankee Fork enters the Salmon River at Sunbeam on U.S. Highway 93, there are several hot springs with a temperature of 107 degrees F that have attracted visitors for more than 100 years. U.S. Highway 93 then follows the Salmon River all the way to Challis. A gravel side road south along Slate Creek ends at trailheads into the gorgeous White Cloud Peaks area. Scenic Castle Rock, at 11,820 feet, dominates the region. A short side road leads to picturesque Bayhorse Bay and its campground.

Another pleasant drive into the southern end of the Salmon River Mountains is along State Route 21. Partly in the Boise National Forest, this route is also the Ponderosa Pine Scenic Route. From Warm Springs in the Boise National Forest, this highway follows Canyon Creek to Summit and Bench Creek rest areas inside the Challis National Forest. The Frank Church–River of No Return Wilderness Area is north of the scenic route. When the highway comes to Marsh Creek, it makes a sharp turn to the southeast. If you elect to stay straight on a gravel road along Beaver Creek, you can drive to Pinyon Peak and Pinyon Lake. The Ponderosa Pine Scenic Route follows Marsh Creek past Stanley Lake to the community of Upper Stanley.

U.S. Highway 93A splits off from U.S. Highway 93 just south of Challis and passes through Thousand Springs Valley and big Lost River Valley to Arco. East of the highway and above the valleys is the Lost River Range. A side road along Lime Creek enters the national forest. From the end of the road is a hiking trail to picturesque Grover Creek Lake. From Dickey, a paved road climbs to Doublespring Pass and then descends into Pashimeroi Valley east of the Lost River Range.

Two miles east of Dickey is a road off U.S. Highway 93A to Birch Spring. From the spring is the trail to 12,655-foot Borah Peak (pl. 34), the highest mountain in Idaho. Vigorous hikers can make it to the top of Borah Peak and back in one day. From the top of the peak are incomparable 360° views. South of the Birch Spring road is a good paved forest highway that crosses another part of the Challis National Forest and leaves the forest at Trail Creek Summit. It is then about 12 miles through the Sawtooth National Forest to Ketchum. Halfway to Trail Creek Summit is a paved road along the East Fork of Big Lost River. This very scenic route passes Big Rocky Canyon, Castle Rock, and the Copper Basin Mining Area. After climbing over Antelope Pass at 8,934 feet, the road drops down to the Antelope Guard Station on the Antelope Creek Road. The right fork from the guard station goes to the Iron Bog Campground where there are primitive roads and trails to Brodie Lake and Iron Bog Lake.

U.S. Highway 93A continues south in the Big Lost River Valley to the tiny community of Leslie. A favorite drive is up Pass Creek Canyon, carved by the turbulent waters of Pass Creek for millions of years. The rock layers in the canyon are tilted upward, and some of them contain caves, including Hidden Mouth Cave. The road passes over Pass Creek Summit and into the area south of Pashimeroi Valley.

To explore the far eastern end of the Challis National Forest in the Lemhi Range, there is a paved highway from Ellis on U.S. Highway 93 through the Pashimeroi Valley to Howe at the southern end of the Lemhi Range. A road from May goes to the pleasantly situated campground along Morse Creek. Another side road, this one paved, between Goldburg and Clyde, heads into the Lemhi Range where there are hiking trails to Mill Creek Lake and Iron Creek Point.

Sawtooth National Forest

SIZE AND LOCATION: Approximately 2.2 million acres in south-central Idaho, with a small part in northern Utah. Major access routes are Interstate 84, U.S. Highways 30S and 93, and State Routes 23, 27, 68, 75, and 77. District Ranger Stations: Burley, Fairfield, Stanley, Ketchum. Forest Supervisor's Office: 2647 Kimberly Road East, Twin Falls, ID 83301, www.fs.fed.us/r4/sawtooth.

SPECIAL FACILITIES: Winter sports areas; visitor centers; boat ramps; swimming beaches.

SPECIAL ATTRACTIONS: Sawtooth National Recreation Area; Ponderosa Pine Scenic Route; Salmon River Scenic Route; Sawtooth National Scenic Byway.

WILDERNESS AREAS: Sawtooth (217,088 acres, partly in three national forests).

If you stand on the shore of handsome Redfish Lake northwest of Ketchum and look westward, you can admire the jagged peaks of the Sawtooth Range towering above you. This is one of the most splendid sights in the country. Redfish Lake is at the northern end of the Sawtooth National Recreation Area, the most visited national recreation area in the United States. The visitor center at the northern end of Redfish Lake has information to help plan your visit to the area. The Sawtooth Mountains are mostly in the Sawtooth Wilderness.

The Sawtooth National Recreation Area is at the northwest corner of the huge division of the Sawtooth National Forest that extends north of State

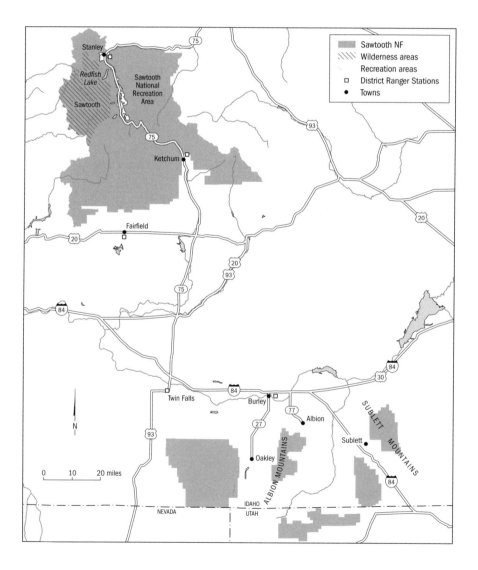

Route 20 and Interstate 84 and between the Boise National Forest to the west and the Challis National Forest to the east. This division of the national forest also includes other mountains—the Pioneer Mountains at the extreme eastern edge, the Boulder Mountains in the central part, and the Soldier Mountains in the southern end. Five small units of the Sawtooth National Forest are south of Interstate 84. These units include the Albion Mountains, the Sublett Mountains, the Black Pine Mountains, the Raft River Mountains, and Deadline Ridge. That part containing the Raft River Mountains is in northern Utah.

The Sawtooth National Recreation Area has 37 campgrounds, six major lakes, and numerous hiking trails. You can get a great overview of the recreation area by driving the Sawtooth National Scenic Byway (pl. 35), which is part of U.S. Highway 93. The road from Stanley will take you to Redfish Lake. After your visit to the lake, continue on the scenic byway south. The road, which is never more than six miles from the Sawtooth Wilderness, passes Alturas Lake and then the remains of the old mining town of Sawtooth City. A road south from Sawtooth City follows Smiley Creek to another old gold-mining townsite called Vienna. Before you get to Galena Summit you pass Galena Overlook where you can get a good view of the Sawtooth Mountains. The forests at the summit are dominated by subalpine fir, whitebark pine, and Douglas fir. Near the southeastern corner of the 756,000-acre recreation area is Boulder Basin where gold was discovered in the 1860s. A four-wheel-drive road goes to the basin and the Boulder Lakes. Boulder Peak, at 10,981 feet, towers over the area. At the Big Wood River Campground is an easy, self-guided nature trail.

Back at Stanley, the Ponderosa Pine Scenic Route proceeds northwest to Stanley Lake in the Sawtooth National Recreation Area. The scenic route climbs to Banner Summit and enters the Boise National Forest.

Two roads from Ketchum go to different parts of the Sawtooth National Forest. To the northeast, State Route 75 follows Trail Creek through the Pioneer Mountains and into the Challis National Forest. The final four-mile section of the road is cut into the side of a scenic canyon. From the top of Trail Creek Summit is a steep one-mile trail to the Trail Creek Falls that drops 60 feet into a catch pool. The Pioneer Mountains have been subjected to glacial activity that resulted in cirques, glacial scoured ridges, and U-shaped valleys. Pioneer Cabin is along the western edge of the mountains overlooking a colorful basin. The rustic cabin has two rooms and was built as a skiing hut in the 1930s. A four-mile trail to the cabin begins from the end of Coral Creek Road.

The road from Ketchum to the southwest is an unpaved forest road that first comes to lovely Dollar Lake and then goes along Warm Springs Creek and into the Smokey Mountains. Warm Springs Creek Canyon has produced considerable amounts of gold. Several hot springs are in the canyon, including the once famous Guyer Hot Springs, which had a resort. A series of switchbacks leads to the abandoned townsite of Carrietown. At the Big Smokey Campground are two other roads to choose from. The north fork follows the South Fork of the Boise River all the way to Featherville through luscious scenery. The south road climbs into the Soldier Mountains where there is a ski area before reaching Fairfield on State Route 20.

Southeast of Twin Falls is the Cassia Division of the Sawtooth National Forest that features the Deadline Ridge on the western side. A forest road penetrates the heart of the division along Rock Creek all the way to Rock Creek Campground. This is a beautiful area that includes Buckbrush Flats, Steer Basin, Magpie Basin, and Elk Basin. This area has many natural springs. Another entrance into the Cassia Division is up the Big Cedar Canyon Road northwest of Oakley. After coming out of the canyon at Oliver Springs, the right fork heads north to Artesian City. West of Oakley is a road along Little Cottonwood Creek. Southwest of Oakley is a long road that provides access to the southeastern quarter of the division and is a highly scenic route that winds its way past beautiful canyons and small mountain peaks.

The Albion Mountains are southeast of Burley, and a part of these mountains is in the Sawtooth National Forest. A paved forest highway between Albion and Connor climbs into the mountains, ending at Mt. Harrison. Several campgrounds and Lake Cleveland are attractive along the way. Between Elba and Oakley is a forest road that cuts across the middle of the Albion Mountains. Scenic hiking trails lead up New Canyon and Flat Canyon.

East of Sublett are the Sublett Mountains in the national forest. A loop road from the Sublett Exit on Interstate 84 winds through much of the area, which has a number of trails into scenic canyons. The loop road eventually returns to Interstate 84, near the Cotterel townsite.

Between the Albion Mountains and the Sublett Mountains are the Black Pine Mountains. Primitive roads off of U.S. Highway 30S go into Eightmile Canyon and Kelsaw Canyon. On the eastern side of the range are the eastern foothills, which are nonforested but support a curl-leaf mountain mahogany–bluebunch wheatgrass community. A road from the Interstate 84 exit at Juniper enters the national forest in East Dry Canyon and leaves the forest at Curlew Junction on U.S. Highway 30S.

The Raft River Mountains are part of a division of the Sawtooth National Forest that is entirely in northern Utah. The Mahogany Peaks above 8,000 feet are the highlights of this region.

NATIONAL FORESTS IN NEVADA

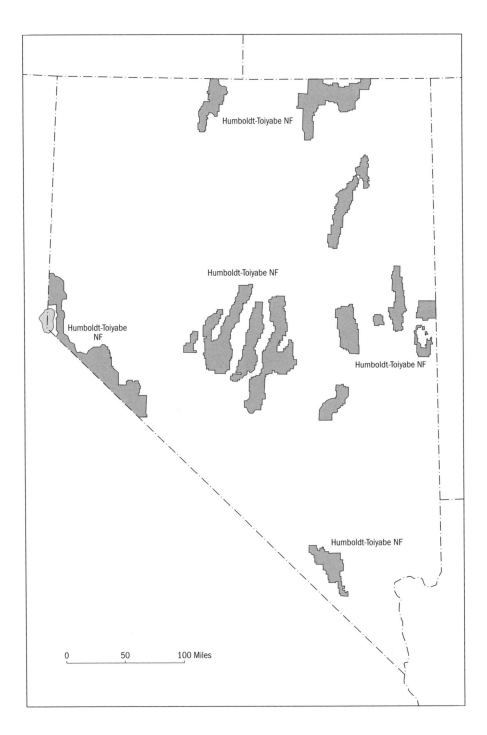

Humboldt-Toiyabe NF

Humboldt-Toiyabe NF

Humboldt-Toiyabe
NF

Humboldt-Toiyabe NF

Humboldt-Toiyabe NF

0   50   100 Miles

The two national forests in Nevada have been combined into one administrative unit. Nevada is in U.S. Forest Service Intermountain Region 4 and has 6.3 million Forest Service acres and 16 wildernesses. The regional office is in the Federal Building, 324 5th Street, Ogden, UT 84401.

Humboldt-Toiyabe National Forest

SIZE AND LOCATION: 6.3 million acres, scattered throughout much of Nevada, with small areas in eastern California. Numerous federal and state highways provide access to the forest. District Ranger Stations: Austin, Carson City, Tonopah, Wells, Ely, Elko, Winnemucca (NV); Bridgeport (CA). Forest Supervisor's Office: 1200 Franklin Way, Sparks, NV 89431, www.fs.fed.us/r4/htnf.

SPECIAL FACILITIES: Boat ramps; swimming beaches; winter sports areas; off-highway vehicle areas.

SPECIAL ATTRACTIONS: Spring Mountains National Recreation Area; Lamoille Canyon National Scenic Byway.

WILDERNESS AREAS: East Humboldts (36,900 acres); Mokelumne (99,161 acres, partly in the Eldorado and Stanislaus National Forests); Carson-Iceberg (161,181 acres, partly in the Stanislaus National Forest); Hoover (48,601 acres, partly in the Inyo National Forest); Currant Mountain (11,513 acres); Arc Dome (115,000 acres); Alta Toquima (38,000 acres); Grant Range (50,000 acres); Jarbidge (113,167 acres); Mt. Charleston (43,000 acres); Mt. Moriah (76,435 acres; 6,435 acres on Bureau of Land Management land); Mt. Rose (28,121 acres); Quinn Canyon (27,000 acres); Santa Rosa–Paradise Peak (31,000 acres); Table Mountain (98,000 acres); Ruby Mountains (20,000 acres).

Most of Nevada's 6.3 million acres of national forest land is encompassed in the vast Humboldt-Toiyabe National Forest. This national forest is scattered over much of Nevada and eastern California. Two of its units reach the Oregon and Idaho borders on the north, one is isolated in the Las Vegas area, two are in the Reno–Carson City area, and several units are across central Nevada. For many years the Humboldt National Forest in central and northern Nevada and the Toiyabe National Forest, mostly in the Reno–Carson City–Las Vegas area, were separate forests, but they have now been combined under one administrative unit.

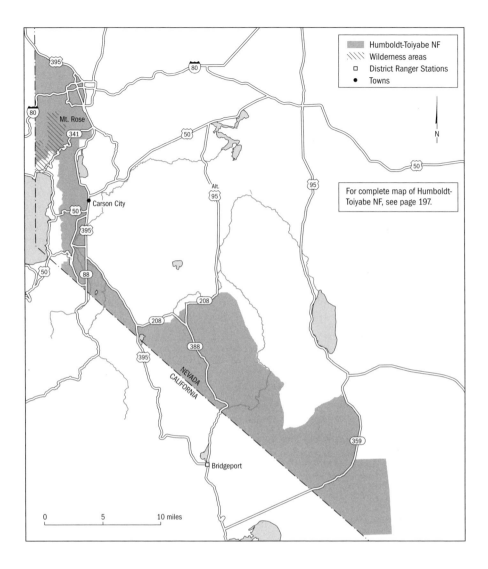

Humboldt-Toiyabe NF
Wilderness areas
District Ranger Stations
Towns

For complete map of Humboldt-Toiyabe NF, see page 197.

Humboldt National Forest

The Humboldt is spread over an area nearly 300 miles from north to south and 200 miles east to west. The forest comprises 11 distinct units with each unit centered around a particular mountain range.

Located near the Nevada-Idaho border, about 80 miles north of Elko, is the wild but extremely remote Jarbidge Wilderness. Legend has it that Shoshone Indians chased the jarbidge, a weird beastly creature, into a cave in Jarbidge Canyon and closed up the entrance with rocks. Although you are unlikely to run across this creature, there are other points of interest to entice you. To give you an idea of the remoteness of this area, not one of the eight peaks that exceed 10,000 feet can be seen from any major roadway. The Jarbidge Mountains form a continuous crest along the spine of the wilderness area, never dropping below 9,800 feet. Alpine habitats prevail on Jarbidge Peak and Cougar Peak. At the foot of Cougar Peak is Emerald Lake, one of the prettiest in the forest. Its name describes its color. The perimeter of the wilderness area around the base of the mountain range is occupied by a sagebrush scrub community. The most popular backcountry trail is the Three Day Creek Trail that meanders across the northern end of the wilderness area. This 11-mile path closely follows Three Day Creek and then Slide Creek, at one point crossing the East Fork of the Jarbidge River. Just west of the wilderness area are several pleasant campgrounds along the Jarbidge River.

If you are not into wilderness hiking, the forest on all sides of Mountain City a few miles west of the Jarbidge Wilderness may be for you. Although there is steep topography that goes from 4,600 feet at Bruneau River to 10,439 feet on McAfee Peak, several easy to moderate trails are available to choose from.

The Ruby Mountains, a narrow range trending southwest to northeast, are only a few miles southeast of Elko. The major landmark is the 11,387-foot Ruby Dome, which is actually a half dome that rises 5,500 feet above the valley below. Wild horses are at the south end of the range. Big and broad Lamoille Canyon (pl. 36), surrounded by rugged mountain peaks, is easily accessible via a 12-mile-long paved scenic byway. The canyon was gouged out through glacial activity. The impressive mountain above the eastern side of Lamoille Canyon is Verdi Peak. The Lamoille Canyon National Scenic Byway follows Lamoille Creek for a while, then climbs above the creek. Look for waterfalls and wet seepage areas on the walls of the canyon. If you want a close-up look at waterfalls, hike the easy trail from Roads End to Island Lake nestled snuggly at the foot of Thomas Peak. In a couple of miles is one of the nicest wildflower meadows in the national forest. Bluebells, columbines, and larkspurs begin the season in June, followed by Indian paintbrushes and lupines, and ending in August with the bright heads of sunflowers. Although you are only at about 8,800 feet, short, twisted bristlecone pines somehow cling to rocky ledges. The highway then climbs 3,500 feet out of the canyon to Road's End. The 40-mile-long Ruby Crest National Recreation Trail may be hiked from here, although the first 1.5 miles will satisfy most visitors. After one mile are serene Dollar and Lamoille Lakes. The trail then climbs to Liberty Pass where you can look back down on these lakes and many others in the U-shaped glacial valley. The ill-fated Donner party of gold rush fame traveled down the eastern side of the Ruby Mountains and crossed over

Overland Pass in the southeastern corner of the national forest. Later the Pony Express used this same route. A poor road still crosses the pass. Across the center of the Ruby Mountains is the Harrison Pass Road that was first used by the Bidwell-Bartleson party in 1841 and later by explorer John C. Fremont.

As you travel east on U.S. Highway 50 from the Utah-Nevada border through the town of Ely to the small community of Eureka, you pass, in order, the Snake Range, the Schell Creek Range, and the White Pine Mountains. Each of these mountainous areas, separated from each other by broad valleys running north to south, is part of the Humboldt National Forest.

The most spectacular area in all of Nevada is at the southern end of the Snake Range where Wheeler Peak, the remarkable Lehman Caves, and other special attractions are located. These wonderful areas were removed from the Humboldt National Forest when the Great Basin National Park was created in 1986. The park is well worth a visit. The north part of the Snake Range, containing Mt. Moriah, is still part of the national forest. Just east of the 12,050-foot Mt. Moriah is a one-square-mile flat-topped plateau known as The Table. Some of the most grotesquely beautiful bristlecone pines live on this flat summit, along with limber pines whose shapes have been influenced by the constant winds. A good diversity of dwarf wildflowers is present during the short growing season in July and August. Much of the area is designated wilderness and is full of prehistoric caves. Enough snow falls during the long winter to support four perennial streams, all containing the Bonneville cutthroat trout. Rocky Mountain bighorn sheep have been seen in some of the steep-walled canyons. Two trails are popular. One follows Hendrys Creek to Mt. Moriah and the other stays close to Hampton Creek before climbing to The Table.

The Schell Creek Range stands tall above Spring Valley to the east and Steptoe Valley to the west. The range is long and narrow, extending for approximately 60 miles north to south but less than 15 miles at its widest point. Many parts of the Schell Creek Range have been mined for ore, and remnants of old mining camps may be found. Two that can be reached at the end of poor roads are Aurum in the northeast corner of the range and Taylor at the southern end. The Success Loop is a 32-mile-long scenic drive along the western edge of the range. Starting at Cave Lake State Park, the road follows Steptoe Creek before climbing over Success Summit. It then drops down into the Duck Creek Basin. A very scenic section of U.S. Highway 50 cuts across the bottom of the Schell Creek Range, rising up over Connors Pass and then dropping into beautiful Connors Canyon. Near the Cleve Crest Campground are many Paiute Indian pictographs.

The White Pine Mountains are 50 miles west of Ely. The valleys surrounding the mountains and the foothills are covered by sagebrush. At mid-slope, the piñon pine–juniper woodland community takes over. As you continue to ascend to greater heights, forests of white fir and limber pine occur. On the highest peaks, such as Currant Mountain, bristlecone pines are found at elevations above 11,000 feet. Desert bighorn sheep are sometimes seen in the area. Currant Mountain is the dominant feature of the Currant Mountain Wilderness. This wilderness is seldom visited because there are few trails in the rugged mountains and no available water. Several back roads are in the northern part of the White Pine Mountains. It is still possible to travel along the Hamilton–Pioche Stage Line route that enters national forest land on the east side of the White Pine Mountains near Willow Grove and meanders in a northwesterly direction past the historical site of Hamilton, which had one of the best silver-producing mines in the area. Other historic sites such as Shermantown, Eberhardt, and Belmont Mill may be visited. The road to Eberhardt passes an impressive natural arch.

Grant Range is another mountainous area in the Ely District of the Humboldt National Forest. Lying south of the White Pine Mountains, this range is very isolated and therefore receives little use. Two wilderness areas are in this range, separated only by the dirt Cherry Creek Road. The northern wild area is appropriately called the Grant Range Wilderness and it soars to 11,295-foot Troy Peak in the center of the wilderness. The summit has a nice stand of bristlecone pines as well as red and white firs. One trail that uses an old road bed goes up Scofield Canyon for about six miles. The Troy Canyon Trail enters the wilderness area from the west and passes a couple of abandoned mines. Be careful if you plan to explore around the old mine shafts. Cherry Creek Road is highly scenic, particularly in the vicinity of Cherry Springs and Little Cherry Springs near the Cherry Creek summit. A campground is situated along Little Cherry Creek.

South of the Cherry Creek Road is the very primitive Quinn Canyon Wilderness. Although scenic Quinn Canyon is worth seeing, it is difficult for the average hiker to get to. Above the canyon is a lengthy crest above 10,000 feet with several east-west canyons leading away from it. Four streams in the wilderness area have flowing water all year.

Shortly after U.S. Highway 6 branches south from U.S. Highway 50 at Ely, the Ward Mountains appear to the east. This small unit of the national forest has excellent winter sports facilities. A very rough but scenic back road runs along the western side of Ward Mountain and passes several bubbling springs. Just before leaving the southern end of the national forest land, the road comes within a short distance of another picturesque natural arch.

The Santa Rosa Range of the Humboldt National Forest touches the Oregon state line. Granite peaks capped by lava in the southern part of the Santa

Rosa Range contrast with the peaks of rhyolite in the northern part. Plenty of minerals are in the rhyolite for rockhounds to seek. Some of the most impressive rhyolite pillars may be seen along the drive up Indian Creek to Hinkey Summit. The tranquil and very cool Lye Creek Campground is just north of the summit.

Numerous trails and back roads permit easy exploration of the many peaks, streams, and old mine sites in the northern part of the Santa Rosa Range. The southern part is very remote and is in the Santa Rosa–Paradise Peak Wilderness.

Toiyabe National Forest

Because the Humboldt National Forest is often in remote areas with only a few small towns nearby, much of it is lightly visited. Some of the Toiyabe National Forest, on the other hand, is in the vicinity of Reno, Sparks, Carson City, and Las Vegas and receives an enormous amount of traffic from sightseers, hikers, backpackers, and campers. A group of four mountain ranges in the center of Nevada, however, is more isolated. U.S. Highway 50 between Eureka and Fallon goes along the northern edge of the Monitor Range, the Toquima Range (pl. 37), the Toiyabe Range, and the Shoshone Mountains. Each range, which is narrow with a north-south orientation, is a distinct unit in the Toiyabe National Forest.

The Toiyabe has a good diversity of plant communities. The Spring Mountains National Recreation Area near Las Vegas has desertlike vegetation that may include Joshua trees, various cacti, creosote bush, sagebrush, and blackbrush. In some of the driest areas are alkaline flats where salt-tolerant species such as saltbush and saltgrass are common. Most of the foothills and lower mountain slopes support fine piñon pine–juniper woodlands. Above this community are stands of ponderosa pine and Jeffrey pine. Higher up, red fir, white fir, and lodgepole pine are the usual dominants. On the highest peaks are mountain hemlock, western white pine, whitebark pine, limber pine, and bristlecone pine.

The easternmost of the central mountains is the Monitor Range. This is high plateau country that is penetrated by several perennial streams that are excellent for trout fishing. A number of difficult roads are in the northern part of the range, most of them proceeding up such scenic canyons as Walters, White Rock, White Sage, Copenhagen, and Ryegrass. Table Mountain at the south end of the range is a remote wilderness area. In this wilderness is an interesting and intricate low-lying stone wall, apparently the work of prehistoric Indians. You can see this wall west of Lower Dry Lake. Trailheads are at Willow Creek, Mosquito Creek, and Green Monster Canyon.

Ten-mile-wide Monitor Valley separates the Monitor Range from the Toquima Range to the west. The only road that crosses the Toquima follows Moore's Creek over Charnock Pass just north of the Alta Toquima Wilderness. The wilderness area features rugged Mt. Jefferson, which has three summits that are connected by a ridge eight miles long and two miles wide. The trail across the ridge is part of the 16-mile trail that originates at Charnock Pass. Mt. Jefferson became a center of interest in 1978 when a forest ranger discovered a previously unknown Indian village on the upper slopes of the 11,949-foot mountain. The village comprised 31 tepees made from pine with rock bases to provide protection from cold mountain winds. Archaeologists date the village back 6,000 years, speculating that it was used as a hunting camp. About 1,500 years ago, Western Shoshone Indian families lived here during the summer. What is remarkable about this village is that it is the highest Indian village known in the Western Hemisphere, easily topping the 8,000-foot Inca city of Machu Picchu in the Andes Mountains.

A huge alkali flat in the Big Smoky Valley is between the Toquima and Toiyabe Ranges. The huge Arc Dome Wilderness in the Toiyabe Range occupies the southern one-third of this rugged mountain range. Thirty miles of the Toiyabe National Recreation Trail crosses the wilderness area from Ophir Summit to the South Twin River. North of the wilderness area is a scenic drive along Kingston Creek, passing a Forest Service campground. The two-mile hike up Straight Canyon, which is curved, is rewarding. Some of the largest mountain mahoganies you will ever see are here.

Mt. Charleston is the dominant landmark of the Spring Mountains National Recreation Area a few miles west of Las Vegas. A total of 42,000 acres around Charleston Peak have been designated a wilderness area. If you plan on hiking here, take water, because there are no streams and few springs. Drive the Sheep Mountain Scenic Loop for panoramic views into the desert below. A side road up Kyle Canyon leads you to the trailhead for a 1.5-mile hike to Cathedral Rock and Little Falls. The trail originates in a beautiful ponderosa pine–white fir forest in Mazie Canyon. After a short side trail to the falls, you will climb a few short switchbacks to the rock. Once there, Kyle Canyon is visible several hundred feet below. Across from Mazie Canyon is the Mummy Springs Trail. Follow the trail for three miles to arrive at Mummy Springs. However, you may wish just to observe a huge bristlecone pine at the trailhead.

Mt. Rose separates the bustling city of Reno from the ever popular Lake Tahoe. State Route 431 crosses the mountain and does not require mountain-driving skills. At the rocky, nearly barren summit is a nice stand of whitebark pine. Part of the Mt. Rose area is designated as wilderness. The smaller northern section includes Hunter Creek Canyon, and there is a pleasant two-mile foot trail here that passes by attractive mountain meadows. Mt. Rose is in the much larger southern section and may be reached by hiking the Mt. Rose National Recreation Trail to the 10,778-foot summit. A less strenuous three-mile trail is along Thomas Creek.

Toiyabe's most spectacular high Sierra country can be seen from the West Walker River Trail. The trailhead is on State Route 108 at Leavitt Meadows. After six miles along the West Fork of the Walker River, essentially following the Duckwell Emigrant Trail, the trail divides. If you choose the left fork, you will eventually climb to Emigrant Pass, as the Duckwell's did. This is the gateway to the Emigrant Basin Wilderness in the Stanislaus National Forest. If you take the right fork, you will pass through tranquil Paiute Meadows and, after eight more miles, come to Kirkwood Pass at the edge of the Hoover Wilderness.

Eighty percent of the Hoover Wilderness, which is entirely in California, is in the Toiyabe National Forest. The region has marvelous U-shaped glacial valleys surrounded by jagged peaks. Remnants of five glaciers are on Sawtooth Ridge, where such landmarks as Matterhorn Peak, Three Teeth Peek, Sawblade Peak, and The Cleavers tower above gemlike Avalanche Lake, Glacier Lake, and Snow Lake. A back road from Conway Summit on U.S. Highway 395 parallels Virginia Creek, culminating at the blue waters of the Virginia Lakes where there is a nice campground at 9,500 feet. Be sure you have cold-weather camping gear if you decide to stay here. From the campground is the trail into the Hoover Wilderness, first circling around Blue Lake before passing in between Black Mountain and Cooney Lake. Black Mountain looms just in front of you. If you ascend any of the mountains here, you will be in forests of lodgepole and Jeffrey pines, red fir, and mountain hemlock. John C. Fremont and his party of explorers crossed this region in the mid-1840s.

Charleston Mountains

Forty miles northwest of Las Vegas, the lofty Spring Mountains rise out of the Mojave Desert, affording welcome relief from summer swelter and, with snowfalls of 14 or more feet, ample opportunity for winter recreation. About 42 miles long and up to 26 miles wide, the range is among the most isolated in North America, its nearest neighbor being the Panamint Range of California, 100 miles away. Dominated by 11,910-foot-high Charleston Peak, which lies within Nevada's Toiyabe National Forest, the Spring Mountains are also commonly referred to as the Charleston Mountains.

On the lower slopes, creosote bush and sagebrush, typical plants of the Great Basin, taper down into the Mojave Desert, where a variety of cactuses and other desert-dwelling plants prevail. In the lower foothills, at about 5,000 feet, piñon pines and Utah junipers take over the drier habitats, whereas from about 6,600 to 8,100 feet ponderosa pines dominate the moist canyons. Bristlecone pines range from as low as 6,900 feet elevation all the way to tree line, which is reached on Charleston Peak at 10,000 feet. At their higher elevations, the bristlecones are joined by limber pines. The treeless summit of Charleston Peak is bordered by small, gnarled specimens of bristlecone pine.

The mountain rock consists of many sedimentary layers of limestone, dolomite, sandstone, shale, and gypsum deposited by the shallow sea that covered the region during the Paleozoic era (590 to 250 million years ago). Geologist Chester Longwell estimated that the deposits are nearly 30,000 feet thick and determined that the mountains themselves were formed about 60 million years ago, when great east-west pressure caused the sedimentary layers to buckle and shear. The violent forces often resulted in older rock formations coming to lie on top of younger ones; this is particularly evident on the south side of Lee Canyon. After a long period of erosion, during which the mountains gradually became lower, renewed pressures thrust some rocks upward and others down.

During the Pleistocene—the epoch of frequent glaciation that lasted from about 1.6 million to 12,000 years ago—southern Nevada was much cooler and moister than it is today. A large lake developed in Las Vegas Valley, and relatives of elephants, camels, and horses lived in the boggy habitat, benefiting from the plant life that must have thrived there. At that time, plant species were able to migrate across the low, wet terrain and colonize the Charleston Mountains.

As the Pleistocene ended and the climate became warmer and drier, the lake in Las Vegas Valley gradually disappeared and the Mojave Desert was formed. The plants that had become established in the Charleston Mountains became isolated, sometimes a long distance from their place of origin. Extended isolation, particularly in high mountain areas, provides an opportunity for plants to develop distinctive characteristics or even evolve into separate species.

The plants of the Charleston Mountains were studied by botanist Ira W. Clokey half a century ago. He discovered that approximately 20 kinds of flowering plants are found in this range and nowhere else, although all but one of them has a close relative elsewhere in the western United States. (The exception is Charleston kitten-tails, a plant whose leaves closely resemble those of alumroot, a totally unrelated species with which it grows.) Because of their uniqueness, these endemic species are monitored by Toiyabe National Forest personnel.

Most of the endemics grow at the higher elevations. Five are found 1,000 feet or so below tree line, where ponderosa pine is the dominant tree. Because environmental conditions in this zone are not as harsh as they are higher up, some of these plants grow quite large. Rough angelica, whose white flower clusters resemble those of Queen Anne's lace, has stems two or more feet tall. The others in this zone are yellow violet, a rosy-flowered sandwort, and two kinds of tiny-leaved milk vetches.

Eight other endemics grow at or above 10,000 feet, either beneath bristle-cone pines at the uppermost level of tree growth or above tree line in the high mountain meadows or rocky crevices of Charleston Peak. Rarest of these alpine species is the arching pussy-toes, a dwarf member of the aster family with clustered basal leaves and short, thick heads of white, fuzzy-looking flowers. Although this species spreads by slender runners as well as seeds, it is uncommon because of its hostile habitat, and the U.S. Fish and Wildlife Service is considering listing it as endangered.

Other dwarf species are nearly as rare, such as the yellow-headed Charleston tansy; the tiny, yellow-flowered hidden ivesia (a member of the rose family whose summer flowers barely peek out around small limestone pebbles); and the inconspicuous, white-blooming Charleston draba, which is found near natural springs in the high mountains. Prettiest of the alpine rarities is Clokey's catchfly, a delicate plant with a single, large, nodding, rosy pink flower. All provide a pleasant, scientifically interesting, and certainly less expensive change of pace from the glitter of nearby Las Vegas.

NATIONAL FORESTS IN OREGON

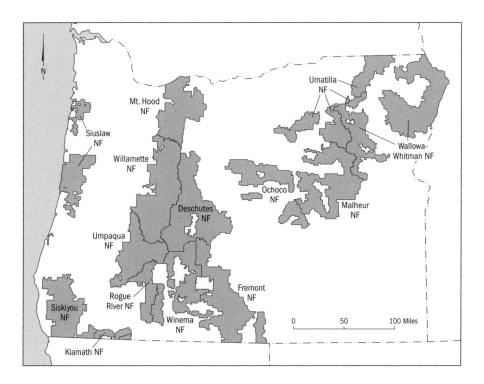

The national forests in Oregon are in U.S. Forest Service Pacific Northwest Region 6, along with the forests in Washington. The 14 national forests in Oregon have 15.9 million acres and 31 wildernesses. In addition, the Forest Service manages the Columbia River Gorge National Scenic Area. The Pacific Northwest Regional office is at 333 SW 1st Avenue, Portland, OR 97208.

Columbia River Gorge
National Scenic Area

SIZE AND LOCATION: 292,500 acres, beginning 25 miles east of Portland and continuing along the Columbia River. Major access route is Interstate 84. Headquarters: 902 Wasco Avenue, Hood River, OR 97031.

SPECIAL FACILITIES: Lodges.

SPECIAL ATTRACTIONS: Waterfalls.

Although the gorge of the Columbia River east of Portland, Oregon, was a part of the Mt. Hood National Forest, it is now the Columbia River Gorge National Scenic Area with a separate administrative unit. The gorge is incredible with its dozens of first-class waterfalls. Interstate 84 between the Columbia River and the Mt. Hood National Forest provides easy access to the trails and numerous waterfalls. Fifteen and thirty million years ago, two

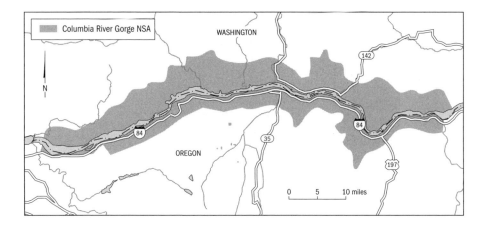

major lava flows covered the area and, as the lava cooled, it formed basalt. The Columbia River was powerful enough to erode a gorge in the basalt on its way to the Pacific Ocean. The small streams from the south that empty into the Columbia River are not powerful enough to carve gorges so that their water drops over the basalt cliffs in the form of waterfalls.

Interstate 84 enters the scenic area 25 miles east of Portland, and immediately there are waterfalls. Mist Falls is the first one, and its water plunges hundreds of feet from Mill Creek. A plunge waterfall is one that drops vertically from a stream, losing contact with the bedrock. Mist Falls may be seen from the highway. At the Wahkeena Picnic Area is Wahkeena Falls. This is a tiered waterfall that descends 242 feet. A tiered falls is one that drops in two or more distinct falls. If you hike the nature trail from the picnic area, the trail crosses in front of the falls and soon passes the east side of Necktie Falls. This falls also originates from Wahkeena Creek and drops 50 feet. It is a horsetail falls, which drops vertically from a stream but maintains some contact with the bedrock. A quarter-mile past Necktie Falls, the trail crosses in front of Fairy Falls with its 30-foot drop from Wahkeena Creek. This is a fan-type falls, which is a horsetail falls whose width increases downward.

A half-mile east of the Wahkeena Picnic Area on Interstate 84 is the pullout for the incredible Multnomah Falls. This is a plunge falls in two steps, the upper step with a remarkable 542-foot drop, and the lower step with a 69-foot drop. A concrete bridge is between the falls and you may thus get a perfect view in front of the falls. A combination visitor center–gift shop–restaurant is near the base of the falls. The Larch Mountain Trail passes Multnomah Falls and, within 1.5 miles, comes to Dutchman Falls and Double Falls. Dutchman Falls descends in three 15-foot drops. Because of their extreme width, they are referred to as block falls. Double Falls is two plunge-type falls, the upper 125 feet high, the lower 75 feet.

Two miles east of Multnomah Falls along Interstate 84 is the Oneonta Falls Trailhead. After hiking this trail for 1.7 miles, you come to Triple Falls. These falls are three falls side-by-side that may drop as much as 125 feet.

In another two miles along the interstate is Horsetail Falls, whose 176-foot descent may be seen from the highway. Horsetail Falls Trail goes behind this remarkable falls. If you continue on the Horsetail Falls Trail for another four-tenths of a mile, you will come to the slightly shorter 125-foot horsetail-type falls appropriately named Ponytail Falls.

Take the Eagle Creek Park Exit on Interstate 84 in another eight miles if you wish to hike the six-mile-long Eagle Creek Trail, or at least a portion of it. It is worth every step. You can see at least seven falls along the way. In order, they are Wauna Falls, a cascade type that drops over a series of rocky steps; Metlako Falls, a 150-foot plunge type; Punch Bowl Falls, only 15 feet high but a classic example of a punchbowl type that has a narrow drop into a catch pool below; Loowit Falls, Skoonichuk Falls; Wy'east Falls; and Tunnel Falls.

Deschutes National Forest

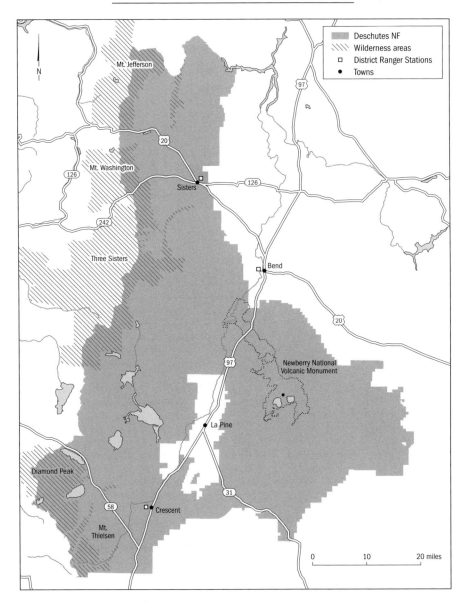

SIZE AND LOCATION: Approximately 1.6 million acres in the Cascade Range in west-central Oregon, north, south, and west of Bend. Major access routes are U.S. Highways 20 and 97, State Routes 58, 126, and 242, and Forest Highway 46. District Ranger Stations: Bend, Crescent, Sisters. Forest Supervisor's Office: 1645 Highway 20 E, Bend, OR 97701, www.fs.fed.us/r6/centraloregon.

SPECIAL FACILITIES: Ski areas; boat ramps; resorts.

SPECIAL ATTRACTIONS: Newberry National Volcanic Monument; Cascade Lakes National Scenic Byway (Forest Highway 46); McKenzie Pass Highway (State Route 242); Santiam Pass Highway (U.S. Highway 20); three Wild and Scenic Rivers—Deschutes, Metolius, Squaw Creek.

WILDERNESS AREAS: Three Sisters (286,708 acres, partly in the Willamette National Forest); Mt. Washington (52,738 acres, partly in the Willamette National Forest); Mt. Jefferson (107,008 acres, partly in the Mt. Hood and Willamette National Forests); Diamond Peak (54,185 acres, partly in the Willamette National Forest); Mt. Thielsen (61,281 acres, mostly in the Umpqua and Winema National Forests).

The Deschutes National Forest is a land of contrasts, with tall volcanoes of the Cascade Range bearing glaciers on their flanks, subalpine forests of firs, parklike ponderosa pine forests, dry scrubby areas of sagebrush, and reminders of past volcanic activities. Everywhere, it seems, are lava flows, cinder cones, lava tubes, volcanic lakes, and pumice flats. Extraordinary views of rugged volcanic landscapes are easily accessible in Newberry National Volcanic Monument (administered by the Deschutes National Forest), along the Cascade Lakes Scenic Highway, and along the McKenzie Pass Highway.

The national monument begins about 10 miles south of Bend and consists of 55,000 acres of the most intriguing and scenic parts of giant Newberry Volcano. At the monument's heart is Newberry Caldera. It holds Paulina Lake and East Lake, lava flows of black, glassy obsidian, the Central Pumice Cone, and Paulina Creek Falls. High on the southwest rim of Newberry Crater is Paulina Peak. It offers superlative views of the crater, central Oregon, the Cascade Range, and the High Desert to the east. Other attractions in the monument are Lava Cast Forest, Lava River Cave, and Lava Butte. Near the northwestern corner of the national monument, the wild and turbulent Deschutes River tumbles along the western edge of the vast lava fields that erupted from Lava Butte. At three places the river goes over impressive falls—Lava Island Falls, Dillon Falls, and Benham Falls. All of these falls may be reached via side roads and short trails off of the Cascade Lakes National Scenic Byway. Both Dillon and Benham Falls cascade about 60 feet.

At the southern edge of the Deschutes National Forest, 20 miles south of the Newberry Caldera, is another fantastic volcanic feature. Hole-in-the-Ground is a large volcanic crater one mile in diameter and several hundred feet deep. Geologists call this type of crater a maar. It develops when molten rock (magma) pushes up into groundwater or a lake. A series of great steam explosions blast a hole in the ground. In 1970, NASA studied Hole-in-the-

Ground in detail to understand if such eruptions could have produced some of the craters of the moon. Six miles to the west is another similar maar called Big Hole. While in the area, take a primitive road two miles north of Big Hole to Katati Butte. On the butte are rock formations called hornitos, which are shaped like beehive ovens. The two- to three-foot openings are in the top of the hornito and open into 12-foot-deep bottle-shaped interiors with glassy walls.

Many lava caves may be reached by a good forest road. These caves formed when a ribbon or sheets of molten lava cooled on the outside but remained molten on the inside. The molten lava inside would drain out the toe of the lava flow leaving a long, open, sinuous tube. Some of these caves are a short distance south of Bend. Lava River Cave, within the monument, is along U.S. Highway 97, and other caves such as Boyd Cave, Skeleton Cave, and the two-layered Wind Cave can be reached from Forest Road 18 (check locally for seasonal closures due to bats). Some caves are several thousand feet long.

Most of the Deschutes National Forest west and south of Bend is readily accessible via Forest Highway 46, the Cascade Lakes National Scenic Byway. This route is open only from June until the first heavy snowfall blocks it. A comfortable 100-mile loop trip can be made along the Cascade Lakes National Scenic Byway, Forest Roads 42 and 43, and U.S. Highway 97.

For nearly 18 miles the Cascade Lakes National Scenic Byway climbs steadily to a saddle between Tumalo Mountain to the north and Mt. Bachelor to the south. Mt. Bachelor is a favorite paradise for skiers and snowboarders. Dutchman Flat is just beyond and is a pumice desert with sparse vegetation. Looking to the north and northwest across Dutchman Flat, you see the Three Sisters and Broken Top, impressive volcanoes that rise as high as 10,350 feet elevation. Soon the Cascade Lakes National Scenic Byway passes a black lava flow from Mt. Bachelor and descends to a meadow where popular Sparks Lake is located. From the boat ramp on the eastern side of the lake you can get a perfect view of South Sister (pl. 38), the largest of the Three Sisters volcanoes. From the Fall Creek Trailhead just beyond Sparks Lake turnoff is a five-mile-long trail to Green Lakes in the Three Sisters Wilderness. From Green Lakes, South Sister is less than three miles away, Middle Sister is about six miles away, and North Sister is about seven miles away.

Between Sparks Lake and Devils Lake is an important pile of 2,000-year-old rock. Astronaut Jim Irwin aboard Apollo 15 placed a piece of this rock on the moon in 1971.

The Cascade Lakes National Scenic Byway makes a sharp bend to the south at Devils Lake, whose sandy pumice bottom gives the water an aqua color. A trail leads to Moraine Lake and up South Sister from here, and another trail winds across the pumice flats of the Wickiup Plains to connect with the Pacific Crest National Scenic Trail at the Sisters Mirror Lake. The highway then passes through a pleasant forest of lodgepole pine to Elk Lake and Lava Lake, both dammed by lava flows from Mt. Bachelor. The historic Elk Lake Guard Station welcomes summer visitors.

Several miles south of Lava Lake, take a side road to expansive Cultus Lake nestled at the base of Cultus Mountain, a lake dammed behind an Ice Age moraine. The mountain has a dense forest of ponderosa pine and firs. In three miles, near the west side of Crane Prairie Reservoir, is a special management area set aside to enhance the nesting of ospreys. At the end of a short trail just south of the Quinn River Campground is an observation point. At the southern end of Crane Prairie Reservoir, Forest Road 42 heads back toward the east for its return to Bend. Pringle Falls on the Deschutes River and Fall River Falls are two minor waterfalls you may want to seek out. A little less than a mile southeast of Pringle Falls is the Turn-of-the-Century Forest. Here, pumice soil supports a forest of pondersa pine and lodgepole pine with an understory of bitterbrush. The area is managed to have the appearance that it did at the end of the nineteenth century. At that time the forests of central Oregon were subjected to natural fires. Small, ground-hugging fires routinely burned the understory of brush and small trees, leaving the larger ponderosa pines to grow tall and unscathed.

Twelve miles west of Bend's Drake Park and Mirror Pond is attractive Tumalo Falls, which drops 100 feet. A picnic area is adjacent.

One of the most exciting drives is on State Route 242 over McKenzie Pass, which forms a narrow corridor between Mt. Washington Wilderness to the north and Three Sisters Wilderness to the south. The western side of the pass is in the Willamette National Forest. The pass is in the midst of a huge field of lava flows. Some are only 1,500 years old. That part of the Pacific Crest National Scenic Trail from the pass to Belknap Crater is about three miles long. As the highway descends from the pass to the east, it passes through an open forest of ponderosa pine.

U.S. Highway 20 crosses Santiam Pass along the south edge of the Mt. Jefferson Wilderness. The large mountain that looms mightily south of the pass is Mt. Washington, a volcano nearly eroded away by glaciers of the last Ice Age. The highway descends rapidly down the east side of the pass to volcanic Blue Lake and glacier-scoured Suttle Lake. Blue Lake fills a volcanic crater blasted out of the ground and is unique in that it is only 54 acres in size but 314 feet deep. A hiking trail is on the cliffs surrounding the lake, but be aware that much of the lake's perimeter is private land. The highway then levels out and passes through more forests of ponderosa pines until the turnoff to the Camp Sherman and Metolius River area. Nearby Black Butte, a distinct, conically shaped volcano, is to the south and is densely forested. A secondary road goes halfway up the butte, and a two-mile trail continues to the top. At the north base of Black Butte, Metolius Springs bubble vigorously and form the headwaters of Metolius River. This is a particularly beautiful area. A short distance north is the Metolius Research Natural Area, which has two very different ecosystems. One of these is on a nearly flat bench along the Metolius River, whereas the other one is on a very steep west-facing slope of Green Ridge. You will find ponderosa pine forests in the natural area, about half of them with an understory dominated by bitterbrush and the other half with an understory of green-leaved manzanita. Several miles northeast of Metolius Springs is the pretty Head of Jack Creek, a group of springs set in a rich diversity of trees, mosses, and other plants.

Lavalands

Nearly every variety of formation caused by volcanism can be found in the 100 miles from the central Oregon city of Bend south to Crater Lake. Known officially as the Newberry National Volcanic Monument and locally as Lavalands and managed by the Deschutes National Forest, this diverse landscape has been shaped by sporadic volcanic eruptions for millions of years. The last eruption was about 1,350 years ago, but geologists consider the region volcanically active, with future eruptions possible at any time.

Rock layers containing fossil corals and sponges show that most of present-day Oregon was once covered by the sea. About 40 million years ago, however, the Cascade Mountains began emerging from the sea, primarily as a result of tectonic activity. Geologist Lawrence Chitwood estimates that about 17 million years ago, the Cascades reached such lofty heights that they blocked moisture coming in from the Pacific Ocean, drying out the area to the east, where Lavalands is now located.

During the early part of this drying-out period, volcanoes east of the Cascades began to spread great amounts of lava, cinder, and ash over the land. Through the years, vegetation has gained a foothold only slowly in these deposits. Low, dry sites are covered with shrubby sagebrush and brittlebush, whereas western juniper abounds on slightly higher terrain. More lush vegetation grows on the volcanic mountains themselves: ponderosa, lodgepole, white, sugar, and whitebark pines; Douglas fir; Engelmann spruce; and mountain mahogany.

One of the volcanoes that especially contributed to the features of south-central Oregon visible today is Newberry Volcano, located south of Bend. It began erupting about 200,000 years ago and has erupted sporadically ever since. Its broad crater, or caldera, apparently formed about 10,000 years ago when lava drained away to feed nearly 200 cinder cones, towers of ash and pumice, on the volcano's outer slopes. As the lava gradually drained out, the top of the mountain collapsed in stages, leaving a five-mile-wide caldera with steplike walls.

A thorough tour of Newberry Volcano could take several days. A few miles south of Bend, a visitors center provides a good orientation to the region. Displays and dioramas depict past geologic events and current topography.

The first stop after leaving the visitors center should be the top of Paulina Peak, which at 7,985 feet affords an overall view from the southwest rim of the caldera. The hundreds of cinder cones that dot the out-slopes of Newberry Volcano often obscure that it is a single landform, with a base 40 miles long and 25 miles wide. The cinder cones were created by the eruption of molten rock containing dissolved gases. Where the erupting lava contained little gas, rounded domes formed instead of cones. One is easily visible near the southeast corner of the rim, about four miles from Paulina Peak.

A hiking trail that stays on the volcano's rim covers 21 miles before returning to the starting point, and a road provides easy access to features within the caldera. Among these are two lakes, kept full from rain and snow that drain from the rim above and by hot springs that enter from below. Geologists believe that originally there was only one lake, which may have been 2,000 feet deep, but that later eruptions divided it and filled the bottom of the lakes with volcanic debris.

The larger of the two lakes is Paulina Lake, two miles long, 1.5 miles wide, and up to 200 feet deep. About a mile to the east and 40 feet higher is East Lake. The two are separated by a porous land mass, and water from East Lake gradually trickles through this "flow zone" into Paulina Lake. Overflow from the west side of Paulina Lake enters Paulina Creek, which at one place plummets thunderously over a 100-foot precipice, forming spectacular Paulina Falls.

South of the flow zone between the two lakes is a huge pumice cone that was formed about 5,000 years ago. An obsidian flow, 1.5 miles wide, begins about a half-mile southeast of Paulina Lake. The flow of shiny, black volcanic glass was formed during the most recent eruption of Newberry Volcano, about 1,350 years ago.

Outside the perimeter of Newberry Volcano are many other volcanic features. Fifteen miles to the north is the Lava Cast Forest, created about 6,000 years ago when lava from the volcano flowed through a stand of trees. Many trees were knocked over or burned, but some of the larger trunks were left standing with a coating of lava, which cooled rapidly when it came into contact with the living wood. The wood eventually rotted away, leaving shells of lava up to 20 feet high. Little soil has accumulated on the lava that covers the ground between these so-called lava casts. As a result, woody species such as ponderosa pine, which grows nearly 100 feet tall on nearby mountains, are dwarfed and grotesquely twisted.

Scattered throughout the region are underground tubes of lava, called lava caves, which formed when basalt, a very fluid type of lava, flowed quickly down the sides of Newberry Volcano and surged into fissures along the earth's surface. Cooled by the air, the upper surface of the lava in each fissure formed a crust, beneath which the rest of the lava continued to flow. When all the fluid lava finally drained out, it left behind a lava-lined tunnel within the earth, with a roof as much as 20 feet thick.

Visitors can proceed through Lava River Cave, walking underground for nearly a mile before the way is blocked by an accumulation of sand. (Rent a lantern from the Forest Service personnel stationed near the entrance and take a sweater—the temperature in the cave is always about 40 degrees F.) The only animals that inhabit the cave are bats, insects, and spiders.

Fragments of the inner lava crust have flaked off and fallen to the floor of the cave in places, whereas elsewhere the cave walls have a smooth, glazed appearance. Although often only three feet wide, the tube widens into rooms, one of which is 50 feet across with a ceiling 58 feet high. Hanging from the ceiling are occasional formations of solid lava, called lavacicles, which resemble the stalactites that are formed in limestone caves from mineral drippings.

Fremont National Forest

SIZE AND LOCATION: Approximately 1.2 million acres in south-central Oregon, extending to the California border. Major access routes are U.S. Highway 395 and State Routes 31 and 140. District Ranger Stations: Bly, Lakeview, Silver Lake, Paisley. Forest Supervisor's Office: 1301 S. G Street, Lakeview, OR 97630, www.fs.fed.us/r6/frewin.

SPECIAL FACILITIES: Hang-gliding areas; boat ramps; swimming beaches.

SPECIAL ATTRACTIONS: Wild and Scenic North Fork of the Sprague River.

WILDERNESS AREAS: Gearhart Mountain (22,809 acres).

The Fremont National Forest is surrounded by high desert country, and the national forest itself is on the east side of the Cascade Range. To get a preview of the transition between high desert and forested mountains, head to

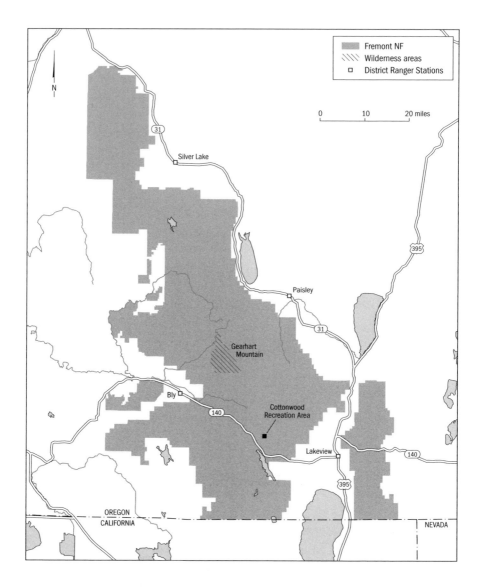

the Goodlow Mountain Natural Area 43 miles from Klamath Falls. This 1,260-acre tract goes from high desert sagebrush steppe to ponderosa pine savanna to ponderosa pine–white fir forest. The topography is gently rolling with a low butte known as Goodlow Mountain at the edge of the sagebrush steppe. Beyond the summit of the butte are the forest communities. The dominant species in the shallow soils of the steppe is low sagebrush, with Sandberg's bluegrass interspersed. With a slight increase in elevation, western juniper appears with big sagebrush and Idaho fescue. The savanna marks the transition from dry steppe and juniper communities to forest. It is recognized by the presence of ponderosa pine above an understory of bitter-brush, curl-leaf mountain mahogany, and Ross's sedge. As elevation increases, Parry's manzanita replaces bitterbrush as the dominant shrub. The highest elevations support old-growth ponderosa pine, white fir, and lodgepole pine.

The Fremont National Forest consists of two major units. The southeastern unit includes the Warner Mountains and the impressive Abert Rim. The larger unit of the forest is to the west and includes the mountains between Lakeview and Klamath Falls.

The Warner Mountains area is easily accessed from forest highways north and south of State Route 140. To the north, the Fremont National Recreation Trail is in the semiprimitive Drake-McDowell Area and offers several marvelous views of Warner Mountain, Warner Valley, and the Abert Rim. However, you can drive to the North Warner Viewpoint and the Drake Peak Lookout for equally splendid panoramic vistas. The latter peak is at an elevation of 8,222 feet, and a view across the high desert is unobstructed. The south end of the Abert Rim is in the Fremont National Forest. It is the largest exposed geologic fault in North America. The top of the rim, at an elevation of 2,500 feet, is covered by lava. Mud Creek Forest Camp, situated in a lodgepole pine forest, is near the Crane Mountain National Recreation Trail. The Warner Mountain winter sports area is located where State Route 140 enters the national forest.

The most pristine part of the mountainous region between Lakeview and Klamath Falls is the Gearhart Wilderness. Gearhart Mountain is a volcanic dome that is 8,364 feet in elevation. The spectacular cliffs of the dome in the southern part of the wilderness are up to 400 feet high and extend for three-quarters of a mile. Palisade Rocks are heavily weathered pillars, columns, pedestals, and toadstool-shaped formations. The Gearhart Mountain Trail from Corral Creek Campground passes the Palisade Rocks and the dome on its way to Gearhart Mountain. It is about six miles to the mountain. The Lee Thomas Campground adjacent to a large wet meadow is another place from which you can enter the wilderness area.

In a ponderosa pine forest 3.5 miles from the Corral Creek Campground is Mitchell Monument, site of a most bizarre incident. It is the only place on the North American continent where death resulted from enemy action during World War II. It happened on May 5, 1945, when Rev. Archie Mitchell, his wife, and five Sunday school children from the Bly Christian and Missionary Alliance Church were on a fishing and picnic outing. When they found the main road they were going to travel on blocked, they proceeded to traipse through the forest. One of the group found a strange-looking object that turned out to be a Japanese balloon bomb, which exploded when it was picked up, killing all but the minister. The Cottonwood Recreation Area southeast of Gearhart Mountain Wilderness Area is another popular campground with a boat ramp on Cottonwood Lake.

From the edge of the cliff at Fremont Point, located at the northern edge of Winter Ridge, is a spectacular view of Summer Lake in the valley below. A historic cabin is near the point. Captain John C. Fremont and his exploration party, which included Kit Carson, stopped at the lake in 1843. A few miles southeast of Fremont Point is a hang-gliding area known as Hadley Butte.

The North Fork of the Sprague River, which penetrates the national forest, is a designated Wild and Scenic River.

Malheur National Forest

SIZE AND LOCATION: Approximately 1.5 million acres in east-central Oregon, from the southern part of the Blue Mountains to Oregon's high desert. Major access routes are U.S. Highways 20, 26, and 395 and State Route 7. District Ranger Stations: Hines, John Day, Prairie City. Forest Supervisor's Office: 431 Patterson Bridge Road, John Day, OR 97845, www.fs.fed.us/r6/malheur.

SPECIAL FACILITIES: Snowplay areas; boat ramps.

SPECIAL ATTRACTIONS: Indian Rock–Vinegar Hill Scenic Area.

WILDERNESS AREAS: Monument Rock (19,650 acres, partly in the Wallowa-Whitman National Forest); Strawberry Mountain (68,700 acres).

One of the great things about the Malheur National Forest is that almost any point of interest is accessible to most forest visitors. Two wildernesses are in the forest. Although Monument Rock Wilderness is wild and rugged, the Strawberry Mountain Wilderness is visited by most people.

Most of the Malheur National Forest is in the southern part of the Blue Mountains. Vegetation types range from grassslands and sage to juniper woodlands to forests of pines and firs. Alpine lakes are present as well as alpine meadows, even though the highest elevation is 9,038 feet.

To start your forest adventure, you may want to head straight for the 4,000-acre Indian Rock–Vinegar Hill Scenic Area at the northeast corner of the national forest in the Greenhorn Mountains. Indian Rock is at the western end of the scenic area, Vinegar Hill at the eastern end, and Summit Butte in the center. This used to be an area of beautiful alpine meadows where dozens of kinds of wildflowers bloom from June until the end of the grow-

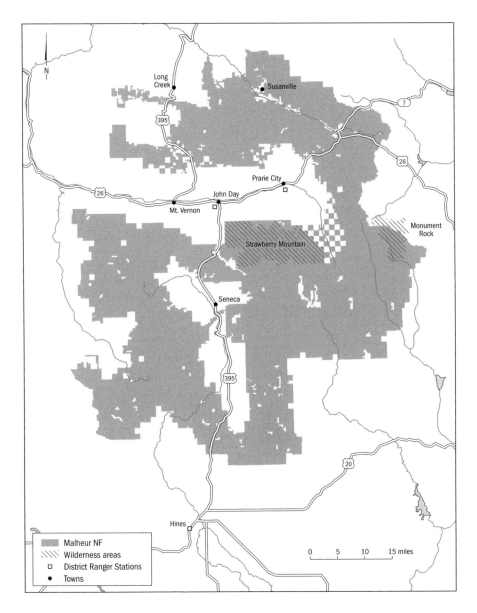

Legend:

- ▨ Malheur NF
- ⧄ Wilderness areas
- □ District Ranger Stations
- ● Towns

0   5   10   15 miles

ing season, but a major forest fire near the summit in 1996 has burned the area. Dispersed camping is available at Indian Rock, which makes the exploration of this high meadow area at 7,353 feet easy to reach. A primitive dirt road goes to the even higher Vinegar Hill. To get to Summit Butte will require a 3.5-mile hike that goes up the south face of the butte. While you are in the area, pay a visit to the old mining community of Susanville near the Middle Fork of the John Day River.

All the rest of the Malheur National Forest is south of the Middle Fork.

Drive south a few miles to the incredible Arch Rock, an ashflow tuff with an opening more than 15 feet high. A little farther to the southwest is Magone Lake and the nearby remarkable Magone Slide. Magone Lake was formed by a landslide that happened in the early 1800s. A trail passes through the area where trees carried down by the slide are growing at angles.

Fifteen miles east of Magone Lake is Dixie Summit, located on U.S. Highway 26. Three miles to the north is Dixie Butte. From the summit you can look down at the old switchbacks of the Sumpter Valley Railroad. An accessible interpretive trail is off of U.S. Highway 26.

Looking across the valley you will see the Strawberry Mountain Wilderness. Two trails are worth taking. A trail north of Canyon Meadows Campground goes for nearly four miles through Buckhorn Meadow to Wildcat Basin. It passes several interesting geologic formations. The 1996 fire burned through the center of the north side of the wilderness. Six miles west of Wildcat Basin is the Tamarack Creek Trail, which circles around the northeastern corner of the Canyon Creek Research Natural Area. The 700-acre natural area has stands of virgin ponderosa pines. Many of the forests consist of ponderosa pines, but as the elevation increases, Douglas firs and grand firs appear. Plants growing beneath the pines are usually whortleberry, lupines, heart-leaved arnica (pl. 39), and woolly-weed. A small dry area dominated by western juniper with curl-leaf manzanita in the understory is also present in the natural area. On rocky outcrops, the shrubby mountain balm clings to life.

The Cedar Grove Botanical Area near Buck Cabin Creek is at the western edge of the Malheur National Forest in the Aldrich Mountains. The cedars are Alaska yellow cedars, some of them more than 300 years old. What is intriguing is that the nearest stands of the Alaska yellow cedar to this one are in the Cascade Mountains of Oregon and Washington. A mile-long nature trail runs through the botanical area, but it is steep and difficult when it gets to a forest of dense firs.

About 10 miles south of the cedars is Rosebud Creek where there is an incredible area of fossilized shells. When you finish looking for fossils, drive about seven miles on back roads to Tex Creek where there is a neat natural bridge.

Fifteen miles north of Burns and just east of U.S. Highway 395 is Devine Ridge. A two-mile-long interpretive trail climbs out of desert sagebrush to an open ponderosa pine forest to a higher forest of mountain mahogany to dry, rocky outcrops on the ridgetop.

Yellowjacket Reservoir is a popular lake with a campground at the southwest corner of the national forest.

Strawberry Mountain Wilderness is 18 miles east to west and one to four miles north to south. A 1.5-mile trail leads from Strawberry Campground outside the wilderness to scenic Strawberry Lake inside the wilderness. Monument Rock Wilderness at the eastern edge of the Malheur, which is shared with the Wallowa-Whitman National Forest, has three impressive peaks—Monument Rock, Table Rock, and Bullrun Rock. Subalpine fir dominates the tops of the peaks, with ponderosa pine, Douglas fir, white fir, lodgepole pine, and quaking aspen a little lower down.

Mt. Hood National Forest

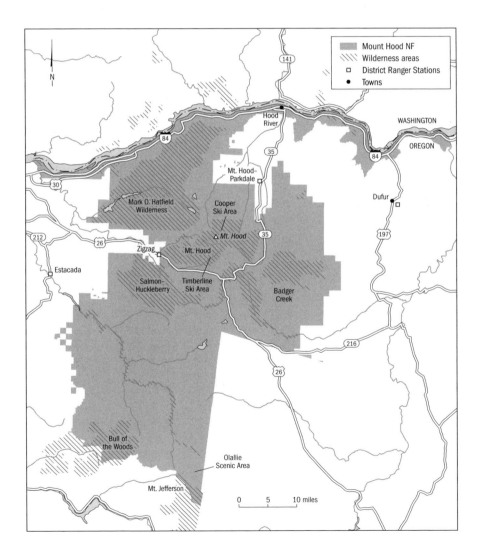

SIZE AND LOCATION: Approximately 1.1 million acres in northwestern Oregon, from the Columbia River to Mt. Jefferson in the Cascade Range. Major access routes are Interstate 84, U.S. Highways 26 and 30, and State Routes 35, 36, 212, and 224. District Ranger Stations: Dufur, Estacada, Mt. Hood-Parkdale, Zigzag. Forest Supervisor's Office: 16400 Champion Way, Sandy, OR 97055, www.fs.fed.us/r6/mthood.

SPECIAL FACILITIES: Winter sports areas; boat ramps; swimming beaches; visitor center; lodge.

SPECIAL ATTRACTIONS: Mt. Hood Loop Highway.

WILDERNESS AREAS: Badger Creek (24,000 acres); Bull of the Woods (34,900 acres, partly in the Willamette National Forest); Mt. Hood (47,100 acres); Mt. Jefferson (107,100 acres, mostly in the Deschutes and Willamette National Forests); Salmon-Huckleberry (44,600 acres); Mark O. Hatfield (39,000 acres).

The majestic snow-capped Mt. Hood (pl. 40) is the centerpiece of this national forest. Although the 11,240-foot summit of Mt. Hood is in a wilderness area, several of the attractions around it are more readily visited. Even if you wish to climb Mt. Hood, you will do so from impressive Timberline Lodge, reached by a paved but curvy mountain road. The climb to the summit of Mt. Hood from the lodge and back requires a minimum of 12 hours. Most climbers use the Hogback route, but other routes are available. Any of the routes up the south face of the mountain come to the Silcox Warming Hut in the Palmer Glacier after about a one-hour hike. Beyond the warming hut, the climb becomes technical, that is, there are no hiking trails over the icy slopes. The Magic Mile chairlift also goes from the lodge area to the 7,000-foot level of the mountain. Perhaps you would just rather relax before the huge fireplaces in the Timberline Lodge and enjoy some refreshments. The lodge, dedicated on September 28, 1937, has 40,000 square feet of floor space and is constructed of native wood and stone. All of the interior was done by local craftsmen. Workrooms were set up where people worked with iron and wood, artists painted in watercolors or oils or carved in wood, a mosaicist worked in tile and glass, and another in wood marquetry. Old railroad tracks were used for andirons, old hinges for doors and gates, and old grills for windows. Upholstery and bedspreads for the guest rooms were woven from wool and flax, draperies were of sailcloth, and rugs were hooked. The high center of the lodge is hexagonal with two levels, and an enormous stone chimney with three walk-in fireplaces at each of the two levels rises through the center.

From the town of Hood River, State Route 35 proceeds south and enters the Mt. Hood National Forest in about 20 miles. The Cooper Ski Area is to the west. The highway keeps pace with the East Fork of the Hood River, passes Mt. Hood Meadows Ski Area, and turns westward. Just beyond the Hood River Meadows Campground is a loop hiking trail connecting Sahalie Falls with Umbrella Falls. Sahalie Falls is a horsetail type that descends nearly 100 feet, whereas Umbrella Falls is a fan type that plunges 60 feet. Where State Route 35 makes a sharp bend to the north and then back to the south is Switchback Falls, whose water cascades for up to 200 feet.

After climbing over Barlow Pass, which was on the old Oregon Trail, the highway comes to a side road that proceeds south to lovely Trillium Lake. Here is a two-mile loop trail with boardwalks where the terrain is exceptionally wet. Just beyond the Trillium Lake Road is the highway to Timberline Lodge and Timberline Ski Area. State Route 35 then follows the Zigzag River with the Salmon-Huckleberry Wilderness to the south. North of Bull Run Lake is the postcard-pretty Lost Lake, especially scenic when Mt. Hood's snow-capped peak is reflected in it. An easy 1.5-mile trail is around part of the lake.

The eastern edge of the Mt. Hood National Forest is approached from Dufur on Forest Highway 44. After a circuitous route over back roads is the Mill Creek Research Natural Area on the east slope of the Cascades. This 815-acre tract exhibits the transition from grassland to oak forests to ponderosa pine–Douglas fir forests. The grasses are bunch grasses and include bluebunch wheatgrass, Idaho fescue, Sandberg's bluegrass, and needlegrass. At an elevation above the grasslands are Oregon white oak woodlands, and above them, stands of ponderosa pine and Douglas fir with some grand fir and western larch usually present as well.

Returning to Forest Highway 44, continue to High Prairie just outside the northern boundary of the Badger Creek Wilderness. In the wilderness is a one-mile trail to the top of Lookout Mountain.

Another great outing is to take State Route 224 and side roads off of it from Estacada. The highway follows the Clackamas River for more than 50 miles. The scenery is superb. At the Ripplebrook Guard Station, you may wish to take the road along the Oak Grove Fork of the Clackamas River to Timothy Lake to enjoy water-based activities or just to camp. A few miles north is the Catalpa Lake Nature Trail. This secluded little lake is surrounded by a forest of firs and western hemlocks (fig. 20). A rocky butte is on one side.

Halfway to Timothy Lake is a forest road along Shellrock Creek. Take it to

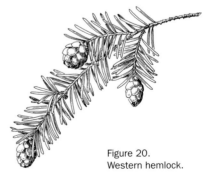

Figure 20.
Western hemlock.

the Frazier Fork Campground in the remote Roaring River headwaters area. Slender subalpine firs tower over the camping area. A 7.5-mile loop trail has stopovers to view Middle Rock Lake, Lower Rock Lake, Serene Lake, Roaring River Canyon, Grouse Point, Cache Meadows, and an old log shelter.

About four miles south of the Ripplebrook Guard Station on State Route 224 is the Collawash River Road, which heads south to the Riverford and Two Rivers campgrounds. In four more miles, take the right fork at Alder Springs. After another six miles, there is a short trail to Pegley Falls, which drops about 35 feet. A fish ladder bypasses the falls. In a couple of miles is the Bagby Hot Springs Trailhead. These three natural hot springs are reached after a 1.5-mile hike. The cabin near the springs was constructed in 1913. The largest of the springs has a flow of four gallons of 138-degree-F water per minute. Fourteen different minerals are in the water, the biggest percentage being silicon, sodium, sulfate, carbonate, and chloride.

On either side of the springs is the Bagby Research Natural Area. About four miles southeast of Bagby Hot Springs is the northern boundary of Bull of the Woods Wilderness.

From the Ripplebrook Guard Station, State Route 224 goes on and on along the Clackamas River until it enters the Olallie Scenic Area at the southern end of the national forest. Several campgrounds are in the scenic area, stately forests, numerous subalpine lakes, including the largest, Olallie Lake, and wildflower-laden Olallie Meadow.

Bagby Research Natural Area

After visiting Bagby Hot Springs, I wanted to explore the Bagby Research Natural Area, which occurs in two sections, one on either side of the trail from Pegley Falls to Bagby Hot Springs. The trail follows the course of the Hot Springs Fork of the Collawash River. The natural area is in the western Cascade Range in the Mt. Hood National Forest. Near the waterway that separates the two sections of the natural area, the elevation is at about 2,200 feet. At the far edge of each section are high ridges with elevations around 3,850 feet. The steep slopes leading to these ridges are interrupted by a number of benches. The 70 inches of annual rainfall received accounts for the relatively lush vegetation. The 5,600-acre natural area was designated to preserve 300-year-old stands of western hemlock. Trails extend into the southern end of each unit of the natural area.

In addition to the fine old specimens of western hemlock, other trees well represented are Douglas fir, western red cedar, noble fir, and grand fir. In most places there are dense entanglements of rose-bay and, in drier places, salal. Here and there are western yew, ocean spray, baldhip rose, red bilberry, and creeping snowberry. Young trees of western hemlock are also plentiful. On the steep slope at the western edge of the natural area, vine maple and Cascade Oregon grape are common. Herbs observed in the natural area are limited to less than two dozen species. Those that appear to be most abundant are Oregon bedstraw, bunchberry, western rattlesnake plantain orchid, trail plant, prince's pine, little pipsissewa, twin-flower, coolwort foamflower, redwoods violet, cutleaf goldthread, vanilla leaf, and deer fern.

Botanists Thomas Furman and James Trappe have noticed an unusually high number of mycotrophic, nongreen flowering plants living on the shaded forest floor of the Bagby Natural Area. Included among these species that have integral relationships between their root system and fungi in the soil are three species of coral-root orchid, candystick, one-flowered Indian pipe, and white-veined pyrola.

Ochoco National Forest

SIZE AND LOCATION: Nearly 900,000 acres in central Oregon, north and east of Prineville at the western end of the Blue Mountain Range. Major access roads are U.S. Highways 20, 26, and 97. District Ranger Stations: Paulina, Prineville. Forest Supervisor's Office: 3160 NE 3rd Street, Prineville, OR 97754, www.fs.fed.us/r6/centraloregon.

SPECIAL FACILITIES: Boat ramps; swimming beaches.

SPECIAL ATTRACTIONS: Steins Pillar.

WILDERNESS AREAS: Mill Creek (17,400 acres); Bridge Creek (5,400 acres); Black Canyon (13,400 acres).

The main mountain range in the Ochoco National Forest is the Blue Mountains, a relatively low range whose highest peak in the national forest is 7,165-foot Snow Mountain. The main attractions in the Ochoco National Forest are several campgrounds, a number of hiking trails, fishing, particularly for rainbow trout, hunting as there are ample deer, and rockhounding, as well as viewing some interesting and scenic rock formations. The national forest consists of three noncontiguous units.

Start your exploration by taking U.S. Highway 26 east from Prineville and parallel Marks Creek into the national forest. On either side of the highway are mineral-bearing areas that are fun for both the amateur and more serious rock collectors. Rockhounds can head to White Rock for red- and white-

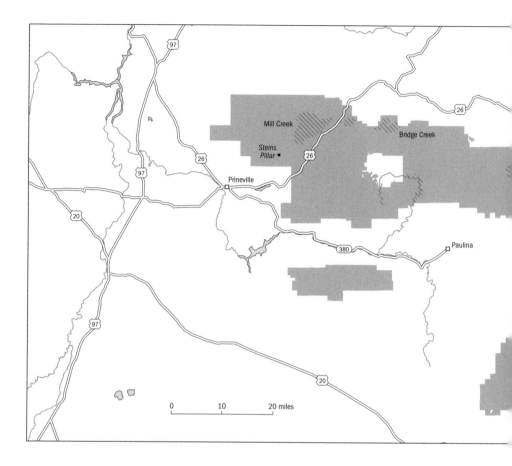

banded rhyolite and for the prize of them all, thundereggs. Thundereggs are geodes whose interior might have opal or agate or nothing. You can also search for green jasper at Coyle Spring and vistaite at Thronsen Creek, or agatized wood at Grant Spring, or red and green jasper at Cow Camp, green moss agate at Johnson Creek, greenish amygdules at Harvey Gap, and more thundereggs at White Fir Spring.

A Forest Service road off of U.S. Highway 26 from Marks Creek Campground leads to a marvelous viewpoint overlooking the northeastern edge of the Mill Creek Wilderness. The road circles around the north side of the wilderness area and comes to another great viewpoint at Hash Rock. When you get to Harvey Gap, try to find greenish amygdules at the Wildcat Campground. This is the trailhead for an 8.5-mile hike to Twin Pillars, an outstanding pair of 200-foot-tall rocky columns. The trail passes through scenic canyons before coming to the twins. A mile north of the pillars is the rugged and desolate Desolation Canyon. This is supposed to be a good thunder-egg area.

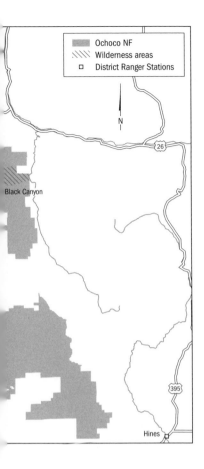

From the Wildcat Campground you may drive along Mill Creek to the trailhead for Steins Pillar, the most outstanding scenic feature in the Ochoco National Forest. Steins Pillar, which is just outside the wilderness area, is a 350-foot-tall rocky column above a base 120 feet in diameter. The pillar appears top-heavy because the upper end is much broader than its pedestal-like base.

Near the middle of the Blue Mountains is Scotts Campground. A primitive forest road runs about six miles to the North Point viewing area where you can look directly into the Bridge Creek Wilderness. North Point stands atop a 600-foot drop-off. Most of the forest here is composed of fir and western larch. At the eastern edge of the Ochoco National Forest in the Blue Mountains is the Black Canyon Wilderness. The South Fork of the John Day River crosses this wilderness and is lined by steep cliffs in its rocky gorge. The 12-mile-long Black Canyon Trail starts in dry sagebrush, climbs through an open forest of ponderosa pine, and enters a forest of lodgepole pine and Douglas fir at higher elevations.

A very small unit of the Ochoco National Forest lies south of the Crooked River. Antelope Reservoir has a campground and is reportedly good for fishing. Two recommended rockhound areas are nearby. At Shotgun Creek look for moss agate and plume agate, and at Friday Creek try finding red and green jasper.

A rather large unit of the national forest is 25 miles south of the Black Canyon Wilderness Area. Delintment Lake has a fine campground, but five miles east of the lake and reached by a Forest Service road is one of the finest wildflower meadows in the national forest.

Rogue River National Forest

SIZE AND LOCATION: Approximately 600,000 acres in southwestern Oregon, with a small area in northern California, west of Crater Lake National Park. Major access roads are State Routes 62, 230, and 238. District Ranger Stations: Jacksonville, Ashland, Butte Falls, Prospect. Forest Supervisor's Office: 333 W. 8th Street, Medford, OR 97501, www.fs.fed.us/r6/rogue.

SPECIAL FACILITIES: Visitor center; boat ramps; swimming beaches; ski areas.

SPECIAL ATTRACTIONS: Rogue-Umpqua National Scenic Byway; Union Creek Historic District; Butte Falls Discovery Loop Tour; Gin Lin Trail; Siskiyou Loop.

WILDERNESS AREAS: Rogue-Umpqua Divide (33,200 acres, partly in the Umpqua National Forest); Sky Lakes (116,300 acres, partly in the Winema National Forest).

Since this manuscript was submitted for publication, the Rogue River National Forest has been placed under the administration of the Siskiyou National Forest.

The Rogue River National Forest lies mostly in the Southern Cascade Range and has several impressive mountain peaks, picturesque rivers and streams, and several historic sites. Many hiking trails including a number of short and easy ones lead to spectacular sites.

You may wish to divide your exploration into three parts—one north of Prospect, one in the middle of the forest in the vicinity of the community of Butte Falls, and one near the California border west of Ashland.

To explore the northern part of the national forest above Prospect, take the Rogue-Umpqua National Scenic Byway at the entrance to the national forest on State Route 62. The scenic byway roughly follows the course of the Rogue River, and several campgrounds and picnic areas are along the route. Just beyond the Mill Creek Campground is a dense forest with a closed canopy over the highway. Many of the trees are more than 200 feet tall and include Douglas firs, grand firs, ponderosa pines, sugar pines, and western hemlocks. Below these giants are small trees and shrubs such as Pacific dogwood, Pacific yew, golden chinquapin, mountain ash, serviceberry, and California beaked hazelnut.

A forest side road to the east between Mill Creek and Ginkgo Creek will bring you to the Frenchman Camp Trail. This two-mile-long trail passes through the Len's Camp Meadow, where small open areas referred to as glades have small but colorful wildflowers during the summer.

You may continue your forest drive for about four miles to the trailhead at the Natural Bridge Campground. A short trail leads to the natural bridge

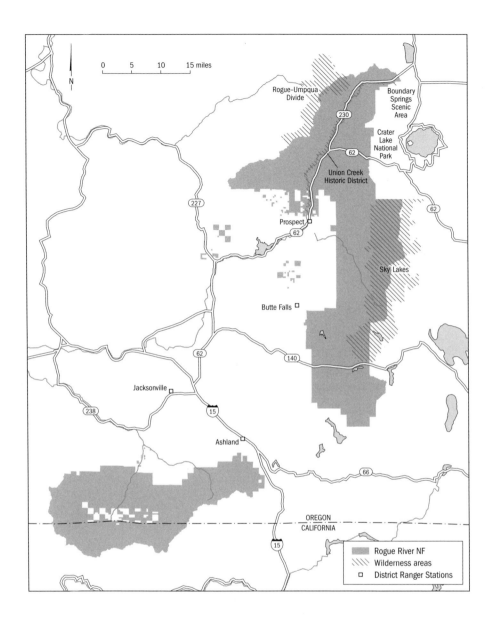

	Rogue River NF
	Wilderness areas
□	District Ranger Stations

where the Rogue River disappears into an old lava tube and continues underground for a while. Some nice wetlands are along the trail that developed because of extensive beaver activity.

A visitor information station and campground are at historic Union Creek, where State Route 62 curves eastward to Crater Lake National Park and the Rogue-Umpqua National Scenic Byway continues north on State Route 230. Union Creek Resort, with its rustic architecture, is listed on the National Register of Historic Places. A part of the Upper Rogue River National Recreation Trail may be picked up from the campground. It is near here, in the Rogue Gorge, where the river literally flows through a cleft in the vertical rock wall. After seeing this phenomenon, take a forest road east to Huckleberry Mountain where there is a stupendous view before you reach the top into Crater Lake National Park whose boundary is only 2.5 miles away.

A forest road off the Mt. Stella Road forms a loop to the north past Rabbit Ears and Hershberger Mountain. Rabbit Ears is a volcanic cone with two vents. A historic lookout tower on Hershberger Mountain allows for a panoramic view of the Cascade Mountains. The Abbott Creek Research Natural Area four miles west of Mt. Stella preserves a beautiful example of a mixed conifer forest. This 2,660-acre tract may be reached by an unimproved dirt road from the Abbott Creek Campground. The Rogue River National Forest part of the Rogue-Umpqua Divide Wilderness is near here, with Abbott Butte and Elephant Head straddling the border between the two national forests.

Proceeding slowly through impressive forests along State Route 230, you will come to a side road to the National Creek Trailhead for a half-mile hike to a waterfall fed by springs high up on Mt. Mazama. You may continue by road or by foot another six miles beyond the falls to the Sphagnun Bog, just inside Crater Lake National Park, where there is an enormous bog filled with bog-loving species.

A side road that follows Lost Creek eventually comes to Hole-in-the-Ground, a once shallow glacial lake that has filled in to become a fine wet meadow. If you take another side road north, you will come to the Alkali Creek Falls Trail. A one-fifth-mile trail uphill brings you to the lower end of the falls where Alkali Creek drops 600 feet down a rock chute.

After returning to the Rogue-Umpqua National Scenic Byway, drive about three miles to the Hamaker Meadows Campground. Just beyond this shady campground is a pleasant trail that follows Minnehaha Creek. Also in the area to the west is a marvelous wildflower meadow. The Hummingbird Meadow Trail winds through the meadow for a half mile.

Soon the Rogue-Umpqua National Scenic Byway makes a turn to head due east. You can see the rim of Crater Lake from the Crater Rim Viewpoint. Less than one mile south of the viewpoint is the Boundary Springs Scenic Area. The springs at this picturesque place form the headwaters of the Rogue River. In another three miles, the scenic byway crosses into the Umpqua National Forest.

To explore the middle part of the Rogue River National Forest, proceed to the village of Butte Falls and drive the Butte Falls Discovery Loop Tour. Heading east, you will come to Whiskey Springs in 10 miles where you may take a mile-long interpretive trail along the edge of a marsh and near a beaver pond. Your next stop on the loop tour is to see the remains of an old logging arch that was used to drag huge cut timber through the forest to the nearest railroad spur. This hydraulic device, which was the forerunner of the Cater-pillar, dates back to the 1920s. To the east you can see the effect of a major forest fire in 1910. The forest still has not fully recovered. From the loop road you will get an occasional glimpse of Mt. McLoughlon in the Sky Lakes Wilderness. This 9,495-foot mountain is an old volcano, and there is a hik-ing trail to its top. A trailhead for hikes into the wilderness area are at Blue Rock. Dozens of high elevation lakes are in the wilderness, but most of them are in the Winema National Forest.

Be sure you take your next hike on the Lower South Fork Trail at Stop No. 7. At the beginning of the trail are two of the largest Pacific yew trees any-where. The Imnaha Guard Station is a rustic Civilian Conservation Corps building dating back to the 1930s. It is on the National Register of Historic Places. It sits in a nice meadow along the creek. The loop highway begins its return to Butte Falls and follows the South Fork of the Rogue River. The South Fork Trailhead is where Green Creek enters the river. We admired a huge sugar pine at the beginning of the trail.

The southern unit of the Rogue River National Forest is more than 20 miles from the rest of the forest and lies south of Medford and west of Ash-land and even includes a part of northern California. The Siskiyou Loop is a driving route through this part of the national forest. It may be started from Medford or Ashland. Leave from Medford and follow the loop, which begins as State Route 238, for several miles to the tiny crossroads of Ruch. State Route 238 continues west past Ruch, but the Siskiyou Loop takes another good highway due south to the national forest boundary. At the Star Gulch Work Center is a 1911 cabin, the oldest building in the national forest. Just beyond, at McKee Bridge, is an old covered bridge.

This was an active gold-mining area in the late 1800s. When news of the gold discovery reached China in 1848, there was an influx of Chinese who came to make their fortune and then return to their homeland. One Chinese mining boss was Gin Lin. After successful mining operations in Applegate Valley, Gin Lin purchased claims along Palmer Creek and began hydraulic mining operations in 1881. From the Flumet Flat Campground near Palmer Creek is the Gin Lin Trail, three-quarters of a mile long with numbered sta-tions along its length that explain the many facets of the mining operation. It is an interesting account of a bygone day.

From the Flumet Flat Campground, the Siskiyou Loop turns east onto Beaver Creek Road. The creek is lined with alders, willows, and Oregon ash. A number of switchbacks are along the gravel road, which narrows near Deadman's Point. From here is a great view into Squaw Creek Canyon. The loop road, which can be rough in spots and is generally not open until late July, passes through old-growth forests of ponderosa pine, incense cedar, white fir, noble fir, Shasta red fir, and mountain hemlock. At Jackson Gap and Dutchman Peak there are breathtaking views of the Siskiyou Crest. A nice stand of mountain mahogany is at the historic Dutchman Peak Lookout at an elevation of 7,418 feet. After Dutchman Peak, the Siskiyou Loop enters the Klamath National Forest on its return to Ashland and Medford.

Applegate Lake is a popular recreation site in the southwestern corner of the national forest and has swimming beaches, a boat ramp, and a visitor center, as well as campgrounds and several trails around the lake.

Two miles south of Ashland is the Ashland Research Natural Area, a 1,408-acre area on the slopes of a rugged canyon along the East Fork of Ashland Creek. The natural area is a fine example of a ponderosa pine–Douglas fir–sugar pine forest. Outside the northern boundary of the preserve is the Ashland Nature Trail.

Siskiyou National Forest

SIZE AND LOCATION: 1,163,484 acres in southwestern Oregon and extending into northwestern California, between the Pacific Ocean and Grants Pass. Major access routes are Interstate 5 and U.S. Highways 101 and 199. District Ranger Stations: Brookings, Grants Pass, Gold Beach, Cave Junction, Powers. Forest Supervisor's Office: 333 W. 8th Street, Medford, OR 97501, www.fs.fed.us/r6/siskiyou.

SPECIAL FACILITIES: Boat ramps.

SPECIAL ATTRACTIONS: National Wild and Scenic Rogue River; National Wild and Scenic Illinois River; Babyfoot Lake Botanical Area; York Creek Botanical Area; Big Craggies Botanical Area.

WILDERNESS AREAS: Kalmiopsis (179,655 acres); Wild Rogue (35,318 acres, partly on Bureau of Land Management land); Red Buttes (19,900 acres, partly in the Rogue River National Forest); Grassy Knob (17,200 acres).

The Kalmiopsis Wilderness is an indication that the incredible diversity of plant species is one of the features of the Siskiyou National Forest. Kalmiopsis is the name of a shrub that resembles a small rhododendron with bright rose pink flowers. The wilderness area and a few small areas nearby are the only places in the world where this plant grows.

The Siskiyou National Forest occupies part of the Coast Ranges and Klamath Mountains in southwestern Oregon and a small part of the Siskiyou Mountains in northwestern California. Not only are there unusual plants in the national forest, but there are two world-class rivers where whitewater rafting is at its best. Throw in some rugged canyons, wildflower meadows, and abandoned gold mines, and you have one of the most interesting national forests in the United States.

The easiest access to the Kalmiopsis Wilderness is from the end of Long Ridge Road at the western edge of the wilderness about 17 miles northeast of Brookings. Because the best colonies of kalmiopsis are on the eastern side, however, you should enter the wilderness area from Chetco Pass. To even drive to Chetco Pass can be a challenge because the last seven miles of the road from Selma on U.S. Highway 199 are rough, twisty, steep, and narrow. From Chetco Pass there are trails north to Pearsoll Peak and Granite Spring, south to Onion Camp and Babyfoot Lake, and west to the kalmiopsis plants. Only the last of these trails is in the wilderness area.

The trail to the kalmiopsis quickly comes to the abandoned Uncle Sam Mine. When the trail crosses Slide Creek, there is a small colony of kalmiopsis. About three miles farther south is another nice colony, and a mile beyond that is the largest patch, extending for 1.5 miles to Bailey Mountain. This wilderness area has very low mountains but very rocky and brushy canyons. To indicate the brushiness of the area, there are 72 different kinds of shrubs known from it. The Forest Service also lists 30 tree species, 14 of them conifers, including lodgepole, sugar, Jeffrey, ponderosa, and western white pine. Knobcone pine grows in the rockiest places. Fourteen different kinds of ferns and about 250 different kinds of wildflowers are also present, 32 of them considered rare. An incredible 32 members of the lily family and 10 kinds of wild orchids are also present.

If you hike north to Pearsoll Peak, a distance of about four miles from Chetco Pass, you will be in an area where most of the rocks have a distinct reddish tint while others are greenish gray. The red ones consist of peridotite, the greenish ones of serpentinite, and both are rocks that contain so much iron and magnesium that only a few kinds of plants can survive on them. Those that do live on them are often rare. The tree that seemingly has adapted for growth here is Jeffrey pine. North of 5,098-foot Pearsoll Peak is Granite Spring. In this area is another rare shrub called the Siskiyou leucothoe, a plant related to kalmiopsis. A nice stand of sugar pines is near the spring.

South from Chetco Pass is a trail that climbs to Eagle Mountain, goes through Eagle Gap where Jeffrey pines grow among the red peridotite, and comes to Onion Camp. It is just a short way then to scenic Babyfoot Lake just inside the Kalmiopsis Wilderness. The lake is in a glacial cirque, and the

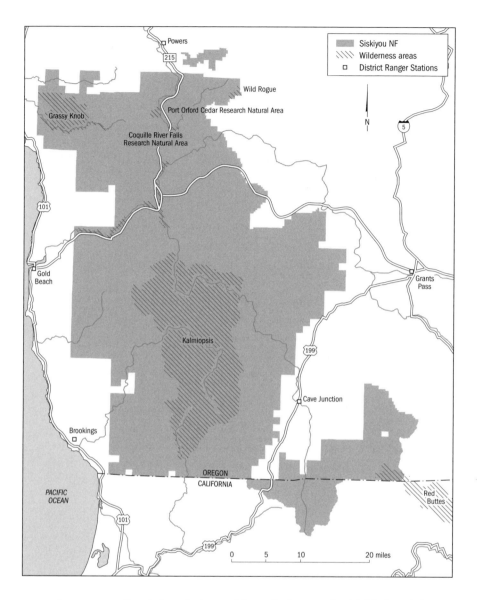

region surrounding it is a designated botanical area. The highlight of the botanical area is the presence of Brewer's weeping spruce around the lake. This handsome species is one of the rarest conifers in the country. Other trees around the lake are Douglas fir, Shasta red fir, sugar pine, incense cedar (fig. 22), and Port Orford cedar (fig. 21). An extremely rough 20-mile-long four-wheel-drive road runs from U.S. Highway 199 to Onion Camp.

From Selma on U.S. Highway 199 is a road that stays near all the twists and turns of the Illinois River as far as Briggs Creek. The Illinois River is a National Wild and Scenic River that tumbles through steep canyons for 29 miles. Of the 150 rapids along this stretch, 11 of them are rated Class IV and one is Class V, the most difficult rating of all. If you hike the two miles along the Illinois River from Briggs Creek to York Creek, you will come to the York Creek Botanical Area, a special area where, among other things, you may see more Brewer's weeping spruce, the Port Orford cedar, and the fascinating insectivorous cobra lily. An interesting trail is along Briggs Creek to Soldier Creek. For five miles the trail goes through some old-growth forests and passes the Courier Mine Historic Site. Fording Briggs Creek is necessary several times. The Big Craggies Botanical Area just outside the northwest boundary of the Kalmiopsis Wilderness is nearly inaccessible.

The National Wild and Scenic Rogue River is another must in the Siskiyou National Forest. Some parts of the river are very serene and tranquil, other parts are wildly turbulent. Thirty-five miles of this river are in the national forest, from the Lobster Creek boat ramp to Marial where the Siskiyou National Forest ends and Bureau of Land Management land begins. The Rogue River empties into the Pacific Ocean about 12 miles from the Lobster Creek. The Rogue is another river with challenging whitewater runs. A commercial jet boat plies the Rogue River from Gold Beach on the Pacific Ocean to Agness, about halfway to Marial. There is a private resort at Agness.

At Lobster Creek is the Frances Schrader Old Growth Trail that meanders for 1.5 miles through an old-growth Douglas fir forest. A trail guide points out 22 species of plants.

A 39-mile hiking trail parallels the Rogue River, the first 20 miles of it being on Bureau of Land Management land. Near Marial at the northeast corner of the Siskiyou National Forest, the trail goes beneath Douglas firs, ponderosa pines, sugar pines, white firs, and incense cedars. As you approach the western edge of the national forest, western red cedar, Port Orford cedar, and Pacific yew become common in more moist locations. If you are observant, you may see Lawson cypress, Oregon white oak, California black oak, Pacific madrone, Oregon ash, golden chinquapin, Oregon myrtle, and bigleaf maple.

Several worthwhile areas are west of Grants Pass. Drive State Route 199 to Riverside Road and then to Limpy Creek Road in the national forest. After three miles on the forest road you come to the trailhead for the Limpy Botanical Interpretive Loop Trail. This easy one-mile loop is in a marvelous serpentine area where a good diversity of plants live. Several rest benches are along Limpy Creek, and there is a fine view over the Rogue Valley. You may also take the Merlin-Galice Exit off of Interstate 5 north of Grants Pass and drive to Taylor Creek Road. Five miles down this road is the Burned Timber Interpretive Loop Trail. The name comes from Burned Timber Creek. This easy two-mile loop has interpretive signs that explain ecology and timber management. A special treat is the Burned Timber Creek Falls. Continue on beyond this trail to Big Pine Campground where there are four loop trails ranging in length from one-sixth to one-half mile. Big Pine Loop passes the world's tallest ponderosa pine.

If you want to see tall trees, however, head to the Redwood Nature Trail nine miles north of Brookings. Here is an inspiring redwood forest with trees at least 250 feet tall and 10 feet in diameter. Other plants you will encounter on this trail are tanoak, black huckleberry, Douglas fir, Pacific rhododendron, red wood sorrel, sword fern, lady fern, deer fern, maidenhair fern, the shrubby salal, bigleaf maple, red alder, and Oregon myrtle. A wheelchair-accessible trail leads through a stand of giant redwoods along Peavine Ridge. It is possible to enter a hollowed-out portion of a redwood.

In the vicinity is Wheeler Creek Research Natural Area, which has been set aside to preserve a great example of a redwood–Douglas fir community. The natural area is on a steep north-facing slope above Wheeler Creek.

Five other special areas are in the Siskiyou. Red Butte Wilderness is at the southern edge of the national forest. The red orange color of the buttes, due to a high content of iron and magnesium, makes this a very scenic region. Wild Rogue Wilderness is north of Marial. Its dominant feature is 4,319-foot Mt. Bolivar. The canyon carved by the Rogue River is 4,000 feet deep. At the northwest corner of the national forest is Grassy Knob Wilderness. The dense vegetation is part of the temperate rainforest. Some plants of the Port Orford cedar are said to be more than six feet in diameter. The trail along Dry Creek goes through the most lush part of the forest. Between the Wild Rogue and Grassy Knob wilderness areas are the Coquille River Falls Research Natural Area and the Port Orford Cedar Research Natural Area. The former is 500 acres of virgin old-growth Port Orford cedar. Although the cedar is the most prevalent tree, there is no shortage of Douglas fir, western hemlock, grand fir, sugar pine, and Pacific yew. Two waterfalls are in this natural area, one along the south Fork of the Coquille River and one along the Squaw Creek tributary. A forest road parallels the southern boundary of the natural area. The Port Orford Cedar Natural Area three miles to the northwest preserves a similar ecosystem, with the addition of western white pine.

Whenever you hike in the Siskiyou National Forest, be careful of poison oak. It is plentiful.

Wheeler Creek Natural Area

California redwoods, the tallest living organisms in the world, live in a narrow band along the Pacific Coast from the Big Sur area of California to southwestern Oregon, extending into that state for a few miles. The dense coastal fog characteristic of this area provides the moisture required for the survival of the redwoods.

The best northernmost stand of redwoods occurs along Wheeler Creek in the Siskiyou National Forest, a few miles east of Brookings, Oregon. The Wheeler Creek Natural Area has been set aside to preserve this northern outpost of redwoods. Some redwoods are about eight miles north of Wheeler Creek, but the Wheeler Creek site has large numbers of old-growth trees.

Wheeler Creek consists of 334 acres of old-growth redwood forests, riparian communities, and dry ridgetop communities. A forest road crosses the southern end of the natural area, providing easy access to the regions. South of the road is a dry ridge at an elevation of 1,771 feet, still a part of the natural area. Wheeler Creek forms the northern boundary of the natural area. It is at an elevation of 200 feet.

Redwoods and Douglas firs occur in all parts of the area, except on the dry ridgetop where tanoak, Pacific madrone, and knobcone pine are the principal woody plants.

In the lush riparian habitats along rocky Wheeler Creek, redwoods and Douglas firs are common, but California laurel and bigleaf maple are more abundant. The understory is almost totally dominated by deer fern, with the absence of rose-bay and salal. Occasional western hemlocks occur in the canopy, with California beaked hazelnut, evergreen blueberry, red bilberry, and vine maple in the shrub layer. Where deer fern has not completely closed in the forest floor, plants of large-flowered fairy bells, Oregon oxalis, rosy twisted stalk, maidenhair fern, inside-out-flower, white trillium, bedstraw, and polypody fern may occur.

Redwoods and Douglas firs thrive on the moist slopes above the creek, although tanoak is also an abundant species. Some of the redwoods are nearly 200 feet tall, with Douglas firs almost as tall. Rose-bay and evergreen blueberry are found throughout the shrub layer, with occasional plants of salal and Cascade Oregon grape. Herbaceous plants are not plentiful in this redwood-dominated forest, and their distribution depends on the amount of shade and moisture available. Those kinds that do occur include beargrass, redwoods violet, vanilla-leaf, deer fern, and bracken fern.

Port Orford Cedar Natural Area

One of the most beautiful coniferous trees in the country, and certainly one of the most uncommon, is the Port Orford cedar (fig. 21). This species is found near the coast from a few miles in northern California to midway up the Oregon coast. The tree's spire-shaped tip contrasts with the flaring base. The flattened branchlets tend to spread or droop.

Figure 21.
Port Orford cedar.

A virgin old-growth stand of Port Orford cedar occurs in a 1,122-acre natural area in the Siskiyou National Forest. The Port Orford Cedar Natural Area is about 35 miles south of the community of Powers in southwestern Oregon. It is not an easy hike to reach the natural area, but if you get there, seeing this fine species and its associated vegetation is breathtaking.

Although most of the natural area is devoted to old-growth Port Orford cedars, a few drier and rockier sites support a fine community of tanoak and Pacific madrone. At the northern tip of the natural area are outcrops of serpentinite. On some of the serpentinite are forests of small trees, whereas on other serpentinite areas, grassy openings prevail. Johnson Creek forms the northern boundary of the natural area, and the South Fork Coquille River forms the eastern boundary. Along both of these waterways are riparian communities.

The significant features of the natural area are the virgin stands of Port Orford cedars. Some of the oldest and largest of these monarchs of the forest are more than 175 feet tall and more than three feet in diameter. They are thought to be somewhere around 500 years old. Even larger in these forests are ancient Douglas firs that top 200 feet and more than four feet in diameter. Other coniferous trees that share the canopy include grand fir, western hemlock, sugar pine, and western white pine. Pacific yew is found sporadically in the forest. Hardwood trees such as tanoak, golden chinquapin, and Pacific madrone are often also present, but these species never reach the size of the coniferous trees. Only where natural seeps occur in the forest is western red cedar found.

Beneath the Port Orford cedars is a dense understory of small trees and shrubs. Thickets of rose-bay, red bilberry, evergreen blueberry, salal (pl. 41), and Cascade Oregon grape are usual. On the ground are deer fern and wildflowers that include Oregon oxalis, redwoods violet, sweet-scented bedstraw, western rattlesnake plantain orchid, twin-flower, Pacific blackberry, and whipplevine.

Where areas are drier and rockier, tanoak and Pacific madrone become the dominant species, although the shrubs beneath them are generally the same species as those beneath the Port Orford cedars. Rather uncommon herbs in this community are gnomeplant and Vancouver groundcone, two nongreen saprophytes, and Alaska rein orchid.

Plant communities that live on serpentinite are usually fascinating because there is a limited number of species that can tolerate serpentinite with its high concentration of elements toxic to certain plants. On some serpentinite in the natural area, there are forests, but the trees are more widely separated and decidedly smaller. Douglas fir and Port Orford cedar grow here with California laurel. Beneath these species are giant fawn lily, snowqueen (also called round-leaved synthyris), western azalea, Bolander's groundsel, wild iris, Piper's Oregon grape, and beargrass. Thickets of California buckthorn and canyon live oak are occasional.

Other serpentinite areas are too dry to support the growth of trees. These openings include two ferns — lip fern and mountain holly fern — and wildflowers such as Fremont's death camas, Tolmie's Mariposa lily, bluedicks brodiaea, wallflower, spatulate-leaved stonecrop, and frosted paintbrush. Coastal range fescue fills in the spaces between the wildflowers.

Along Johnson Creek and the South Fork Coquille River are red alder, Oregon ash, and bigleaf maple. This is also a well-documented area for amphibians, with Del Norte salamander, Dunn's salamander, Pacific giant salamander, clouded salamander, Oregon salamander, northwestern salamander, rough-skinned newt, and Pacific tree frog (pl. 42) recorded from here.

Siuslaw National Forest

SIZE AND LOCATION: Approximately 650,000 acres in western Oregon, along the Pacific Ocean from Tillamook to North Bend, and up to 35 miles inland in the Coast Ranges. Major access routes are U.S. Highways 20 and 101, and State Routes 18, 22, 34, 36, 38, 126, and 229. District Ranger Stations: Waldport, Hebo, Florence; Oregon Dunes National Recreation Area, Reedsport. Forest Supervisor's Office: 4077 Research Way, Corvallis, OR 97339, www.fs.fed.us/r6/siuslaw.

SPECIAL FACILITIES: Visitor centers; boat ramp; off-highway vehicle areas.

SPECIAL ATTRACTIONS: Oregon Dunes National Recreation Area; Cape Perpetua Scenic Area; Heceta Head Lighthouse.

WILDERNESS AREAS: Drift Creek (5,798 acres); Cummins Creek (9,173 acres).

The Siuslaw National Forest is a study in contrasts. The three ecosystems in the forest are as different as night and day. The lower stretch of national forest along the Pacific Ocean is a huge complex of sand dunes, whereas the northern stretch is rocky coastline with picturesque promontories and wild, turbulent water. Immediately east of these coastal ecosystems is the temperate rainforest of the Coast Ranges with old-growth forests.

We began with the sand dunes, which are in the special Oregon Dunes National Recreation Area. This extensive area of sand dunes parallels the Pacific Ocean for 40 miles, most of it in the national recreation area. Part of the area south of Winchester Bay was originally the Umpqua Dunes Scenic Area. U.S. Highway 101 provides easy access to all points of interest.

The south end of the recreation area from North Bend to Winchester Bay has a number of campgrounds and hiking trails. One of the best hiking trails is the Lakeside Trail from Elk Creek Campground, but there are trails through the sand at Bluegill Campground and Horsfall Dune. Horsfall and Beale are two interesting lakes in this area.

North of Winchester Bay, the recreation opportunities increase. A splendid hiking trail leads from Tahkenitch Campground and there is a boat ramp for sailing on Tahkenitch Lake. At the Oregon Dunes Overlook on U.S. Highway 101 are magnificent overviews of the vast sand dunes from three viewing platforms that are wheelchair accessible. Trails are at Carter Lake and Driftwood Campground. From the several parking areas along South Jetty are numerous places in a wetland to observe a large variety of birds. You can also climb the face of a dune here. The northern tip of South Jetty marks the end of the national recreation area, although there are still more dunes north at Sutton Lake.

North of Heceta Beach is the Sutton Recreation Area where there are six miles of trails and two campgrounds. Holman Vista has a platform from which there are great views of the beach. Across U.S. Highway 101 at the lower end of the Sutton Recreation Area is the state-owned Darlingtonia Wayside. A boardwalk over the boggy terrain lets you have a close-up look at the incredible insectivorous cobra lily.

Just beyond Lily Lake, U.S. Highway 101 comes near the ocean. You do not want to miss Sea Lion Caves, Sea Lion Point, and Heceta Head Lighthouse. Although Sea Lion Caves and Sea Lion Point are not in the Siuslaw, which is less than a half-mile away, this is an opportunity to see the only mainland home for wild Steller sea lions. In spring and summer, look for gray whales out in the ocean.

Heceta Head Lighthouse with its red top is on a low promontory with a steep cliff behind it. The lighthouse, which is still a working lighthouse, and the light-keeper's house were built in 1894. It is 250 feet above the ocean. Guided tours are given five days a week from Memorial Day to Labor Day. Although the lighthouse and two-story Victorian style keeper's house are

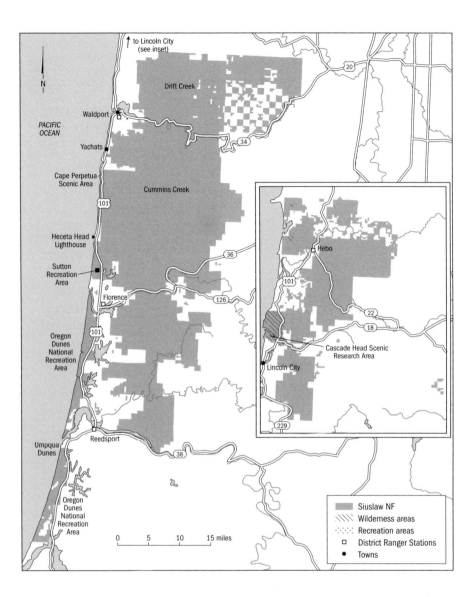

owned by the Forest Service, the house is operated as a bed and breakfast by a private concessionaire. A trail leads from the keeper's house to the lighthouse and to the beach below.

From Heceta Head north, the Pacific coastline becomes rocky and often wild, with water forming huge sprays as it rushes against the rocks. Rock Creek Campground is situated where Rock Creek enters the ocean. After U.S. Highway 101 crosses Cummins Creek, the wild area to the east is the Cummins Creek Wilderness. As you climb along the creek that is lined by willows and alders, you enter an old-growth temperate rainforest where there are

Sitka spruces up to nine feet in diameter. Other huge trees in the forest, all draped with mosses and lichens, are Douglas fir, western red cedar, and western hemlock.

Just north of Cummins Creek is the Cape Perpetua Scenic Area, 2,100 acres of unmatched scenery. The visitor center is a must stop if you want to learn about the phenomena along the coast and to get a panoramic view of the ocean. A trail to the rocky coast lets you examine tidepools and see the remarkable Devil's Churn and spouting horns where water erupts from cavities in the rock, then slowly drains back into the cavities, only to erupt again. At the tidepools look for sea anemones, starfish, and spiny purple sea urchins.

North of the community of Yachats, the Forest Service has a pleasant campground at Tillicum Beach. Then, between Waldport and Lincoln City, the area adjacent to the Pacific Ocean is not in the Siuslaw National Forest. Five miles north of Lincoln City, and back in the National Forest, is the Cascade Head Scenic Research Area. A six-mile trail follows Falls Creek through beautiful forests. At the northern edge of the Cascade head area is the Neskowin Crest Research Natural Area in the national forest. This is a most remarkable area because this is a 686-acre tract of temperate rainforest whose western edge plunges abruptly to the Pacific Ocean. The tops of the ocean cliffs are only 250 feet above the water but support a rainforest. A hiking trail leads through the natural area, beginning at Chitwood Creek. The natural area is dominated by Sitka spruce and western hemlock, with some red alders present. Beneath the trees are rusty menziesia, red bilberry, deer fern, false lily-of-the-valley, western spring beauty, and two varieties of foamflower.

The northern nine miles of the Siuslaw National Forest between Cape Kiwanda and Cape Lookout feature the Sand Beach Campground and the nearby Sand Beach Research Natural Area.

After exploring the Siuslaw National Forest land adjacent to the Pacific Ocean, you should next see that part of the national forest that is inland. Hebo Lake is southeast of Sand Beach and has a nice campground with a nature trail that completely encircles it. Flynn Creek five miles southeast of Toledo is a forested area preserved as the Flynn Creek Research Natural Area.

At the eastern edge of the national forest and only 13 miles from Corvallis is Mary's Peak. Because of the unusual assemblage of plants on the 4,097-foot peak, the peak is a designated botanical area. Amidst a forest of noble fir are meadows with a great diversity of wildflowers including the uncommon frigid shooting-star.

Only five miles northeast of Waldport is the small Drift Creek Wilderness. This wilderness is home to giant Sitka spruces and western hemlocks.

Drift Creek is lined by steep cliffs, you get the feeling that you are in a remote mountain canyon. A 1.5-mile trail leads to Drift Creek Falls where the 75-foot drop may be observed from a 240-foot-long suspension bridge.

If you follow a fine paved road from Florence that proceeds northeast along the North Fork of the Siuslaw River, you encounter a beautiful old-growth forest surrounding the North Fork Siuslaw Campground. Three miles southeast is the Mapleton Hill Pioneer Trail, a self-guiding trail that includes the site of a covered bridge that collapsed during a snowstorm and fell into the North Fork. The trail follows part of the old pioneer route between Florence and Eugene.

About five miles south of Mapleton and the North Fork of the Siuslaw River is Sweet Creek. If you hike the 3.5-mile trail along this picturesque creek, you encounter 11 waterfalls. Two miles southeast of the eastern end of Sweet Creek are the marvelous Upper and Lower Kentucky Falls.

Umpqua Dunes

Almost 40 percent of Oregon's 340-mile Pacific coastline consists of sandy shores and dunes that form an ever-changing system of ridges and depressions. An example of this habitat, known as Umpqua Dunes, lies in the Siuslaw National Forest, between Heceta Head and Coos Bay. A popular place for swimming, boating, fishing, gray whale watching, and other seaside activities, most of Umpqua Dunes has been set aside as the Oregon Dunes National Recreation Area.

Alfred Wiedemann, a biologist at Evergreen State College who has been studying the Oregon dunes for nearly 25 years, has pieced together their geological history. About 53 million years ago, he reports, there was downward bending of the earth's crust all along the Pacific Coast. The land was submerged under the sea, and began to be covered by sediments of sand, silt, clay, and volcanic material that were washed into the ocean. In particular, along what is now the Oregon coast, feldspar and quartz sand were deposited by erosion of an uplift of sandstone to the south, in the region of northern California.

Thirty million years into this continuous process of deposition, there was an uplift inland that formed the Coast Ranges (Cascade Mountains), visible east of the dunes today. Rivers and streams flowing from these mountains have swept additional materials into the sea. This action was especially intense during Ice Age times (1.8 million to 10,000 years ago), when periodic glaciation caused the sea level to rise and fall, encouraging rivers to cut trenches across the continental shelf and carry sand and other erosional products to the ocean. These and earlier deposits of sand are now being washed up by wave action and blown by the predominantly landward winds into the various types of dunes we see today.

A short distance inland, parallel to the shoreline, is a ridge of sand known as the foredune, which may be 25 feet high. Foredunes were absent along this part of the Oregon coast until half a century ago, when people introduced European beach grass in an effort to stabilize the sand. With its extensive root system, beach grass is able to remain anchored in the sand under the assault of the wind, and in fact, this grass grows best in windy places. As the wind encounters the beach grass, sand is deposited, forming mounds around each clump. As a result, the movement of sand inland from the beach is slowed, and the foredune slowly increases in size.

When winds are severe or in areas where there is little or no foredune development, sand is blown rapidly inland, often reducing the sand behind the foredune down to the water table. From the crest of the foredune, the sandy slope usually drops gradually into this "deflation plain," where the small degree of sand movement and the abundance of water are conducive to the growth of some plants.

Along the higher, drier edges of the plain, lupines, seashore bluegrass, and European beach grass grow in widely separated tufts to form a community that Wiedemann calls a dry meadow. Lower down, where the water table in the deflation plain lies only two or three feet below the surface of the sand, and where water stands for nearly two months during the winter, a meadow community of red fescue grass, false dandelion, and coast strawberry replaces the dry meadow. The vegetation is so thick that the dominant grasses may form a sod.

Still lower lying areas, 12 to 18 inches above the water table, where water stands for three or four months during the winter, support a dense growth of rushes, golden-eyed grass, springbank clover, and California aster. The bright pink heads of the clover and the purple blossoms of the aster provide vivid color in this rush meadow community. A marsh, where water may stand for six months of the year, occupies the wettest part of the deflation plain. Slough sedge, the most common plant, is joined by the yellow-flowering silverweed and creeping buttercup in this habitat.

The sand that has been scooped out by the wind during the formation of the deflation plain piles up eastward into a series of low, undulating ridges, about 75 to 150 feet apart. These ridges are oriented at right angles to the northwesterly summer winds and thus generally run at a transverse angle to the coast. Because the sand is continuously moving, no vegetation can become established. When the wind shifts during the winter and comes in from the southwest, the ridges themselves are partly obliterated, only to re-form the following summer.

The terrain gradually rises eastward until it comes in contact with forest, 2 to 2.5 miles inland. Here the sand is deposited in dunes known as precipitation ridges, which may be as much as 165 feet tall. They are created as the sand-laden wind strikes the trees of the adjacent forest and is deflected upward, losing its velocity and depositing sand on the side of the ridge toward the forest. When the lee slope reaches an angle of 33°, the sand slips down this steep face and invades the forest, burying part of it.

A stable plant cover sometimes becomes established on the precipitation ridges. Seashore bluegrass and large-headed sedge, two species with large seeds that are readily spread by the wind, are often the pioneers. Once germinated, these plants anchor themselves with extensive roots, slowing up the movement of the sand. Shade from their leaves lowers the temperature of the sandy surface and raises the moisture content, allowing for the invasion of other species. Not long after, colorful plants such as the seaside tansy, seashore lupine, and others become established. As the decaying parts of these herbaceous plants build up an organic soil over the sand, the seedlings of woody plants are able to survive. Coast pine, Douglas fir, and Sitka spruce seedlings are common, as are the young plants of the shrubby salal and evergreen huckleberry, two members of the heath family.

As these woody plants grow to maturity, a Douglas fir forest with an entangled understory of salal, huckleberry, and western rhododendron dominates the once barren locations. Eventually, even the Douglas fir gives way to a forest of western hemlock and western red cedar. The complete succession of plants takes more than 100 years. Meanwhile, more sand is washed up on shore, creating new expanses of beach and dune west of the encroaching forest.

Umatilla National Forest

SIZE AND LOCATION: Approximately 1.4 million acres in northeastern Oregon and southeastern Washington, including the northern Blue Mountain Range. Major access routes are Interstate 84, U.S. Highways 30 and 395, and State Routes 204 and 341. District Ranger Stations: Heppner, Ukiah, Pomeroy (WA), Walla Walla (WA). Forest Supervisor's Office: 2517 SW Harley Avenue, Pendleton, OR 97801, www.fs.fed.us/r6/uma.

SPECIAL FACILITIES: Boat ramps; ski areas.

SPECIAL ATTRACTIONS: Blue Mountain National Scenic Byway; Wild and Scenic North Fork of the John Day River.

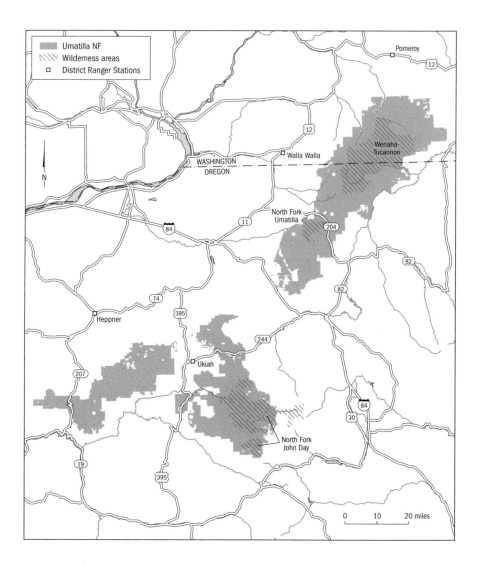

WILDERNESS AREAS: Wenaha-Tucannon (177,465 acres); North Fork John Day (121,352 acres, partly in the Wallowa-Whitman National Forest); North Fork Umatilla (20,144 acres).

The northern reaches of the Blue Mountains are the dominant features of the Umatilla National Forest. The wildest and wooliest part has been set aside as the Wenaha-Tucannon Wilderness. The Wenaha River has carved a deep and scenic gorge along the southern boundary of the wilderness, and there is a rough hiking trail that attempts to follow the course of the river. Along the northern edge of the wilderness are four areas that are interesting and fairly accessible. Just south of the Misery Springs Campground on Ray Ridge is the best viewpoint for looking into the wilderness area. Farther west, from the Teepee Campground, there is a three-mile trail to Oregon Butte, the highest point in the wilderness at 6,387 feet. Continuing west, from the Godman Campground, there is a two-mile trail to Rainbow Creek Research Natural Area in the wilderness. If you can get to the natural area, you will be inspired by a virgin forest of grand fir, western white pine, Douglas fir, ponderosa pine, and western larch. Beneath the trees is a shrub layer of blueberry, Pacific yew, and thinleaf alder. Signs of elk are everywhere. Outside the northwest corner of the wilderness is the primitive four-wheel-drive Kendall Skyline Road. If you get as far as Table Rock, you will have a panoramic view of the entire area. South of the wilderness area are two areas that will appeal to some outdoorsmen. The artificial Jubilee Lake provides a place for all kinds of water-based activities. The Jubilee Lake National Recreation Trail is here. It is a wheelchair-accessible trail encircling the 97-acre lake. A little farther south is the Spout Springs Ski Area, which is popular in winter.

From the Mt. Emily Exit on Interstate 84 is a circuitous forest road that eventually comes to Sacajawea Springs. Nearby is a grand viewpoint overlooking Meacham Canyon. This back road is part of the historic Whitman Route that Marcus and Narcissa Whitman took over the Blue Mountains long ago. Whitewater enthusiasts will be interested in the 45-mile stretch of the Wallowa and Grande Ronde Rivers between Minam and Troy. About half of the river course is along the border between the Umatilla and Wallowa-Whitman National Forests.

The remainder of the Umatilla National Forest is south of Interstate 84. From the community of Heppner, you may begin your exploration by taking the Blue Mountain National Scenic Byway. This is a 130-mile route that merges with the Elkhorn National Scenic Byway at the Wallowa-Whitman National Forest boundary. As soon as the scenic byway enters the Umatilla National Forest, take a side trip to the south and walk the 1.25-mile trail through pastoral Martin Prairie. The trailhead is at the Ditch Creek Guard Station.

Back on the scenic byway, it is a pleasant drive to Ukiah through forests of Douglas fir, grand fir, and ponderosa pine. Here and there are wildflower meadows, basalt columns, rock outcrops, and cliff-lined canyons. After a 15-mile side trip to Potamus Point, you may look down on a cliff-lined canyon along the North Fork of the John Day River.

After getting refreshments and relaxing from mountain driving in Ukiah, you may continue on the Blue Mountain National Scenic Byway east of town where the road climbs back into the national forest. The highway goes through part of the 50,000-acre Tower Fire, which burned in 1996. A side road north to the top of Tower Mountain will reward you with another panoramic view.

The scenic byway twists and turns until it reaches the North Fork John Day Campground. From the campground you may either begin the Elkhorn National Scenic Byway, which makes a loop through the Wallowa-Whitman National Forest, or you may continue south past Crane Flats to the gold-mining town of Granite. As you drive toward Granite, observe the rock walls in the stream bottoms. The boulders in these walls were hand carried by Chinese gold miners at the beginning of the twentieth century. The boulders were piled up in order to expose the sand and gravel where the gold was deposited. Although Granite is the termination of the scenic byway, you may continue west for five miles to the Fremont Powerhouse and then another six miles to Olive Lake. When the California–Pacific Utilities Company abandoned their electric generating plant in 1968, they gave the entire powerhouse complex, which had been constructed in 1908, to the Forest Service. You may walk around this assemblage of buildings, pipelines, and machinery. In the vicinity are ghost towns such as Greenhorn and several abandoned mines.

Umpqua National Forest

SIZE AND LOCATION: Approximately 1 million acres in southwestern Oregon on the western flanks of the Cascade Range. Major access routes are State Routes 62, 138, 227, and 230. District Ranger Stations: Cottage Grove, Idlewyld Park, Glide, Tiller. Forest Supervisor's Office: 2900 N. W. Stewart Parkway, Roseburg, OR 97470, www.fs.fed.us/r6/umpqua.

SPECIAL FACILITIES: Boat ramps; swimming beaches; winter sports areas.

SPECIAL ATTRACTIONS: North Umpqua National Wild and Scenic River; Oregon Cascades National Recreation Area; Rogue-Umpqua National Scenic Byway.

WILDERNESS AREAS: Boulder Creek (19,100 acres); Mount Thielsen (61,281 acres, partly in the Deschutes and Winema National Forests); Rogue-Umpqua (33,000 acres, partly in the Rogue River National Forest).

The Umpqua National Forest is a land of old-growth forests, tumultuous waterfalls, turbulent rivers, and rugged mountains on the western edge of the Cascade Range.

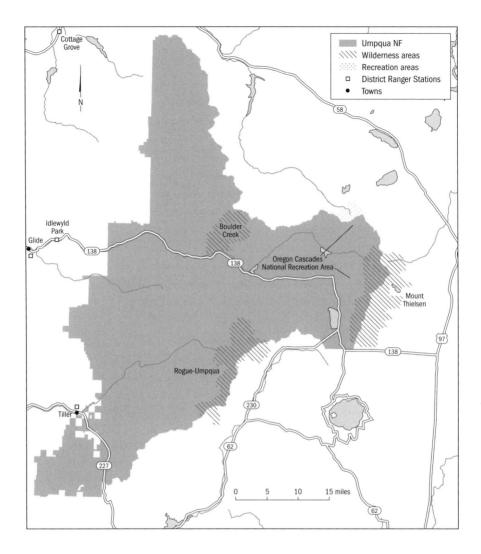

The northern part of the Rogue-Umpqua National Scenic Byway, State Route 138, crosses the heart of the Umpqua National Forest. It is also called The Highway of Waterfalls. From Roseburg, the scenic byway proceeds eastward, entering the national forest just before the Fall Creek Trailhead as it follows the North Fork Umpqua River. This is a Wild and Scenic River, great for whitewater rafting all the way to Soda Springs Dam. This is a major stop because the Falls Creek National Recreation Trail begins here and ends in one mile at the falls, which are actually double, with falls of 35 and 50 feet. The jumble of rocks along the trail is referred to as Job's Garden and, at one point, the creek you are following narrows through a crack in a large boulder.

Near Steamboat Lodge, a side road goes to Steamboat Falls Campground and a short trail to Steamboat Falls. Although the falls are only about 25 feet high, you can usually see fish trying to jump it or use the fish ladder next to it.

Three and one-half miles from Steamboat along State Route 138 are Jack Falls: three falls in one with drops of 70, 40, and 30 feet. At Eagle Rock Campground you are directly below the prominent volcanic Eagle Rock Dome. The wild area north of Eagle Rock is in the Boulder Creek Wilderness. In about one mile is Old Man Rock. This is an amazing formation. You barely get going on the scenic byway when you come to Toketee Falls, which has two plunges. The upper falls drops about 40 feet, the lower about 80 feet. Continue past Toketee Lake to the Umpqua Hot Springs Trailhead. Hot bubbling spring water fills a tublike basin in the rocky surface.

More waterfalls are ahead on the scenic byway. Watson Falls is the most impressive as it cascades for 272 feet, the highest waterfall in southern Oregon. The quarter-mile trail from the Watson Falls Picnic Area brings you face to face with the falls. Whitehorse and Clearwater Falls drop 15 and 30 feet, respectively.

Before continuing on the scenic byway, take the forest highway to Lemolo Lake where there is a resort with boating facilities. Five miles north of the lake is mighty and thunderous Lemolo Falls along the North Umpqua River. It drops nearly 100 feet. Just beyond is Warm Springs Falls, reached by a quarter-mile trail through a dense forest of Douglas fir and lodgepole pine. The falls are about 70 feet high.

The scenic byway turns abruptly southward and skirts the eastern edge of Diamond Lake in the Oregon Cascades National Recreation Area. A forest service information station is near the lake. Three-thousand-acre Diamond Lake was formed when a lava flow from Mt. Thielsen blocked a broad valley. Three miles directly across from the information station is Mt. Thielsen with its spirelike peak, nicknamed "The Lightning Rod of the Cascades." Ten miles west of Diamond Lake is the Incense Cedar Grove, a majestic stand of some of the largest incense cedars (fig. 22) in the country. From the south end of the lake, the scenic byway follows State Route 230 and enters the Rogue River National Forest.

From the town of Glide is a highway that follows Little River and enters the Umpqua National Forest at the Wolf Creek Campground. You need a good forest map to find the Emile Big Tree Trail, Grotto Falls, Yakso Falls, Hemlock Falls, and the Youtlkut Pillars. Lake-in-the-Woods Campground is right in the midst of all of these points of interest.

The Emile Big Tree Trail is only a tenth of a mile long and is in an old-growth ponderosa pine forest. The prize specimen here is the Bill Taft Tree, which is 235 feet tall and nearly 10 feet in diameter. Most trees in the

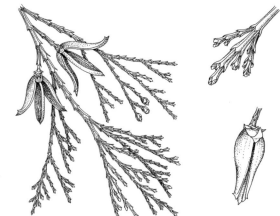

Figure 22.
Incense cedar.

grove, including the Bill Taft Tree, are more than 400 years old. Nearby Grotto Falls plunges 100 feet along Emile Creek, with the trail passing behind the falls. Hemlock Falls is south of Lake-in-the-Woods Campground. A half-mile trail is near this 80-foot falls. Yakso Falls a little north of the campground has a 70-foot drop. Youtlkut Pillars is an impressive site with the columnar basalt pillars strongly tilted.

From the town of Tiller at the far western edge of the Umpqua National Forest is a forest highway that parallels the South Umpqua River. Take this road to the Umpqua Falls Campground where there is another scenic waterfall. If you branch off the South Umpqua River Road, another side road will bring you to the former world's tallest sugar pine that is more than 200 feet tall.

Other isolated areas in the southern part of the Umpqua National Forest are of interest. A few miles southeast of Glide is Shadow Falls. The Cavitt River roars through a natural grotto and plunges nearly 100 feet in three cascades. Downstream are unusual rock formations. Ten miles southeast of Tiller near the Devil's Flat Campground are the Cow Creek Gorge and Falls. They are breathtaking.

In the northern part of the Umpqua National Forest east of Cottage Grove are Swordfern Trail, Parker Creek Falls, and the old Bohemia Mining District. The Swordfern Trail begins at the Rujada Campground and follows Layng Creek for nearly 1.5-miles. The lush swordferns that live beneath the Douglas firs are incredible. Parker Falls is a series of cascades surrounded by a grove of huge Douglas firs. This 1.5-mile-long trail is steep in places.

The site of the Bohemia Mines is south of the Mineral Campground. The Bohemia National Recreation Trail follows a wagon road built in 1864 and 1865 from the Bohemia Mines to Oakland, Oregon. That part of the trail near Canton Point is well preserved.

Wallowa-Whitman National Forest

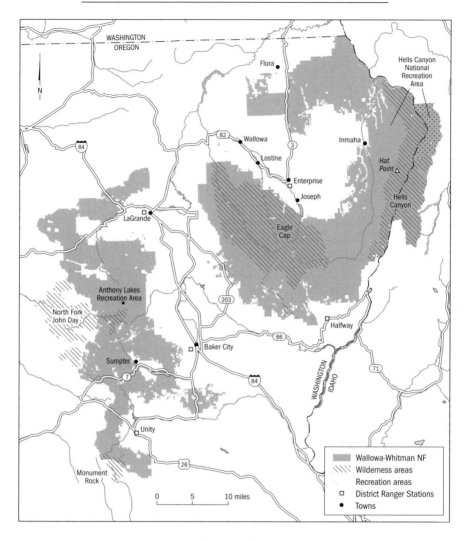

SIZE AND LOCATION: Approximately 2.3 million acres in northeastern Oregon, including the Wallowa Mountains and a part of the Blue Mountains. Major access routes are Interstate 84 and State Routes 3, 7, 82, 86, 203, and 240. District Ranger Stations: Baker City, Enterprise, LaGrande, Halfway, Unity. Forest Supervisor's Office: 1550 Dewey Avenue, Baker City, OR 97814, www.fs .fed.us/r6/ww.

SPECIAL FACILITIES: Boat ramps; swimming beaches; winter sports areas.

SPECIAL ATTRACTIONS: Hells Canyon National Recreation Area; Anthony Lakes Recreation Area; four Wild and Scenic Rivers.

WILDERNESS AREAS: Eagle Cap (354,290 acres); Hells Canyon (204,814 acres, partly in the Payette and Nez Perce National Forests); Monument Rock (19,650 acres, partly in the Malheur National Forest); North Fork John Day (121,352 acres, partly in the Umatilla National Forest).

The Wallowa Mountains and part of the Blue Mountains are in the Wallowa part of the national forest, whereas the Snake River and Hells Canyon are in the Whitman part of the national forest. Until 1954, these were two separate national forests, but they are now united into one administrative unit.

The Snake River has carved the 8,000-foot deep Hells Canyon. It is the deepest gorge in the United Sates, topping the Grand Canyon by 1,000 feet. The river forms the boundary between Oregon and Idaho, and Hells Canyon is in both states. The gorge and its surrounding areas are in the Hells Canyon National Recreation Area, a massive region of 652,877 acres. Within the national recreation area, the most inaccessible parts are further designated as the Hells Canyon Wilderness. Privately run boat tours up the Snake River are given from Hells Canyon Dam at the end of State Route 86. Whitewater rafting trips also take off from here.

For the best view of Hells Canyon and for an exciting drive, take a secondary road from either Enterprise or Joseph to Inmaha at the edge of the Hells Canyon National Recreation Area. From Inmaha, follow a narrow, gravel, 24-mile road to Hat Point which, at 7,000 feet, has a spectacular view into the V-shaped Hells Canyon. The route edges around vertical canyon walls, past alpine meadows, and through forests of ponderosa pine. After the first five miles, there is a marvelous view of the Inmaha River Valley and the Wallowa Mountains in the distance. At 11 miles, the Horse Creek Observation provides views of Inmaha Canyon to the west and Horse Creek Canyon to the east. A little farther along, there is a one-eighth-mile loop trail to Granny View for more great panoramic scenes. After another 1.5-miles, the Western Rim National Recreation Trail leads south to Lookout Mountain. Saddle Creek Viewpoint gives you the first view of Hells Canyon down Saddle Creek and across into the mountains of Idaho. In five more miles you arrive at Hat Point for a breathtaking look into Hells Canyon.

A few miles west of Inmaha, State Route 3 north from Enterprise to Flora passes through a pretty part of the Wallowa-Whitman National Forest. Four miles before reaching Flora is a superb viewpoint into Joseph Canyon.

Fine scenic drives give an overview of the Blue and Wallowa Mountains. The 106-mile loop around the Blue Mountains is a National Scenic Byway. Start it in Baker City and head south on State Route 7 to Salisbury before heading west, following the Powder River. Phillips Reservoir has swimming beaches and a boat ramp. Camping is available at the Union Creek Campground. The Shoreline Trail follows the boundaries of the reservoir. Look for ospreys that nest here. Gold was discovered near the western end of where Phillips Reservoir is today. The scenic byway is now in gold country, and there are abandoned mines everywhere. The byway leaves the national forest boundary temporarily as it makes its way in the Powder River Valley to the old gold-mining town of Sumpter. You might want to take a side road north from Sumpter to Bourne in the national forest. Bourne is a ghost town, and there are still a few buildings left.

The scenic byway west from Sumpter is back in the Wallowa-Whitman National Forest. The McCulley Fork Campground is set beneath a forest of ponderosa pine, Douglas fir, and western larch. After twisting for four miles, the highway arrives at the Blue Spring Snow Park. Granite, another gold-mining town, still has a few residents. North from Granite, the scenic byway enters the Umatilla National Forest for a few miles until it comes to the North Fork John Day Campground. The byway then heads due east to the Anthony Lakes Recreation Area. Although this is a popular ski area in winter, there is a pleasant half-mile Trail of the Alpine Glacier to Hoffer Lakes that can be enjoyed during the summer. The 23-mile Elkhorn Crest National Recreation Trail also begins at Anthony Lake.

The elevation of Anthony Lake is 7,100 feet, but the adjacent high country rises to 8,650 feet. The Trail of the Alpine Glacier climbs through a sub-alpine forest and a small subalpine meadow. The first stop on this interpretive trail is at a 60-ton granite boulder that was carried down from higher on the mountain by glacial action. Gunsight Mountain can be seen from Stop Two. It gets its name from the notch at its summit. The small meadow seen next was a glacial lake at one time that has now filled in. The two Hoffer Lakes developed in a glacial cirque as a single lake, but an accumulation of soil provided a barrier that separated the single lake into two. At Stop Nine, you get a grand view of a hanging valley. As the main canyon was being gouged out by a glacier, it was formed more quickly than the side canyons that were left hanging above the main canyon. The return trip from the Anthony Lakes Recreation Area to Baker City is about 35 miles.

The Wallowa Mountains lie betwen the town of LaGrande and the Hells Canyon area. Much of the mountain range is in the Eagle Cap Wilderness, the largest wilderness area in Oregon. Thirty-one mountain peaks are above 8,000 feet, including Matterhorn Peak, the highest, at 9,845 feet. Many alpine lakes are present, including Legore Lake, Oregon's highest lake at 8,880 feet. The National Wild and Scenic Eagle Creek flows through a part of the wilderness area, as does the National Wild and Scenic Minam River, the National Wild and Scenic Inmaha River, and the National Wild and Scenic Lostine River.

Extending between two lobes of the Eagle Cap Wilderness is the Lostine Canyon Auto Tour that follows a part of the Lostine River. After 18 miles, the auto tour ends at the Two Pan Campground where there are trails into the wilderness area. Along the route to Two Pan Campground, you will see the Pole Bridge Gorge and the Lostine Gorge (pl. 43), both beautiful canyons. The Lostine Guard Station cabin was built in 1933 and is still in use. From the Lostine Auto Tour route are trails to Bowman Lake and Frances Lake.

Northwest of LaGrande, off of Interstate 84, are the Oregon Trail Interpretive Park, the Bird Track Springs Interpretive Trail, and the California Gulch Trail. The Oregon Trail Interpretive Park is on a grassy ridge where you can see wagon-track depressions. From the Interpretive Park is the two-mile California Gulch Trail through an old-growth forest. Nearby is the 1.5-mile Bird Track Springs Interpretive Trail that goes through the floodplain of the Grande Ronde River.

Willamette National Forest

SIZE AND LOCATION: Approximately 1.7 million acres along the western flanks of the Cascade Range in west-central Oregon. Major access routes are U.S. Highway 20 and State Routes 20, 22, 26, 58, 126, and 242. District Ranger Stations: Lowell, Mill City, Westfir, McKenzie Bridge, Sweet Home. Forest Supervisor's Office: 211 E. 7th Avenue, Eugene, OR 97440, www.fs.fed.us/r6/willamette.

SPECIAL FACILITIES: Winter sports areas; boat ramps; swimming beaches.

SPECIAL ATTRACTIONS: McKenzie-Santiam Loop National Scenic Byway; Hoodoo Ski Bowl.

WILDERNESS AREAS: Bull of the Woods (34,900 acres, mostly in the Mt. Hood National Forest); Diamond Peak (54,185 acres, partly in the Deschutes National Forest); Menagerie (4,800 acres); Middle Santiam (7,500 acres); Mt. Jefferson (107,108 acres, partly in the Deschutes and Mt. Hood National Forests); Mt. Washington (52,738 acres, partly in the Deschutes National Forest); Three Sisters (286,708 acres, partly in the Deschutes National Forest); Waldo Lake (39,200 acres).

You might think that with all or part of nine wilderness areas in the Willamette National Forest there would be nothing left for people wanting nonwilderness experiences. But such is not the case. Although the marvelous scenery in the wilderness areas are off-limits to the physically handicapped and the elderly, there is scenery elsewhere in the national forest. It is even possible to nibble around the edges of some of the wilderness areas.

From the west are four major access points to the national forest. State Route 22 from Salem enters the northern part of the Willamette at the west end of Detroit Lake. Piety Island is near the eastern side of the lake, and the Forest Service has a campground on the island. Piety Knob stands 600 feet above the campground. A boat ramp is at Breitenbush Campground along the north side of Detroit Lake. North of State Route 22 is scenery galore in the form of Elephant Rock, Needle Rock, and Dome Peak. At the northern tip of Detroit Lake at the crossroads community of Detroit is a three-way junction. The road to the northwest along French Creek winds and wriggles over Knutson Saddle before ending. A short trail from the road's end goes to Phantom Natural Bridge, an impressive rock arch. Tiny and picturesque Cedar Lake is nearby. A second road from the junction at Detroit follows the Breitenbush River northeast past Eagle Rock, eventually coming to Breitenbush Hot Springs on private property. Less than two miles east of the springs is a trailhead for the South Breitenbush Gorge National Recreation Trail. The entire trail is only 2.5 miles long, but the most impressive feature is the river gorge reached by a short side trail. At the gorge the river squeezes through a narrow passage of basalt for 300 feet but only 3 to 10 feet wide. Where Roaring Creek enters the gorge from the east, there is a wooden footbridge over the creek. This beautiful area is completely shaded by old-growth Douglas firs and western hemlocks. The third highway at the junction at Detroit is the continuation of State Route 22. You may wish to take several trips off this highway. One road makes its way for about six miles to scenic little Rainbow Lake. Another road is to Whispering Falls Campground where there is a short trail to a picturesque waterfall. Whispering Falls is where Misery Creek drops 60 feet into the North Santiam River. After State Route 22 curves to the south, it continues to follow the Santiam River. Parmelia Creek Road eastward dead-ends at the Mt. Jefferson Wilderness. A trail into the wilderness stays alongside Parmelia Creek to lovely Parmelia Lake. The trail is beneath a beautiful stand of Douglas fir. The Pacific Crest National Scenic Trail may be joined here at Parmelia Lake. In another four miles along State Route 22, Independence Rock looms less than one mile to the east. From the Marion Fork Campground, you will come in four miles to Gooch Falls, which plunges 100 feet. Two miles south, another forest road turns back north, passing Bachelor Mountain and coming to Bruno Meadow. The meadow is one of the prettiest wildflower areas around, made even more beautiful by a forest of large Engelmann spruces that surrounds it.

State Route 22 continues south for another 15 miles, eventually going

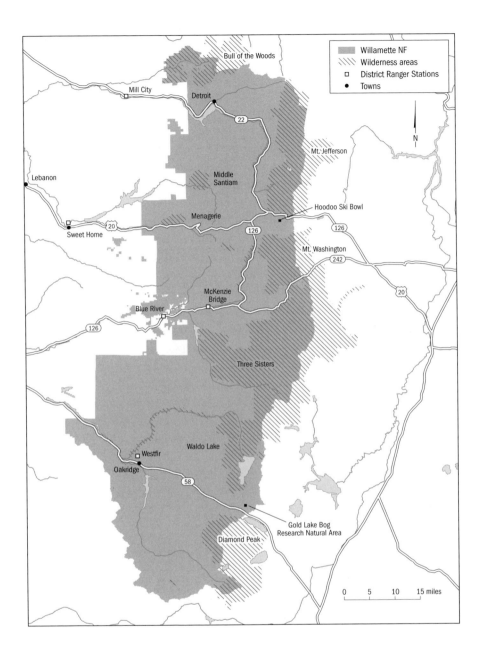

through a lava flow just before it ends at U.S. Highway 20 at Little Nash Crater.

U.S. Highway 20 east from Corvallis passes through Lebanon and Sweet Home and around the south side of Foster Lake before entering the Willamette National Forest just beyond Cascadia State Park. If you want to do some wilderness hiking, drive a different road away from Foster Lake and follow the Middle Santiam River to the Quartzville Work Station just before reaching the national forest. Side roads to trailheads are at the northern edge of the Middle Santiam Wilderness. The primary feature in this wilderness is Chimney Rock, a 4,965-foot volcanic lava plug. Most of the trees in this wilderness are old-growth Douglas fir, western hemlock, and western red cedar, several of them more than 200 feet tall. The highway stays near the South Santiam River as far as the House Rock Campground. Before reaching House Rock, Fernview Campground is situated at the river's edge. North of this campground is the Menagerie Wilderness, named for several unusual rock formations that have the form of animals. Although Rooster Rock, at 3,567 feet, is the most popular for climbing and the easiest to get to because it is only one mile from the campground, other formations are Rooster Tail, Chicken Rock, Hen Rock, Turkey Monster, and South Rabbit Ears. These strange formations are in a dense forest of Douglas fir, western hemlock, and western red cedar.

From the House Rock Campground, State Route 20 begins to climb on a very curvy route, gaining more than 2,000 feet in elevation. About 10 miles east is Civil Road to the north. This side road ends after three miles at the western edge of Iron Mountain. Iron Mountain's rocky peak is above timberline and is a local landmark. The nearly mile-long trail to the summit goes through wet meadows, moist meadows, dry meadows, and various forest communities. In the rocky crevices on the summit are mosses, lichens, and tiny wildflowers. The wet meadows have deep organic soils and stay wet all year.

In the vicinity is the Wildcat Mountain Research Natural Area where you can see more meadows and rocky habitats, but these are surrounded by an ancient stand of noble firs. With the noble firs are impressive specimens of Pacific silver fir, Douglas fir, western white pine, and Alaska cedar.

On the south side of Hackelman Creek is the Hackelman Creek Old Growth Trail that meanders through a forest of 500-year-old Douglas firs, western red cedars, and western hemlocks. The creek contains an unusual subspecies of cutthroat trout known as the Hackelman trout.

Just as the highway comes to a great lava flow before climbing to Santiam Pass, head south on State Route 126. This is the beginning of the western side of the McKenzie-Santiam National Loop Scenic Byway. For the next seven miles you will come to Sahalie Falls, Koosah Falls, and Tamolitch Falls, the first two of these incredible cascades on the McKenzie River. Sahalie Falls plunges 140 feet and Koosah 120 feet, each hitting bottom with thunderous explosions. At Tamolitch Falls, the falls are dry because the river was diverted three miles upstream at Carmen Reservoir and then diverted again by tunnel to Smith Reservoir and power-generating facilities. Springs still fill the pool at the base of the dry falls. State Route 126 continues south to Belknap Springs where it abruptly heads west, eventually exiting the Willamette National Forest beyond Blue River Lake. If you take the highway east from Belknap Springs, it becomes State Route 242 and continues the McKenzie-Santiam National Loop Scenic Byway up over McKenzie Pass and into the Deschutes National Forest. Approaching McKenzie Pass, the highway enters a huge lava flow. The highway is the only feature separating the Mt. Washington Wilderness to the north and the Three Sisters Wilderness to the south. Just before the scenic byway begins its ascent up Deadhorse Gulch to McKenzie Pass, there is a trailhead for a half-mile trail into the Three Sisters Wilderness Area to Proxy Falls. This is a very impressive 200-foot falls. A fine viewpoint overlooks the falls. At McKenzie Pass, in the heart of a huge lava flow, dwarf trees of ponderosa pine, western red cedar, hemlock, Pacific silver fir, and noble fir grow among the lava. Despite their short stature, these trees are several hundred years old.

North where U.S. Highway 20 continues east over Santiam Pass is Sawyers Cave, an interesting lava tube with ice in it. Near Santiam Pass is the Hoodoo Ski Bowl.

If you return to State Route 126 between Belknap Springs and Blue River Lake, take the forest road south to Horse Creek Campground. At the campground the road splits. Both forks head into the notch between the left thumb and the remainder of the Three Sisters Wilderness Area. The left fork follows Horse Creek to the western edge of the wilderness for some of the most startling scenery in the national forest. The right fork twists its way through the Olallie Ridge Research Natural Area to Larch Butte and Larch Butte Scenic Area. The natural area is 720 acres of subalpine forests and mountain meadows, flanked on either end by O'Leary Mountain and Horsepasture Mountain. Here are marvelous forests of hemlock, subalpine fir, and Douglas fir with many meadows interspersed.

The remainder of the Willamette National Forest may be accessed from State Route 58, which crosses the southern half of the national forest. From the town of Lowell just east of Eugene, take the North Shore Road that curves around Fall Creek Lake. From the Dolly Varden Picnic Area, take the forest road north for 13 miles to the Tall Trees Trail. Although the trail is only a quarter-mile long, many giant Douglas firs and western hemlocks tower high above you. Interpretive signs are along the trail..

State Route 58 goes along the southern edge of Lookout Point Lake until it reaches the Willamette River. At Buckhorn Creek is a half-mile Buckhorn Nature Trail through a woods of black cottonwoods and various species of conifers. The highway leaves the community of Oakridge and follows Salt Creek for many miles to Salt Creek Falls where the creek plunges with ear-shattering noise for 286 feet. Short trails lead to Fall Creek Falls in the vicinity. Diamond Peak Wilderness is south of these falls.

A good forest highway goes to Waldo Lake, the second largest lake in Oregon. Waldo Lake has 6,700 acres of surface water surrounded by forests of Douglas fir, western hemlock, and ponderosa pine. The east side of the lake is accessible by road, whereas the north and west sides are in the Waldo Lake Wilderness. Before reaching Waldo Lake (pl. 44), the forest highway passes Gold Lake Bog Research Natural Area. The 500 acres in this natural area contain subalpine bogs at the head of Gold Lake. Forests of Engelmann spruce and subalpine fir are interspersed among the bogs. The sphagnum bogs contain several rare bog species, including five carnivorous plants—two kinds of sundews and three kinds of bladderworts. Other bog species present are cotton grasses, bogbean, alpine laurel, and a plant known only as *Scheuchzeria*.

Wildcat Mountain Natural Area

When I asked a ranger in the Willamette National Forest where I could see a fine stand of noble firs (fig. 23), I was directed to the Wildcat Mountain Natural Area east of Eugene, Oregon. A forest road parallels the southern edge of the natural area, going between Wildcat Mountain to the north and Bunchgrass Mountain to the south. At the western end of the natural area is a hiking trail that climbs to the summit of Wildcat Mountain, going from an

elevation of 4,400 feet at the beginning of the trail to 5,353 feet at the summit. The trail follows a moderate grade to the summit, but the north face of Wildcat Mountain is steeply precipitous and rocky. The natural area is considered subapline, with wet, cool climate. Snow depths during the winter may be as much as nine feet, and snow usually lingers on the ground in protected areas until early July.

The natural area that has been designated occupies 1,000 acres, with 70 percent of it forested. The remainder of the natural area consists of wet meadows, mesic meadows, dry meadows, shrubby thickets, and seemingly bare, rocky areas.

The trail enters the natural area in a beautiful forest of noble fir and Douglas fir.

Figure 23. Noble fir.

Botanists estimate the noble firs in this grove to be more than 175 years old, with some of the huge Douglas firs nearly 500 years old. In addition to the noble firs and Douglas firs, other trees in this forest are Pacific silver fir, mountain hemlock, western white pine, and western hemlock. The wet soil beneath the trees supports a very rich flora. Common in the understory are thin-leaved blueberry, bunchberry, bead lily, and starry false Solomon's-seal. Scattered plants of three-leaved windflower, little pipsissiwa (pl. 45), prince's pine, Oregon bedstraw, stream violet, redwoods violet, coolwort foamflower, white-veined pyrola, one-sided wintergreen, dwarf bramble, vanilla-leaf, and bracken fern are also present.

As the trail climbs higher, western hemlock is no longer present. Where the soils are rockier and not as rich, vine maple becomes very common. Bead lily occurs rarely in this habitat, with the dominant herb being vanilla-leaf.

If you continue past the summit of Wildcat Mountain to the eastern side of the natural area, 350-year-old stands of noble fir occur.

On the cooler, north side of Wildcat Mountain is a forest dominated by mountain hemlock and Pacific silver fir beneath which grow considerable amounts of beargrass and thin-leaved blueberry.

Three types of meadows have developed because of the differences in available moisture. Wet meadows are wet all year, with thickets of Sitka alder often found throughout the wetlands. In openings between the thickets are green false hellebore, arrowleaf groundsel, and Sitka valerian. In openings where there is an adequate supply of moisture only until mid-July, a mesic meadow community develops. Eastwood's fleabane, broad-leaved lupine, poke-leaved smartweed, mountain thistle, and American vetch are the dominant herbs. Most of these mesic meadows are being encroached upon by noble fir and other tree species.

Where the soil is shallow and rocky and dries out by the middle of June, dry meadow communities occur. The vegetation is more sparse than in the other meadow types, but there is a good diversity of species, including scarlet gilia, spreading groundsmoke, mountain navarretia, microsteris, small-flowered blue-eyed Mary (pl. 46), and mountain owl-clover. Alaska cedar sometimes invades these rocky openings.

Crevices of south-facing cliffs of Wildcat Mountain support a fascinating flora. Among the many species growing from the cliffs are harsh paintbrush, bastard toadflax, woolly sunflower (pl. 47), few-flowered desert parsley, northern buckwheat, rough wallflower, virgate phacelia, small-flowered penstemon, cliff penstemon, Menzies' larkspur, matted saxifrage, and the succulent wormleaf stonecrop and spreading stonecrop. Low shrubby mountain juniper and manzanita are usually present, as well.

Olallie Ridge Research Natural Area

To see an example of a subalpine complex of mountain meadows and fir–hemlock forests on ridgetops in the western Cascades of Oregon, I headed to the Olallie Ridge Research Natural Area. From U.S. Highway 126, I followed the South Fork Forest Road and then the North Fork Forest Road to a trailhead and then hiked about one mile to the natural area in the Willamette National Forest.

The 720-acre natural area consists of two mountains—O'Leary at 5,530 feet on the western side of the natural area and Horsepasture at 5,660 feet on the east side. The valley between the mountains is at an elevation of 4,400 feet. Most of the mountain slopes in the natural area are steep and rocky, but there is a switchback trail, Horsepasture Mountain Trail, that leads to the summit of Horsepasture Mountain. Rocks that outcrop are either basalt, dacite, breccias, or volcanic tuffs. During the winter, as much as nine feet of snow may accumulate on the summits of the mountains. Total precipitation of all kinds for the year averages 80 inches. About 60 percent of the natural area is forested, the remainder either occupied by open meadows or shrubby communities.

Although the forests are magnificently beautiful, it is the nonforested areas with a rich flora of species, many of them endemic and others disjunct from their nearest location, that make the area special. On the north-facing slopes of Horsepasture Mountain, where snow depth is greatest, the short summer growing season has such wildflower beauties as western spring beauty, Oregon bluebells, pale fawn lily, turkey-peas, and partridge-foot, which is a trailing member of the rose family. On the south-facing side of Horsepasture Mountain's summit, where freezing and thawing have created crevices in the rocks, several species have found their home in these crevices. Among them are Jeffrey's shooting-star, three-leaf lewisia, Brewer's monkey-flower, yellow monkey-flower, western saxifrage, and dwarf groundsmoke, the latter a tiny-flowered member of the evening primrose family.

On all of the mountain slopes are scattered openings that are essentially devoid of trees but which have a continuous cover of vegetation. These meadows differ in species composition depending on the amount of moisture in the soil during the growing season. Meadows that are continuously wet and sometimes nearly soggy have deep layers of soil. Dominant herbs are green false hellebore, arrowleaf groundsel, and Sitka valerian, all robust species. Other species present include Brewer's mitrewort, Gray's lovage, Fendler's waterleaf, Pacific waterleaf, and the shrubby blue currant and salmonberry.

In meadows where moisture in the soil is usually gone by late July, a different community of plants is present. This mesic meadow supports black-head, Eastwood's fleabane, broad-leaved lupine, poke-leaved smartweed, tall bluebells, fireweed, dwarf groundsmoke, mountain thistle, and American vetch. Thimbleberry and sticky currant are present in most of the mesic meadows.

Just below the summit of Horsepasture Mountain but above the mesic meadow community are openings that support dry meadow species. Rocks and boulders of various sizes are strewn across this community, whose primary species are scarlet gilia, narrow-leaved collomia, spreading ground-smoke, mountain owl-clover, tongue-leaved luina, mountain navarretia, small-flowered blue-eyed Mary, sticky cinquefoil, mountain Mariposa lily, and myrtle boxwood.

One of the more interesting communities is called the outcrop ridge community where patches of exposed rock have broken down around the edges into finer rock particles. Growing in this finer substrate are harsh paintbrush, Menzies' larkspur, small-flowered penstemon, wormleaf stonecrop, spreading stonecrop, woolly sunflower, pinemat manzanita, rough wallflower, pearly everlasting, and Cardwell's penstemon.

As we looked down the vertical rock faces, we could see matted saxifrage, cliff penstemon, Cascade daisy, showy polemonium, tufted saxifrage, and small-flowered alumroot hanging from the crevices.

Hiking up the switchbacks on the dry, south-facing slopes of Horsepasture Mountain, we encountered a nice forest community dominated by Douglas fir and white fir. Beneath these canopy trees are western snowberry, vine maple, thin-leaved blueberry, and baldhip rose. On the forest floor are starry false Solomon's-seal, vanilla-leaf, prince's pine, and bracken fern.

On more moist slopes, the forest consists mostly of Pacific silver fir, noble fir, subalpine fir, mountain hemlock, western hemlock, and western white pine. The understory in this type of subalpine forest is made up of bunch-berry, vanilla-leaf, one-sided wintergreen, redwoods violet, mountain sweet cicely, and bead lily, with occasional shrubs of dwarf bramble and thin-leaved blueberry.

Between the wet meadows and forested communities, there is often a shrubby thicket dominated by Sitka alder and, in somewhat drier sites, vine maple.

Winema National Forest

SIZE AND LOCATION: Approximately 1.1 million acres, north, east, and south of Crater Lake National Park in south-central Oregon. Major access roads are U.S. Highway 97 and State Routes 58, 62, and 140. District Ranger Stations: Chemult, Chiloquin, Klamath Falls. Forest Supervisor's Office: 1301 S. G Street, Lakeview, OR 97630, www.fs.fed.us/r6/frewin.

SPECIAL FACILITIES: Boat ramps; swimming beaches.

SPECIAL ATTRACTIONS: Desert Forest hiking and driving tours.

WILDERNESS AREAS: Mountain Lake (23,071 acres); Mt. Thielsen (61,281 acres, partly in the Deschutes and Umpqua National Forests); Sky Lakes (116,301 acres, partly in the Rogue River National Forest).

Since this manuscript was submitted for publication, the Winema National Forest has been placed under the administration of the Fremont National Forest.

The Winema National Forest is on the eastern slopes of the Cascade Range. The best known feature of this national forest is Mt. Thielsen, whose slender, spirelike peak is a landmark for miles around. The upper 2,000 feet of the mountain stands above timberline. Just below timberline are stands of mountain hemlock, fir, and whitebark pine. The dominant tree on the lower slopes is lodgepole pine.

Immediately east of Crater Lake National Park, the Winema National Forest has developed the Desert Forest Journeys hiking and driving tours. Three half-mile hiking trails and a 3.5-mile driving route are off of U.S. Highway 97. The Ties Through Time Trail shows several aspects of past logging operations. Open Forest Trail is a leisurely trail through an old-growth ponderosa pine forest. Pines in the Pumice Trail allows you to see a forest of ponderosa pine growing with apparent difficulty on deep pumice that was laid down 7,000 years ago when Mt. Mazama erupted. Along the Ponderosa Postcards Drive, there are numbered posts that describe ecology in a ponderosa pine forest. A short distance south of the Desert Forest Journeys are the Sand Creek Pinnacles. These pointed peaks above Sand Creek are impressive.

The Sky Lakes Wilderness stretches south from Crater Lake National Park for 22 miles. The wilderness is shared with the Rogue River National Forest. Although both sides of the wilderness have alpine lakes, there are more in the Winema National Forest. Unusual geological formations known as Goose Egg and Goose Nest at the upper end of the wilderness area are named for their shapes. A trail off of State Route 62 goes to these formations.

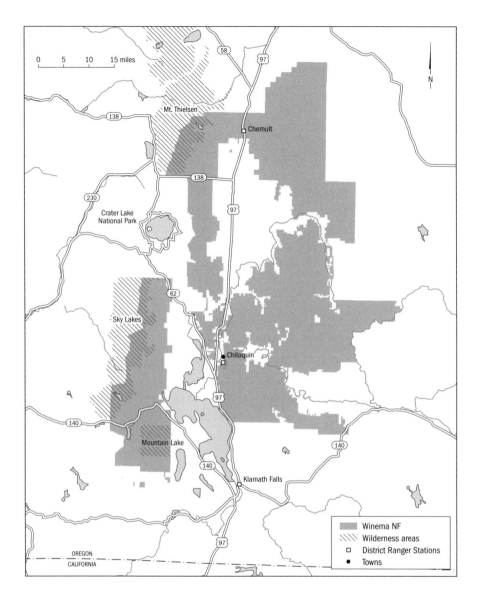

One of the strangest things you will see is Mare's Egg Spring six miles north of Crystal Spring Campground. You will see perfectly round green globules at the bottom of the clear springs. These are colonies of the blue green alga *Nostoc pruniforme*. Very few colonies like this exist in the world. Please do not disturb them.

East of Lake of the Woods is the Mountain Lakes Wilderness. This was originally a 12,000-foot mountain that erupted and became transformed into a large caldera. After glaciation, numerous small alpine lakes were left behind. The wilderness area has eight peaks on the rim of the caldera.

A forest road leads east along the Sprague River out of Chiloquin. In about 12 miles, a good forest side road winds south past Saddle Mountain to the Devil's Garden. This an interesting jumble of rocks with unbelievable shapes. It is believed that these rocks have formed from volcanic eruptions that occurred under water.

Another road east off of U.S. Highway 97, about 20 miles north of Chiloquin, goes through the heart of Klamath Forest National Wildlife Refuge and then enters the national forest. Six miles to the northeast is Bluejay Research Natural Area. This is a 210-acre tract with two pristine plant communities on a plateau covered with 10 feet of pumice. The eastern three-fourths of the natural area is a community of ponderosa pine with an understory of bitterbrush and needlegrass. In the western one-fourth of the natural area is a lodgepole pine community with bearberry, bitterbrush, and needlegrass beneath.

NATIONAL FORESTS IN UTAH

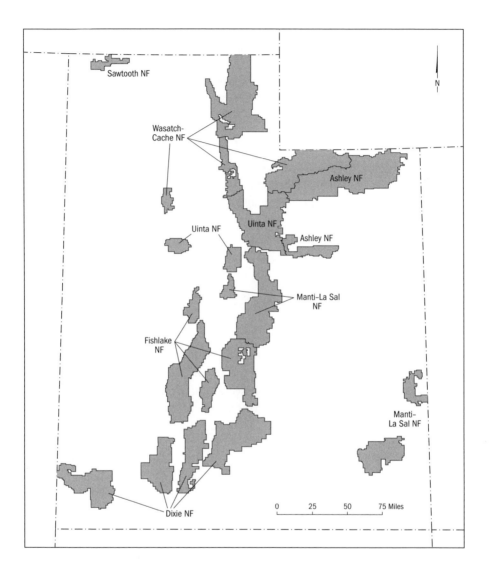

The eight national forests in Utah are in U.S. Forest Service Intermountain Region 4. The regional office is in the Federal Building, 324 5th Street, Ogden, UT 84401. The Utah forests contain 9.1 million acres and 13 wildernesses.

Ashley National Forest

SIZE AND LOCATION: 1,384,132 acres in northeastern Utah and southwestern Wyoming. Major access routes are U.S. Highways 6, 50, 60, and 191 and State Routes 32, 43, 44, 87, 121, and 208. District Ranger Stations: Duchesne, Roosevelt, Vernal; Flaming Gorge National Recreation Area, Manila and Dutch John. Forest Supervisor's Office: 355 N. Vernal Avenue, Vernal, UT 84078, www.fs.fed.us/r4/ashley.

SPECIAL FACILITIES: Visitor center; boat ramps; swimming beaches; lodges.

SPECIAL ATTRACTIONS: Flaming Gorge National Recreation Area; Flaming Gorge–Uintas National Scenic Byway (U.S. Highway 191); Indian Canyon Scenic Byway (U.S. Highway 191); Red Cloud–Dry Fork Loop Scenic Backway; Wildlife Through the Ages.

WILDERNESS AREAS: High Uintas (356,705 acres, partly in the Wasatch-Cache National Forest).

Most visitors to the Ashley National Forest will head straight for the incredibly scenic Flaming Gorge National Recreation Area (pl. 48). The wildly colorful canyons were seen in 1869 and 1872 by explorer Major John Wesley Powell who gave the canyons their descriptive name. The National Recreation Area extends into Wyoming and is administered entirely by the Forest Service. You may reach the National Recreation Area from Green River and Rock Springs, Wyoming, at the northern end, or from Vernal, Utah, at the southern end.

U.S. Highway 191 and State Route 44 make up the Flaming Gorge–Uintas National Scenic Byway. U.S. Highway 191 connects Vernal with the Recreation Area, a distance of 43 miles. The route is not only scenic, but very educational. The theme for the byway is Wildlife Through the Ages, and there are pull-outs that describe the edges of exposed rock layers that you pass. The exposed rocks get progressively older as you approach Flaming Gorge. For example, about three miles north of Vernal are the youngest exposed clays, which settled to the bottom of a sea that covered the area. On the other hand, near Flaming Gorge are layers that date back 1 billion years, with no signs of primitive life. In between are other interesting formations. One formation along the way has yielded fossilized bones of dinosaurs and crocodiles, and at another formation, the layers have dinosaur footprints.

Begin your exploration of the Flaming Gorge National Recreation Area at the Flaming Gorge Dam Visitor Center near the dam for the reservoir. State Route 44 heads west at the entrance to the National Recreation Area; the

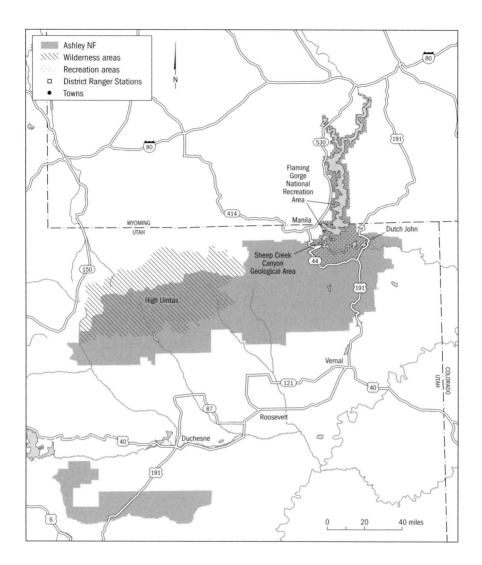

road north proceeds to the visitor center. Beyond the dam, the reservoir has filled several narrow canyons to the west before broadening into a wide reservoir to the north. Below the dam, the Green River is a popular fishing and rafting stream, with one stretch recommended for beginners. Two lodges are in the National Recreation Area, as well as several roads, boat ramps, and numerous campgrounds and hiking trails.

After you have completed your visit to Flaming Gorge, proceed west on State Route 44 to the Sheep Creek Gap Campground. Within a few miles the highway enters the spectacular Sheep Creek Canyon Geological Area. This 3,600-acre site has myriad rock formations such as Tower Rock. Many of the rock strata are bent drastically upward, and there are numerous fossil trilobites, corals, sponges, and other sea creatures. Sheep Creek Cave is along the west wall of the canyon.

One of the more fascinating driving tours is the Red Cloud–Dry Fork Loop Scenic Backway over mostly unpaved roads. This backway branches off U.S. Highway 191 about 14 miles north of Vernal and proceeds through scenic sandstone canyons, forests of lodgepole pine and aspen, and several open meadows. Among the points of interest are Iron Springs, Lily Pad Lake, Sims Peak, and the Dry Fork Overlook. The loop exits the national forest through Brownie Canyon.

Another scenic drive follows the Uinta River north from State Route 121 through Uinta Canyon to its end at a campground. From the campground is a four-mile trail to the High Uintas Wilderness. Once in the wilderness area, this trail winds and climbs past numerous springs and lakes and ultimately comes around the north side of Kings Peak, whose 13,526-foot summit is the highest in Utah. From State Route 87 is another route that terminates at a trailhead at the edge of the wilderness area. This road follows the Yellowstone River to the Swift Creek Campground.

Moon Lake is a large natural lake that touches the southern edge of the High Uintas Wilderness. State Route 87 north from Duchesne goes to the lake, where there is a nice resort. A trail follows the western shore of Moon Lake and then proceeds into the wilderness.

Twenty miles west of Duchesne, a dirt and gravel road heads south off of U.S. Highway 40 and parallels Red Creek to the Strawberry Pinnacles. At the Pinnacles, take the fork to the west that passes through Timber Canyon. Eventually this back road turns abruptly to the north and comes to colorful Strawberry Peak where there are panoramic views.

U.S. Highway 191, between Duchesne and Price, is the Indian Canyon State Scenic Byway, which brings you face to face with colorful rock formations. That part of the scenic byway that passes through the Ashley National Forest goes for miles through Left Fork Indian Canyon. The Avintaguin Campground is on the southern boundary of the national forest. Instead of continuing on U.S. Highway 191 to Price, you may take the rough back road west from the campground. For 20 miles this road across Reservation Ridge does not stray from the southern boundary of the national forest. The road eventually exits onto U.S. Highways 6 and 50.

Dixie National Forest

SIZE AND LOCATION: Approximately 2 million acres in southwestern Utah on the divide between the Great Basin and the Colorado River. Major access routes are Interstate 15, U.S. Highway 89, and State Routes 12, 14, 18, 20, 22, 56, 62, and 123. District Ranger Stations: Cedar City, Escalante, St. George, Panguitch, Teasdale. Forest Supervisor's Office: 1789 North Wedgewood Lane, Cedar City, UT 84720, www.fs.fed.us/r4/dixie_nf.

SPECIAL FACILITIES: Winter sports areas; boat ramps; swimming beaches; visitor center.

SPECIAL ATTRACTIONS: Markagunt National Scenic Byway (State Route 12); Brian Head–Panguitch Lake National Scenic Byway (State Route 143); Highway 12 National Scenic Byway; Red Canyon Recreation Area.

WILDERNESS AREAS: Pine Valley Mountains (50,232 acres); Ashdown Gorge (7,043 acres); Box–Death Hollow (25,751 acres).

The hot climate of the Dixie National Forest around St. George reminded early Mormon settlers of the Deep South they had left behind, and the region has been referred to as Dixie ever since. It can also be called Color Country because some of the most vivid limestone, sandstone, and siltstone rocks are now intensely red or pink. By contrast, there are also white cliffs. Combined with deep green coniferous forests, Color Country is appropriate. Although these colorful and often fanciful rock formations are well known for nearby national parks such as Zion Canyon and Bryce Canyon and for national monuments such as Cedar Breaks, Capitol Reef, and Grand Staircase–Escalante, they are just as intensely colored and fantastically shaped in the Dixie National Forest.

Much of the Dixie is high plateaus that rise up from surrounding sagebrush deserts. The highest elevation forests usually consist of Engelmann spruce and subalpine fir, with ancient bristlecone pines in a few places. Farther down the plateaus are extensive forests of ponderosa pine. Various oaks and quaking aspen may be interspersed. Still lower are piñon pine–juniper woodlands. Some areas such as the summit of Brian Head Peak are above timberline.

The Dixie National Forest comprises five districts. West of Interstate Highway 15 and north of St. George is the Pine Valley Ranger District. The Pine Valley Mountains dominate this area, and 50,000 acres of this mountainous region are in a wilderness area. Signal Peak, at 10,365 feet, has beautiful stands of Engelmann spruce on its upper slopes. Summit Trail follows

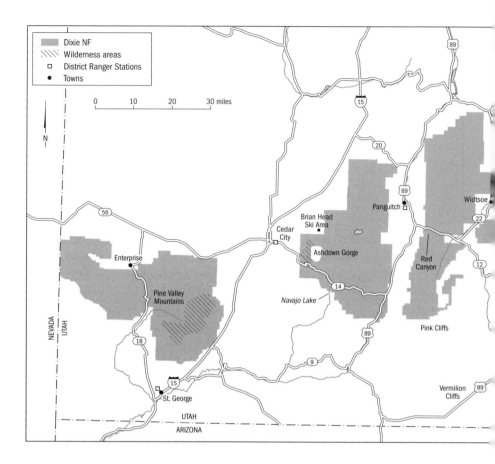

the full 35-mile length of the Pine Valley Mountains, although the most popular trail is the six-mile Whipple Trail from the Pine Valley Campground to scenic Whipple Valley.

From the community of Enterprise at the northern edge of the district, Shoal Creek Road travels for six miles outside the national forest to the Hebron Historical Site. The road south from here along Little Pine Creek brings you in four miles to the fascinating Honeycomb Rocks in the national forest. West from the historic site are some outstanding white rocks just before the road leaves the national forest.

The Cedar City Ranger District lies east of Cedar City and includes the incredible Markagunt Plateau. The Markagunt National Scenic Byway crosses the southern part of the plateau. The byway goes through scenic Cedar Canyon before reaching the national forest. State Route 143 turns off the scenic byway north to colorful Cedar Breaks National Monument, administered by the National Park Service. Between the western boundary of the national monument and Cedar City lies the Ashdown Gorge Wilderness. The color-

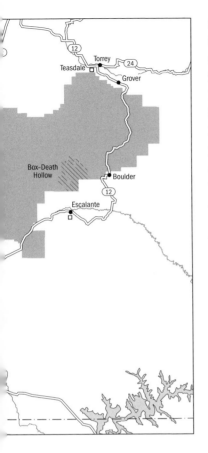

ful red and yellow and purple rock formations in Cedar Breaks are also to be seen in the wilderness area. The color is due to manganese and iron oxides in the rocks. Continue driving through Cedar Breaks. Just beyond the monument's northern boundary is the road to the summit of Brian Head. At 11,307 feet, this is one of the highest points in the Dixie National Forest. The plant communities that are encountered all the way to the top are worth studying. The Brian Head Ski Area is adjacent. Three miles west of the ski area is one of the finest stands of ancient bristlecone pines in the forest.

Back on the scenic byway, the highway comes along the northern shore of deep blue Navajo Lake. This is part of the Markagunt Plateau that was covered by lava from a volcanic explosion eleven million years ago. Numerous colorful wildflowers are in the meadows adjacent to lava fields. The natural outlet of Navajo Lake was blocked during volcanic activity. The water in the lake drains out in two lava tubes. Water emerges in one of the tubes from the pink cliffs on the south side of the lake in the unforgettable Cascade Falls. Water in the other tube appears four miles east on the scenic byway in Duck Springs and then enters Duck Creek Pond. The pond eventually drains into the Sevier River, which, in turn, drains into Sevier Lake. A visitor center is at Duck Creek Pond. From Navajo Lake, the three-mile Virgin River Rim Trail leads to the spectacular Cascade Falls, or you may drive to within a half-mile of the falls on a forest road and hike the short trail to the falls.

About five miles past Duck Creek Pond is a narrow gravel road to Strawberry Point. The point is one of the most spectacular places in any national forest. From the overlook, which is at an elevation of 9,016 feet, the valley below is lined with pink and red cliffs. The scenic byway ends at Long Junction.

Between Brian Head and Panguitch Lake is another scenic byway that is also State Route 143. This 55-mile route goes along the northern end of the Markagunt Plateau. Six miles east of the Brian Head summit road is a trailhead for the six-mile Hancock Peak Trail. In two miles, after coming to Hancock Peak, the trail passes between the Red Desert to the east and the Lava Desert to the west. A side road to Mammoth Springs is through a forest of ponderosa pine with aspens here and there. Meadows are plentiful. After the scenic byway turns abruptly north, there is a forest road to Buck Springs Knoll with fascinating rocks and wildflowers.

Panguitch Lake is nestled in a basin that was not covered by the lava flow. Lodges and resorts are in the area. From Panguitch Lake, the scenic byway descends into a piñon–juniper woodland and ends in Panguitch.

If you choose to drive north from Brian Head toward Parowan, there is a short drive to the breathtaking Vermilion Cliffs (pl. 49).

Five miles east of Panguitch is the western edge of the Powell Ranger District. As soon as State Route 12 reaches the national forest, you are in the unbelievable Red Canyon. It is impossible to describe the beauty in this part of the national forest. The East Fork of the Sevier River runs through the Paunsaugunt Plateau. At the south end of the plateau are the Pink Cliffs. From the top of 9,375-foot Pink Cliff, reached by a half-mile trail, is a marvelous panoramic view. Bryce Canyon National Park is in the Pink Cliffs just to the east. Along the road to the Pink Cliffs is the Tropic Reservoir.

Johnson Valley and State Route 22 separate the Powell Ranger District from the Escalante and Teasdale Ranger Districts. The Escalante and Teasdale are contiguous and contain the Escalante Mountains, the Aquarius Plateau, and Boulder Mountain.

The Clay Creek Forest Road exit off of State Route 22 crosses the southern end of the Escalante Mountains. Pine Lake and its adjacent campground are lovely places. East of the campground, the forest road passes through Pine Canyon and climbs to the Table Cliff Plateau. At the southern end of the plateau is Powell Point. From this 10,188-foot viewpoint, all of southern Utah spreads out before you. On a clear day you can see five states. Bristlecone pines cling to the point.

Another forest road off of State Route 22 from Widtsoe goes through Escalante Canyon, climbs over the crest of the Escalante Mountains, and descends through Main Canyon before eventually arriving at the town of Escalante. From Main Canyon is a side road north to the Barker Reservoir area.

An exciting drive north from Escalante begins along Pine Creek. At the first major intersection, the left fork goes to Posey Lake and crosses the Aquarius Plateau. The right fork takes you on a hairy-scary road that winds down to Hell's Backbone Bridge. The bridge was constructed in 1933 to cross the narrow gap between Sand Creek and Salt Creek. From the bridge are spectacular views of the narrow ridges on both sides and of the Box–Death Hollow Wilderness to the south. The wilderness is in slick rock country where the slick Navajo sandstone rocks are a gray orange color. Pine Creek forms a deep canyon known as the Box.

From the bridge, it is a long, narrow, winding road through magnificent scenery to the settlement of Boulder.

State Highway 12 is paved from Boulder to Grover. It stays along the east side of Boulder Mountain. Trails into this mountainous area lead from the highway. From Lookout Point, there is a fine view of the Circle Cliffs to the east as well as the multicolored desert.

Brian Head

In southwestern Utah, a series of north-trending river valleys divide the terrain into seven distinct plateaus. Brilliant red and pink limestone is exposed in the valleys, and the plateaus are capped by lava from volcanoes to the north and east that erupted between 3 and 4 million years ago (the present Tushar Mountains are what remain of these volcanoes). Among the highest features of the plateaus is Brian Head, in Dixie National Forest, whose 11,307-foot summit in the Markagunt Plateau provides a breathtaking view in all directions across southern Utah and into adjacent Arizona.

Five miles south of Brian Head, State Highway 143 skirts the rim of Cedar Breaks, a colorful exposure of Claron limestone. At one time managed as part of the Dixie National Forest, this horseshoe-shaped gorge is now a national monument under the jurisdiction of the U.S. Park Service. The road here runs along at about 10,400 feet, which happens to be the treeline on the Markagunt Plateau. Trails leading down from the road into Cedar Breaks pass through subalpine coniferous forests containing bristlecone pine, subalpine fir, and Engelmann spruce.

One and a half miles north of Cedar Breaks, a forest road branches off State Route 143 and begins a circular ascent up Brian Head. The last well-formed trees are soon left behind, giving way to three overlapping plant communities. At the lower elevations, usually between 10,600 and 10,900 feet, are mountain meadows, whereas above is the bleak alpine tundra. Occasionally encountered in somewhat protected areas, usually near 10,800 feet, are small clusters of dwarfed and gnarled subalpine firs and Engelmann spruces.

Prominent among the volcanic rocks on Brian Head, according to geologist Richard Kennedy, is rhyolite, a vary acidic material that consists of welded tuff and ash. Flows of basalt also lapped up onto the mountains to form thick deposits. The mountain meadows exist where enough soil has built up to cover the rock thinly. Tufted hair grass and purple reed grass, with their dense growth of leaves and their extensive underground stem systems, form continuous mats in which other plants may find a foothold. Dwarf scarlet paintbrush, two bright beardtongues, three kinds of white daisy fleabanes, and the two-foot-tall blue larkspur bloom during the summer, heralded a week or so earlier by the rosy-flowered pussy-toes. The blue thick-stem aster closes the meadows' flowering season a few days before the first autumnal snow.

Above 10,900 feet and continuing to the summit is the treeless tundra, where the surface is strewn with countless rock fragments. The churning effect of frost action year after year heaves rocks out of the ground and grinds them to pebble size. Close inspection reveals that most of the rocks are etched by lichens of various colors. A number of flowering plants grow in the crevices and among the rock fragments, hugging the ground in this zone of constant strong winds.

Small grasses and sedges are particularly adapted to high altitude because their extremely narrow leaves offer little surface to the wind and thus resist shredding. The plants are anchored by extensive root systems that spread out beneath the surface layer of rocks. If poor conditions prevail during the year so that the grasses and sedges are unable to flower and form seeds, these plants may still expand through branching of the underground stems. Common grasses on Brian Head, according to botanist James Bowns, who has inventoried the area, are Patterson's blue grass, alpine fescue, alpine bent grass, and mountain timothy. Also present is oniongrass, which takes its name from the onionlike bulb that forms just below the surface of the soil.

Some plants are able to survive at the summit because they grow in a small, mounded form that tends to deflect the strong wind, with their leaves clustered toward the center of the plant, where they gain a little extra warmth. Among these is the cushion phlox, whose flowers are pale lilac or pink. Other dwarf plants find shelter from the wind amid the pebbles scattered on the ground. These include a mountain dandelion, a low-growing goldenrod, the handsome Parry primrose, rock jasmine, woolly cinquefoil, Drummond's catchfly, and a creamy white windflower, or anemone, with deeply divided leaves. Although sometimes no more than two inches high, these plants are anchored by extensive underground stem systems or by deep taproots.

Because snow persists on Brian Head from mid-September through June, the vegetation usually has less than three months to reproduce and manufacture food stores for the next growing season. Many plants burst into flower as soon as the snow melts and the first favorable warm period occurs. This rapid blooming is possible because the flower buds are formed during the previous year and food reserves are present in underground roots and stems.

The dwarf subalpine firs and Engelmann spruces, which form a third type of plant community on Brian Head, grow only a tiny amount each year because of the extremely harsh environment. Although seldom more than three or four feet tall, they may be a few hundred years old. The biting wind usually shears off the branches on the windward side of the trees, giving them a flaglike appearance. In the moist, shaded niche provided beneath the small clusters of flag trees may be found an occasional colorful wildflower, such as the tall bluebell, the yellow, daisylike Uinta groundsel, and the delicate blue Jacob's ladder. Elkweed, a green-flowered member of the lily family, may actually grow taller than the flag trees that offer it initial protection.

Pink Cliffs

The rough mountain terrain in the southwestern corner of Utah, a few miles east of Cedar City and Saint George, is called the Land of Color. Visitors flock to the region's national parks, such as Zion, to see the brilliant exposures of the Wasatch Formation, a mass of limestone deposited during Eocene times, some 40 million years ago. The park lands surround the lesser known Dixie National Forest, whose precipitous Pink Cliffs, tinted by small quantities of iron and manganese oxides, are among the most spectacular parts of the formation.

I got my first glimpse of the Pink Cliffs one summer morning as I made my way on back roads northeast of Zion National Park. Following a sign that pointed down a gravel road to Cascade Falls, I soon came to a scenic overlook. Below me, at an elevation of 8,000 feet, flowed the North Fork Virgin River, whose erosive action had shaped the cliffs. Perforated by honeycomb depressions lined with calcite crystals, the iridescent cliffs towered 1,700 feet. From the overlook, a narrow trail led to the Cascade Falls, which emerge midway down the cliffs and tumble into the river below.

Plant life is sparse on the exposed cliffs because of the scarcity of water, the absence of all but a few soil microorganisms, extremely poor aeration, and extensive fluctuation of soil temperatures. The most picturesque plants are the bristlecone pines. The shrubby mountain mahogany, whose feathery fruits are conspicuous in autumn, is a frequent associate of the bristlecone. Although the rocky soil is often bare of vegetation, occasional flashes of color are provided by the dwarf blue columbine and the daisylike goldenweed.

The Pink Cliffs form the south wall of the Markagunt Plateau, a 25-mile-wide block that was uplifted 12 million years ago through violent shifting of the earth. The western edge of the block was elevated nearly 3,000 feet higher than the eastern edge, so the plateau tilts sharply. After the uplift, perhaps as recently as 1,000 years ago, volcanic eruptions covered much of the plateau with lava. Scanning the horizon from the Cascade Falls overlook, you can see a few scattered cinder cones. Lava flows define much of the topography, but where the soft, sedimentary rocks of the Wasatch Formation are exposed, streams have modeled the surface into a series of long ridges and broad, flat-floored valleys.

Navajo Lake, 714 acres of clear blue water, is the focal point of the plateau. It owes its existence to volcanic eruptions that dammed up broad Duck Creek Valley. The lake is a geological phenomenon, for it has no surface outlet nor is it filled from streams on the surface of the plateau. Instead, rainfall, snowmelt, and several small perennial springs supply its water. Navajo Sinks, a series of sinkholes formed at the eastern end of the lake by hundreds of years of the water's dissolving action on the limestone, provides the lake's only outlet.

Through a series of experiments, in which they added fluorescent dye to the lake, geologists M. T. Wilson and H. E. Thomas were able to show that water flowed from Navajo Lake through the series of sinkholes and into an interconnected underground reservoir and system of channels. About two-fifths of the excess water from Navajo Lake meanders through these underground channels until it bursts through an opening in the Pink Cliffs as Cascade Falls. The remainder appears about four miles to the east at Duck Creek Springs. From these springs, the water flows 2.5 miles along Duck Creek and Duck Lake before disappearing into Duck Creek Sinks, eventually emptying into the Sevier River to the east.

The features of the Markagunt Plateau illustrate the concept of biological life zones, developed in 1894 by Clinton Hart Merriam. An American biologist who had served as naturalist on the Hayden surveys in Utah, Idaho, Montana, and Wyoming, Merriam divided the biological communities of North America into east-west belts. He distinguished the different belts by their plant and animal composition, which, he proclaimed, was controlled by temperature. Thus Merriam's Boreal Life Zone is found in the far north, whereas the Sonoran Zone occupies the semiarid southwest. Merriam also maintained that in mountainous regions, such as southwestern Utah, zones similar to these can be found at differing elevations—the summits of the highest mountains are Boreal, their bases are Sonoran. A very general rule of thumb is that every increment of 1,000 feet above sea level is equivalent to 600 miles in a south-north direction. By ascending a mountain you encounter life zones in the same sequence that you observe driving hundreds of miles from south to north.

Between approximately 7,500 and 8,500 feet on the Markagunt Plateau is the Transition Life Zone. Ponderosa pine is the dominant species, usually occupying gentle slopes. These trees allow enough sunlight to reach the forest floor to encourage growth of manzanita, bitterbrush (fig. 24), and mountain ceanothus. Hayle Buchanan, a botanist at Weber State University in Ogden,

Utah, who has spent more than two decades studying the plants in Utah's Land of Color, has noted that occasional depressions in the Transition Life Zone are home to a community dominated by sagebrush and a wide diversity of grasses and wildflowers. Buchanan noted that when young ponderosa pines and Rocky Mountain junipers from nearby forests attempted to invade these treeless areas, they were unable to survive because cold air drainage from higher mountains caused severe minimum winter temperatures. Daily temperature fluctuations were a further obstacle to woody plant growth.

The Canadian Life Zone occupies forested slopes from 8,500 to 9,500 feet. Ponderosa pine and Rocky Mountain juniper

Figure 24.
Bitterbrush.

grow where the slopes are more exposed and therefore drier, but on moist slopes, white fir and Douglas fir are the dominant species. Because of the intense shade, few wildflowers are able to grow beneath these trees. Above 9,500 feet, along the high road between Navajo Lake and Duck Creek Springs, treeless meadows have developed over old lava flows. These meadows, which are assigned to the Hudsonian Life Zone, are covered by snow for as much as eight months of the year, so that the wildflowers and grasses that live there must complete their life cycle in a very short time. Penstemons, paintbrushes, orange sneezeweeds, long-plumed avens, goldenrods, and asters are vivid from late June to mid-August.

Although Merriam's life zone concept was widely accepted at one time, it has fallen into disfavor, partly because the various zones cannot be recognized across the continent on the basis of uniform plant and animal communities. Nevertheless, biologists continue to use Merriam's life zone designations in those areas where they seem to apply.

Red Canyon

Most tourists driving between Utah's Zion and Bryce Canyon National Parks pass through the heart of Red Canyon in the Dixie National Forest. No one can miss the brilliant rocks that make up Red Canyon, but most people, hell bent on getting to either Zion or Bryce, will pause along the road near a short tunnel only long enough to snap a photo from the car window.

Red Canyon lies at the upper end of the Paunsaugunt Plateau, one of the most colorful areas in the United States. The vivid pink and scarlet rocks of Red Canyon, the same type seen at Bryce Canyon, are primarily limestone, a part of the Wasatch Formation deposited in a freshwater lake during the early Tertiary period some 60 million years ago. The Wasatch is the uppermost—and therefore youngest—rock layer found in the Colorado Plateau region, a geologic uplift that covers more than 100,000 square miles. Unweathered Wasatch limestone may be pale pink, red, gray, or even white, but according to geologist Herbert Gregory, weathering produces a strong pink color, which is due to the oxidation of iron and manganese.

The top of the plateau is dominated by coniferous forests and sagebrush and rimmed by spectacular cliffs, called breaks by the local residents. The red Sunset Cliffs form the western border of the canyon, and on the southeastern edge are the contrasting, precipitous White Cliffs.

Much of the dry, exposed red limestone has a sparse cover of vegetation, with a scattering of shrubby plants. Fragments of coniferous forests have developed only in a few, more protected sites. Despite this apparent paucity of vegetation, numerous plants lie scattered across the gravelly limestone. Almost a dozen of them are endemic, known only from Red Canyon and vicinity and found nowhere else in the world. Endemism has always fascinated naturalists. (Some areas, like Hawaii, have more than their share of endemic organisms; others, like my home state of Illinois, have no species of flowering plant to call their own.)

Although botanists are unable to explain precisely why Red Canyon has an unusually high number of endemic plants, it seems safe to say that the answer has to do with the red limestone, as it is only on this particular geological formation that they are found. These plants have formed some sort of link with the rock that permits them to survive and reproduce where many other herbs cannot. These endemics are so well adapted to the locality that they cannot survive in areas where no red limestone is found.

The 10 endemic wildflowers found in Red Canyon are all dwarfs, three to 10 inches tall, perhaps a response to the arid conditions that prevail there during the growing season. For four of these endemics, Red Canyon is the type locality, that is, the first place in the world where each was discovered. These are yellowish cryptanth, a dwarf perennial belonging to the borage family; breaks bladderpod, a miniature yellow-flowered mustard; Jones' locoweed, a three-inch-tall, pink-flowered member of the pea family; and Red Canyon beardstongue, a showy species that stands only six inches tall but bears blue flowers nearly two inches long.

The other six endemics are equally rare, being confined to the colorful limestone of the Wasatch Formation. They are spring parsley, breaks whitlowgrass, Widtsoe buckwheat, daggerleaf phlox, plateau catchfly, and least townsendia, a three-inch-tall plant with daisylike flowering heads two to three inches in diameter.

Although Red Canyon is largely devoid of trees, small coniferous forests of ponderosa pine and Douglas fir exist in areas where more moisture is available. In many regions of the West these species often reach more than 100 feet, but because Red Canyon is so dry they rarely attain a height of 60 feet. Joel Tuhy, public lands coordinator for The Nature Conservancy in Utah, has studied the region extensively and noted that on the cooler, steeper, north-facing slopes, Douglas fir is dominant, whereas on the top of the plateau, ponderosa pine predominates. In both areas, green-leaved manzanita is the prevailing shrub.

Tuhy has discovered a very small but distinct plant community that has developed where groundwater is forced up near the surface by a shallow zone of white sandstone. Plants that require more moisture than is usually available in other parts of Red Canyon have found a home here. Among them are the shrubby cinquefoil (a species also found in the northeastern United States) and the death camas, an elegant wildflower belonging to the lily family.

Although most of Red Canyon is carved out of vivid red limestone, Black Mountain on the western edge provides a stark contrast with its steep slopes of dark volcanic basalt. The northern and eastern slopes support a forest of ponderosa pine and Douglas fir, but piñon pine and Rocky Mountain juniper are also important species. These trees tend to grow taller and more densely than those on the nearby limestone formation, probably because more moisture is available and because organic nutrients do not leach away from basalt as readily as from limestone. Beneath the overstory of trees is a shrub layer consisting primarily of curl-leaf manzanita and Martin's ceanothus.

The southeastern slope of Black Mountain resembles the open woods of piñon pine and Utah juniper that are found in much of the intermountain region. Beneath them is a rocky soil, suitable for only a very few herbaceous species. On the upper sides of the mountain, there are a few nearly barren slopes of basaltic boulders. Specimens of wax currant and red raspberry can be found here, but only in crevices between the boulders, where soil has accumulated over the years.

Recently, 460 acres of Red Canyon, lying to the north of Utah State Highway 12, have been designated a Research Natural Area by the U.S. Forest Service, a status that should help preserve the canyon's special features.

Fishlake National Forest

SIZE AND LOCATION: Approximately 1.5 million acres in central Utah, including the Tushar Mountain Range. Major access routes are Interstates 15 and 70, U.S. Highways 50 and 89, and State Routes 10, 24, 25, 29, 30, 62, 72, 125, 137, and 153. District Ranger Stations: Beaver, Fillmore, Loa, Richfield. Forest Supervisor's Office: 115 East 900 North, Richfield, UT 84701, www.fs.fed.us/r4/fishlake.

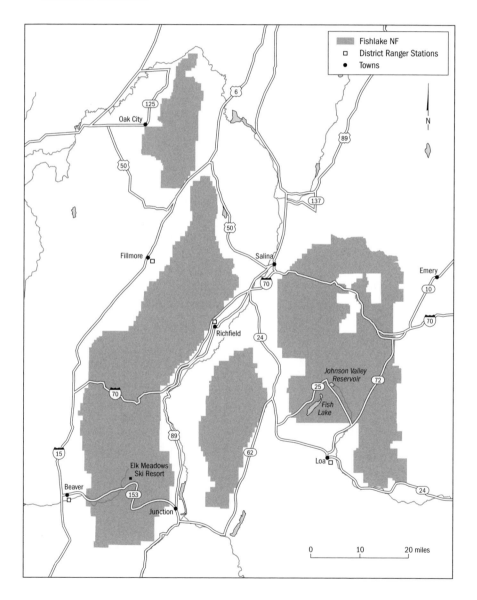

SPECIAL FACILITIES: Boat ramps; swimming beaches; winter sports areas; Paiute ATV Trail.

SPECIAL ATTRACTIONS: Beaver Canyon National Scenic Byway; Fishlake National Scenic Byway (State Route 25); scenic backways.

Fish Lake (pl. 50) is the primary attraction in the Fishlake National Forest for many people. This beautiful 2,600-acre lake has the outline of a giant fish. Personally, I think it looks more like a whale. The depression that the lake has filled was caused by geologic faulting. When glacial deposits dammed the northern end of the depression, Fish Lake was formed. The lake is at an elevation of 8,500 feet and the weather is always cool. Most of the crystal-clear water is around 90 feet deep or more, with a maximum depth of 120 feet. Fishing is great for Mackinaw and rainbow trout. Pelican Point protrudes into the water, causing a notch near the northern end of the lake that helps differentiate the tail from the remainder of the fish. A very scenic trail leads from the Pelican Point Overlook through Pelican Canyon. Doctor Creek Trail also climbs into the Fish Lake Hightop Plateau, an area covered by lava. Boat ramps and resorts are around the lake. Lake Shore Trail makes a 15-mile loop around the lake and climbs Mytoge Mountain. Campfire programs are presented at the amphitheater from June to Labor Day.

The Fishlake National Scenic Byway is State Route 25 and it begins at its junction with State Route 24 west of the town of Loa. The scenic byway goes from sagebrush to high-mountain aspen forests. You will probably see deer and elk and, if you are lucky, a moose or two. Before reaching Fish Lake, the rather primitive road to Mytoge Mountain has some striking views of the lake. At the north end of the lake is a huge marsh where you may see an abundance of waterfowl. If you find Fish Lake too crowded, you might drive a few miles farther north to Johnson Valley Reservoir. The highway beyond Johnson Valley Reservoir follows Sevenmile Creek. Mt. Terrell is off to the east. The road swings around through beautiful scenery to the Gooseberry Guard Station. A little past the guard station is an assemblage of strange-shaped formations known as The Rocks.

A second major feature of the Fishlake National Forest is the Tushar Mountain Range with Delano Peak the highest at 12,168 feet in elevation. The Beaver Canyon National Scenic Byway goes into these mountains. Beginning at the town of Beaver, the 17-mile byway, which is State Route 153, follows the Beaver River through a marvelous rocky canyon whose steep sides rise abruptly above the road. The lower elevation is desertlike with sagebrush, rabbitbrush, bitterbrush, and Mormon tea (fig. 25) common. Growing precariously on the slopes are piñon pine and Utah juniper. The river is lined by black cottonwood and an occasional willow. A forest road from Little Reservoir Campground winds past Kents Lake and Upper Kents Lake, circles around Anderson Meadow and Anderson Reservoir where there is a campground, crosses Dry Hollow, straightens out across Betenson Flat, and rejoins State Route 153 five miles west of Junction.

Just beyond the Little Reservoir Campground, the byway climbs past Mahogany Cove via a series of switchbacks. You can easily tell that you are climbing higher and higher as you pass forests of subalpine fir, white fir, and Engelmann spruce. Hiking trails go into the Tushar Mountains. A good trail is at Puffer Lake. The highway then

Figure 25. Mormon tea.

heads due south and eventually ends at the Elk Meadows Ski and Summer Resort area. The Skyline National Recreation Trail begins near the Big Flat Guard Station. Much of it is above timberline.

Three and one-half miles east of Oak City is the Oak Creek Picnic Area. A half-mile trail south brings you to Partridge Mountain Research Natural Area. Because this area has had little grazing in the past, it exhibits several typically undisturbed plant communities. A few miles north of the Partridge Mountain Research Natural Area are 1,000-year-old petroglyphs in Fool Creek Canyon.

Three scenic backways are in the Fishlake National Forest. The Kimberly–Big John Scenic Backway climbs over the Tushar Mountains. The road begins at the town of Junction and turns onto State Route 153. After passing Elk Meadows, the scenic backway turns north on a dirt road to Big John Flat. As the backway goes over the Tushar Mountains, it passes many high mountain meadows that occur between fields of lava rocks. Above 11,000 feet are tundralike habitats.

The Cove Mountains Scenic Backway begins at Koosharem on State Route 62 and crosses Cove Mountain on a dirt road. Views into Sevier and Koosharem valleys are superb, particularly during the autumn. The historic Koosharem Guard Station was built in 1910. After 25 miles, the backway ends in Glenwood.

The Gooseberry–Fremont Scenic Backway is paved but is very steep and narrow in places. It begins two miles north of Fremont and heads north for 40 miles past mountain meadows and clear streams, skirting the north end of Johnson Valley Reservoir.

Bullion Canyon was the site of gold-mining activity beginning in 1865. A driving tour follows the canyon, with nine numbered stops. From Stop One you can see the old toll road 15 feet above you. Other stops permit you to see the old log buildings of the Witt Tate Mine and the Dalton Mill. At Stop Five is a short trail to the edge of a creek where an arrastra is well preserved. Next is the townsite of Bullion City and the burned-out remains of the Gibbs Cabin. The final stop is at Miner's Park where you may hike a quarter-mile trail past displays of mining equipment and a refurbished cabin.

The Paiute ATV Trail is the longest ATV trail in the country, an incredible 230-mile loop over three mountain ranges, through rocky canyons, and in desert country.

Partridge Mountain Research Natural Area

Partridge Mountain Research Natural Area is located in west-central Utah in the Fishlake National Forest, approximately 35 miles north of Fillmore and two miles from the community of Oak City. The top of Partridge Mountain is at an elevation of 7,985 feet, but drops precipitously on the south end to Rattlesnake Canyon and on the north end into Dry Creek Canyon. Oak Creek follows along the north side of Partridge Mountain, with the Oak Creek Picnic Area situated along the creek. From the picnic area, we followed the gravel road that parallels Oak Creek for 1.5 miles until South Walker Canyon came in from the south. We then hiked through South Walker Canyon until it came to a saddle that separates South Walker Canyon from Rattlesnake Canyon. Climbing the saddle to the west provides the best access to Partridge Mountain.

Most of Partridge Mountain is rugged and broken. At the northwest corner are precipitous limestone cliffs. All sides of the mountain are dissected by steep, narrow canyons. The significance of the natural area is in the diverse communities of trees and shrubs in an area so isolated that it has received little grazing pressure and is therefore undisturbed.

As we climbed up onto the saddle, we were in a dry, treeless community dominated by junegrass with low sagebrush, prickly pear, and cushion cactus interrupted throughout. Higher on the saddle, the grassy cover gradually is replaced by mountain mahogany and Gambel's oak. Low sagebrush is still prominent in this community. The area is so dry that Gambel's oaks are only two to four feet tall at maturity. Weetches Canyon drops away from the saddle to the north, with Gambel's oak and bigtooth maple growing above a shrubby layer of black elderberry and mountain snowberry. The Gambel's oaks in this canyon may reach a height of nearly 20 feet. Aspens grow along the intermittent stream in the bottom of Weetches Canyon.

As we continued along the saddle, where the soil appeared to be a little deeper, Letterman's needlegrass becomes the dominant species with Vasey bigsage interspersed. On top of the first ridge we encountered a piñon–juniper community with widely scattered trees of single-leaf piñon pine and Utah juniper. Low sagebrush dominated the understory.

Near the top of Partridge Mountain is a moderately dense stand of concolor fir and Rocky Mountain Douglas fir above rather thick groves of mountain snowberry. Other plants we observed in this community included alder-leaved serviceberry, wood rose, wax currant, snowbrush ceanothus, and green-leaved manzanita. Common wildflowers of summer are showy goldeneye, leafy polemonium, sweet anise, and waterleaf. Where there were more rocky streams across the forest, creeping barberry took on a dominant role beneath the concolor fir.

The steep limestone slopes at the northeast corner of the 1,200-acre natural area are home to bigtooth maple, mountain snowberry, scattered Gambel's oak, and Vasey bigsage, the latter in more xeric habitats. Here and there are specimens of the rather uncommon shrub known as rock spiraea.

At the western edge of the natural area, where the exposed rocks are mostly quartzite, the Gambel's oaks reach their tallest, some of them more than 30 feet tall. Sweet vetch is the common herbaceous plant on the quartzite.

Manti-LaSal National Forest

SIZE AND LOCATION: 1,238,419 acres in central and southeastern Utah and encompassing three distinct and separate districts. A small portion of the forest extends into western Colorado. Major access routes are Interstate 70, U.S. Highways 6, 89, and 191, and State Routes 10, 29, 31, 46, 95, 96, 128, 211, and 264. District Ranger Stations: Ferron, Moab, Monticello, Price, Ephraim. Forest Supervisor's Office: 599 W. Price River Drive, Price, UT 84501, www.fs .fed.us/r4/mantilasal.

SPECIAL FACILITIES: Winter sports areas; boat ramps; swimming beaches.

SPECIAL ATTRACTIONS: The Energy Loop: Huntington and Eccles Canyons National Scenic Byway; Skyline Drive (State Routes 31 and 264); Great Basin Experimental Range; LaSal Mountain Loop Road.

WILDERNESS AREAS: Dark Canyon (47,116 acres).

The Manti-LaSal National Forest is three forests in one. That part of the forest in central Utah is in the Wasatch Plateau. Until 1949, it was the separate Manti National Forest. In the southeastern corner of Utah, the national forests include the LaSal Mountains east of Moab and the Abajo Mountains west of Monticello. These two ranges made up the LaSal National Forest. In 1949, the Manti and the LaSal were united into one administrative unit.

Manti National Forest

The Manti Mountains is the local name for the Wasatch Plateau in central Utah, which rises nearly 2,000 feet above Castle Valley (pl. 51) to the east and Sanpete Valley to the west. The top of the plateau is not flat but consists of rolling ridges covered by forests and meadows. The Manti National Forest extends 60 miles north to south, wedged in between the Uinta National Forest to the north and Fishlake National Forest to the south. East to west most of it is between State Route 10 and U.S. Highway 89. The eastern side of the plateau has vertical cliffs. The western cliffs above Sanpete Valley are steep but usually not straight up and down.

The Energy Loop: Huntington and Eccles Canyons National Scenic Byway bisects the national forest, using State Route 31 as its major route. Beginning in Fairview, the scenic byway climbs along Cottonwood Creek through Fairview Canyon and onto the Wasatch Plateau. It is a great route to observe the nice scenery and to lead to fishing in several lakes and reservoirs. The route passes Fairview Lake and three reservoirs—Beaver Dam, Huntington, and Cleveland. For scenery, there are the lush forests along Huntington Creek with stands of Engelmann spruce, Douglas fir, and quaking aspen on the slopes above the creek. Where the Left Fork of the Huntington River joins the Huntington River, there is the four-mile Left Fork National Recreation Trail that goes through forests of subalpine fir and Engelmann spruce. Two miles south of this trail is a two-mile trail that proceeds up Tie Fork Canyon. In a few miles the scenic byway leaves the national forest on its final stretch to Huntington. The scenic byway heads east past numerous wildflower meadows. Back in the national forest, a side road drops down to Electric Lake, popular for fishing. When the byway reaches Huntington Creek, it abruptly turns north and ascends into fabulously scenic Huntington Canyon. After passing a very active coal-mining operation on a high ridge, the byway drops into Eccles Canyon. The scenic byway now heads west, past numerous wildflower meadows, before ending at the Huntington Canyon Scenic Byway near Fairview Lake.

A forest road from Mt. Pleasant through Sealey Canyon brings you to Scad Valley Creek and the interesting Scad Valley Botanical Area. Although

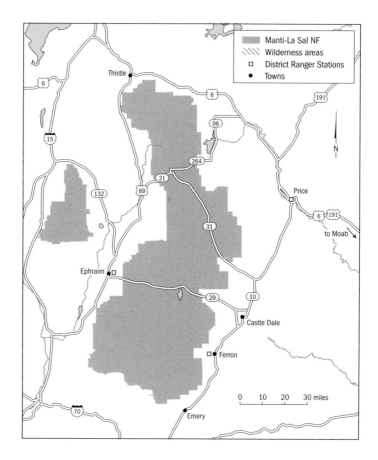

only 50 acres in size, this area near the junction of Scad Valley Creek and the Left Fork of the Huntington River is a wetland formed from the waters of Cold Water Spring. Here are several common wetland grasses, sedges, and marsh marigold, but rare species usually found in more alpine or Arctic sites are present as well. Some of these are spreading gentian, alpine meadow rue, and two sedges.

West of Castle Dale just as State Route 29 enters the national forest, Mahogany Point at the edge of the North Horn Mountains has a nice natural area. Trees you will encounter on the point include ponderosa pine, limber pine, Douglas fir, and Rocky Mountain juniper. The shrubs in the understory consist mostly of manzanita and snowberry. Treeless openings have black sagebrush and wild rye, with many plants of curl-leaf mountain mahogany. A rare daisy known as *Hymenoxys depressa* and a rare mustard called *Arabis demissa* var. *languida* also occur here.

In 1934, in the Black Dragon Canyon area of the North Horn Mountains,

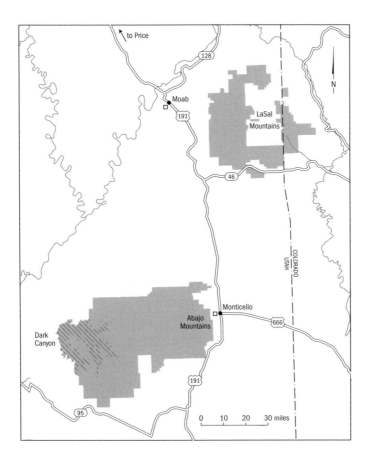

the fossilized remains of a dinosaur were discovered. Additional finds included specimens from five different orders of reptiles, one order of fish, and several large mammals.

State Route 29 from Ephraim is the road to the national forest's Great Basin Experimental Range in Ephraim Canyon. This 4,600-acre range was designed to study the cause of summer floods from water originating in the high-mountain watersheds. An auto tour booklet guides you on a range tour from Alpine Station to Major's Flat. The headquarters, where you can find out more about the range, is about halfway along the route.

South of the Experimental Range above Loggers Fork to the north and Lowry Creek to the south is the 40-acre Elks Knoll Research Natural Area.

From Mayfield is the Mayfield-Ferron Scenic Backway, which comes to the Grove of Great Aspens in Twelve-Mile Canyon after about 13 miles. At the time of its designation as a significant area in 1951, there were 20 giant aspens with diameters greater than three feet. The largest of these trees and a few others have now fallen, but it is still a grove worth visiting. This gravel scenic backway continues on for 37 more miles to Ferron, passing through stands of piñon pines, junipers, aspens, and high-mountain meadows.

Nelson Mountain, at 9,061 feet, is at the southern edge of the Wasatch Plateau five miles southwest of Ferron. This mountain has undisturbed plant communities that are worth the rather long hike to it. Pristine communities found here are piñon pine–juniper, with an understory of black sagebrush, mountain mahogany, and snowberry, curl-leaf mountain mahogany with western rye grass beneath, and Douglas fir–white fir forests with snowberry in the shrub layer. Dotted throughout the mountain slopes are quaking aspens and big mountain sagebrush.

If you want a full day of driving excitement, preferably with a four-wheel-drive vehicle, take the Skyline Drive. Because this gravel or dirt road is often difficult in places, it would be well to check with the Forest Service Supervisor's Office to check on the route. If you can take this road, you will be rewarded by superb scenery. Skyline Drive follows the entire length of the Manti National Forest. The northern end of the road begins at the Tucker Rest Area on U.S. Highways 6 and 50, 18 miles east of Thistle. On the way, it follows the left fork of Clear Creek and passes several areas already described. You can pick up part of the route from State Route 31 from Fairview, from the forest roads from Mt. Pleasant and Ephraim, up the Manti Canyon Road from Manti, and from the Grove of Giant Aspens Road from Mayfield.

LaSal National Forest

The LaSal National Forest has two distinct units in southeastern Utah. The LaSal division east of Moab contains the LaSal Mountains with peaks above 12,000 feet, deep canyons, and red rock formations similar to those in Arches and Canyonlands National Parks. The LaSal division west of Monticello includes the Abajo Mountains and much evidence of prehistoric civilizations.

Two routes through the LaSal Mountains will bring you close to several points of interest as well as trailheads for hiking in these rugged mountains. From Utah Route 128 west of Cisco, take the Castle Creek paved road south toward the LaSal Mountains. You will soon see the spirelike, colorful pinnacles of the Fisher Towers. Just after passing the townsite of Castle Valley, the road splits. For a slight diversion, take the left fork and you will come to one of the largest two-needle piñon pines in the world. It is about 35 feet tall with a circumference of 11.5 feet. A half-mile trail leads to the tree.

The right fork from Castle Valley is the main route to take. The road winds through Pinhook Valley where there are occasional views of Round Mountain and Castle Valley to the northwest and Harpole Mesa to the east. A rough, rocky, three-mile road to Miner's Basin leads to a large grassy meadow with the ruins of several mining cabins.

The main road leads through a picturesque aspen grove, then climbs to the top of a mesa where there are outstanding views westward toward red rock country around Moab.

Be sure to take the five-mile side trip up to Warner Lake. The road is rough and very narrow at times. Warner Lake is nestled among forests of aspens. The 11,612-foot Haystack Mountain has a perfect reflection in the lake (pl. 52). The mountain is snow capped the year round. You may take several hikes from the Warner Lake Campground. A two-mile trail north to Miner's Basin descends to the Schuman Gulch Meadows where there was a mining community in the 1800s. After crossing Schafer Creek, the trail climbs steeply through a dense forest to Miner's Basin. Another two-mile trail (Oowah Lake Trail) leads south to Oowah Lake. After crossing a meadow, the trail heads downhill, crosses a small stream, and then descends steeply through a scrub oak forest to Oowah Lake.

After rejoining the main road, the route becomes very narrow and rocky as it proceeds treacherously downward. At the foot of the ridge and the head of a scenic canyon, a side road from Mill Creek bridge leads to Oowah Lake. The main road then makes its way down Brumley Ridge to a paved road that returns to Moab. Brumley Ridge vegetation is mostly of the piñon pine–juniper type.

The LaSal Mountain Loop Road begins by climbing the same route up Brumley Ridge, but 3.5 miles after entering the national forest, the loop road goes east to Geyser Pass. This is a scenic, high-mountain road that is very rough in places. At Geyser Pass, you are in the heart of the LaSal Mountains, with 12,000-foot mountains to the north and to the south. On Mt. Peale, the highest of these peaks at 12,271 feet, there are forests of subalpine fir, rock glaciers, and bare-rock alpine areas with rare plants such as the Mancos fleabane. After leaving Geyser Pass, the road forks and leaves the national forest temporarily on poor roads. The left fork, which is not the loop road, eventually crosses Geyser Creek and follows along Pine Flat, crossing the Colorado state line just as the road reenters the LaSal National Forest. Buckeye Reservoir, more easily reached from Paradox, Colorado, is a popular area in the Colorado part of the national forest. The right fork is the loop road, and it leads back into the national forest past Blue Lake, eventually coming to the town of Moab on U.S. Highway 191.

West of Monticello is the other division of the LaSal National Forest. The major features are the Abajo Mountains and, farther west, the Dark Canyon Wilderness. Forest Road 105 from the center of Monticello heads west toward the Abajo Mountains. One mile west of Monticello, the Abajo Loop

Backway heads south and then curves back north to the summit of 11,360-foot Abajo Peak. If you continue due west on the road from Monticello, you can drive for miles into the heart of this national forest division. The forest road then crosses Cooley Pass, works its way across Dickson Ridge, then drops into Cooley Gulch. The picturesque Red Bluffs are to the west. From here on, the road goes through areas where the remains of a prehistoric civilization may be seen. It is illegal to remove or in anyway disturb these remnants. The road crosses Dry Wash, and there is a cliff dwelling one mile north by trail.

After turning abruptly north, the forest road comes to a large box canyon, surrounded by semidesert habitat, known as Cliff Dweller's Pasture. Cliff dwellings as well as a beautiful natural bridge and natural arch are here. Because the area has not been disturbed by grazing, the plant communities are pristine. The upper elevations support ponderosa pine forests with snowberry beneath. Drier areas either have scrub oak forests or piñon pine–juniper forests. Large yucca plants are often seen beneath the piñon pines.

West of the Cliff Dweller's Pasture, the road continues through exquisite scenery including The Causeway and the Chippean Rocks. Soon you are at the eastern edge of the Dark Canyon Wilderness. This wilderness has numerous Anasazi sites as well as natural arches and colorful sandstone cliffs. Anasazi pictographs and petroglyphs are common. In the canyon along the creek are large cottonwoods as well as small waterfalls and deep, clear pools.

The area contains phenomena known as hanging gardens where moist plant communities hang from the sides of cliffs. Two of the hanging gardens are in Woodenshoe Canyon and Dark Canyon, both in the wilderness area. Arch Canyon, however, is much more accessible. Turn north off of State Route 95, the highway to Natural Bridge National Monument, and drive into the national forest on the Elk Ridge Scenic Backway (Forest Road 088). After passing a landmark known as Bears Ears, the forest road begins a sweeping curve to the northeast along the spine of Elk Ridge. You will pass through stands of piñon pine, juniper, ponderosa pine, and aspens. From the road there is a trail into Arch Canyon where there not only is a hanging garden but a fine natural arch. It is a beautiful area.

Particularly along the forest road north of Arch Canyon that goes across Elk Ridge, keep on the lookout for the very rare Abert squirrel. This tassel-eared squirrel, a subspecies of the more common tassel-eared squirrel, is found only in the ponderosa pine forests of Elk Ridge and in the Dark Canyon Wilderness.

Brumley Ridge

Known locally as the P-J zone, piñon and juniper woodlands make up nearly one-third of the forested land in the Rocky Mountains, generally occupying the elevations between 5,000 and 7,500 feet. Falling comfortably within these margins is Brumley Ridge, a 6,500-foot foothill in Utah's Manti-LaSal National Forest. To the west of the ridge, the land dips down to desert grasslands, and aspen and dark ponderosa pine forests cloak higher terrain to the east, leading up to the snow-capped 12,000-plus-foot peaks of the LaSal Mountains.

The species of the two small conifers that dominate the P-J zone—juniper and piñon pine, whose oily seed, the piñon nut, was a staple food for prehistoric inhabitants of the Southwest—differ from place to place in the Rocky Mountains. Three types of piñon pines (true, single-leaf, and Mexican) and four kinds of junipers (Rocky Mountain, Utah, one-seed, and alligator) are known to exist in Utah. On Brumley Ridge, the true piñon pine and the Rocky Mountain juniper prevail.

The soil in the P-J zone is shallow and rocky; trees are scattered, and other vegetation is sparse. An occasional shrubby sagebrush, snakeweed, or Gambel's oak breaks up the monotony, and patches of grasses—mutton grass, galleta, needlegrass, Indian ricegrass, and blue grama—grow where moisture accumulates.

Several kinds of cactuses dot the terrain. Once, while enjoying a leisurely hike east of Brumley Ridge, I chanced upon an 18-inch-wide clump of bluish green, succulent cactuses, dozens of two- to three-inch-tall plants crowded together. The most notable feature of this clump was that not a single spine was evident on any of the stems. Having studied the endangered species of the United States for several years, I knew immediately that this was the spineless hedgehog cactus, one of the rarest plants in the country.

First discovered in the LaSal Mountains about 1895 by German botanist Joseph Anton Purpus, and also known in adjacent Colorado, the spineless hedgehog cactus has long puzzled botanists. Despite the plant's striking lack of spines, its flowers (the main feature that botanists examine for reliable distinctions) resemble those of the red claret hedgehog cactus, a common species that grows from California to Colorado to Texas and south to Mexico. The red claret hedgehog cactus is known to be a highly variable species (cactus authorities recognize eight or nine different varieties). Occasionally, too, a clump of spineless hedgehog cactus is found in which one or more stems bear a few spines, clouding the plant's only notable feature.

An important clue that two plants may differ genetically is that they grow in different places. As pointed out by cactus authority Lyman Benson, however, the spineless plants are scattered throughout populations of spined varieties in west-central Colorado and east-central Utah. And, although most varieties of the red claret hedgehog cactus grow in the open sun, whereas the spineless variety often grows in the shade of piñons and junipers, this too is not a clear-cut difference.

Although at first glance spiny and spineless hedgehog cactuses appear to occupy the same areas, recently a team of botanists has noted that, on closer examination, they may actually grow in different microhabitats. The Colorado Plateau is an area of sedimentary rocks, such as sandstones and shales, whose erosion exposes many layers. As it turns out, where there is a series of ledges, the spineless plants grow on level tops and spiny plants occur on side slopes. Because the two varieties occur in nearly pure populations, these botanists feel that they are adapted to different conditions and to a degree are genetically isolated.

Several studies still need to be done to establish the taxonomic status of the spineless hedgehog cactus. A possibility yet to be excluded is that it is an "ecological form" of a spiny cactus—a growth pattern the plant follows wherever it is subjected to certain environmental conditions. To resolve this, attempts must be made to transplant some spiny cactuses to where the spineless types flourish, and vice versa. If such transplants fail, the plants are more likely to be distinct varieties. Analysis of the contents of the cells of the two types could also be made to see if the two have any chemical differences.

The reason for knowing the exact status of the spineless hedgehog cactus revolves around the Endangered Species Act of 1973. In that year, Congress authorized the U.S. Fish and Wildlife Service to begin listing plants, as well as animals, that were on the verge of extinction. To date, more than 700 plants in the United States and its possessions, including many cactuses, have made the list. Prized by plant growers, some cactus species have been reduced to near extinction by poaching.

For a plant, being on the list means that it is to be protected wherever it is found on federal lands. In addition, a listed species may not be transported across state lines. To be protected, however, a plant must be a recognized species, subspecies, or variety. If a plant is only an ecological form, it has no recognition under the law. The continued existence of the spineless hedgehog cactus may well depend on the ability of botanists to quickly and accurately determine its taxonomic status.

Arch Canyon

East of Natural Bridge National Monument, in southeastern Utah, a narrow, dirt Forest Service road leads through Manti-LaSal National Forest, snaking over Maverick Point and between the two prominent Bears Ears peaks. Several overlooks here, in the Abajo Mountains, afford views of the vast Grand Gulch Plateau some 2,000 feet below. From this distance, the few defiles that dissect the plateau cannot be detected. In fact, hikers traversing that arid, sparsely vegetated terrain come upon the rims of the canyons without much warning.

One of the canyons that penetrate Grand Gulch Plateau is Arch Canyon, formed during thousands of years as a tributary of the San Juan River carved a 1,000-foot-deep trench. The canyon walls, often sheer in places, reveal colorful patches and layers of red, iron-bearing shales, red cherts, mudstones, siltstones, and blue-gray limestones in the predominantly white sandstone. Each sandstone stratum is made up of layers, which often slope at an angle to the horizontal—a phenomenon known as crossbedding. Vertical joints also intersect the horizontal strata.

Because shale weathers more quickly than sandstone, the canyon walls are grooved where the shale has eroded. The deeper grooves sometimes serve as animal trails, and some have been sites for ancient cliff houses. Elsewhere in the canyon, huge blocks of sandstone have become separated from the main cliffs and stand isolated in the gorge. Where the blocks are crossbedded with vertical joints, the loss of chunks and sheets of rock has left openings, or windows, which are as large as 100 feet wide and 140 feet high. Among these spectacular formations are Cathedral Arch, Keystone Arch, and the nearly all-white Angel Arch.

Most of the cliffs and the floor of Arch Canyon reflect the arid conditions of the region, and green vegetation is sparse except for an occasional willow or cottonwood where water accumulates in the old streambed. But here and there on the faces of the cliffs in Arch Canyon, and many of the other canyons along the tributaries of the Colorado and San Juan rivers, are wet alcoves that support wet-habitat flora. Many of these species cling precariously to the rock face, whereas others cascade down the cliffs. These pockets of vegetation are called hanging gardens, and they are unique to southeastern Utah and adjacent northeastern Arizona.

John Wesley Powell discovered these green zones in 1869 while exploring the canyons along the Colorado River several miles west of Arch Canyon. He noted that "sometimes the rocks are overhanging; in other curves narrow glens are found [where] oaks grow, and other rich vegetation is seen. We call these Oak Glens." Much of this area, designated Glen Canyon, is now covered by Lake Powell.

The first botanist to explore the region was Alice Eastwood, who traveled there in 1895 with a guide. Riding mules and accompanied by one pack animal, they traversed 150 miles in eight days, from Mancos, Colorado, to wilderness beyond Bluff, Utah, and back. On encountering the hanging gardens, Eastwood described them as "a boreal oasis in the middle of a Sonoran desert."

Botanist Stanley Welsh has studied the plants that grow in these hanging gardens and has suggested how these habitats are formed. As water percolates through softer rock layers, it eventually comes in contact with a more impervious layer of sandstone. Because the water cannot readily penetrate this layer, it spreads out laterally to the edge of the exposed cliff, where a dripline is formed. Eventually, a few small plants begin to grow in the dripline, sending out roots and root hairs into minuscule crevices.

The spreading water dissolves the material that cements the particles making up the sandstone. At the same time, the growth of plants on the rock surface speeds up deterioration of the rock through the production of mild acids. As the sandstone is weakened, it gradually falls off in thin layers. In addition, some of the clinging plants become too heavy for their precarious perches and fall, often pulling some of the sandstone with them. Eventually, large pieces of rock fall from the cliff, creating alcoves. A typical alcove will have a roof, a back wall that is often flat, and a sloping floor built up from sandy detritus and fallen plant materials.

The alcove roof impedes desiccating winds and blocks direct sunlight for much or all of each day. As a result, these gardens are shielded from the very conditions that limit the growth of vegetation in the surrounding arid lands. The species in hanging gardens vary in response to altitude and possibly other factors. The wet, protected back wall may harbor such showy species as the cave primrose, cardinal flower, alcove columbine, and Eastwood's monkey-flower, whereas bog orchids and other species may flourish on the sloping floor near the mouth of the alcove. Downslope from the mouth are zones of grasses, sedges, and rushes, followed by a community of shrubs and, ultimately, small trees. As the trees attain maturity, they too provide shade that helps retain the moisture in the alcoves.

According to Bob Thompson, of the Manti-LaSal National Forest, a few species, such as the dainty Kachina daisy, are native to the hanging gardens. Welsh has found that most of the plants, however, are more familiar elsewhere in North America. Many of them, such as the cave primrose and bog orchid, are plants with boreal affinities. Several of the grasses and sedges are from the prairies and plains. Most likely, all these plants originally colonized the hanging gardens through long-range dispersal of seeds from their usual habitats.

Uinta National Forest

SIZE AND LOCATION: Approximately 958,000 acres in north-central Utah on the front of the Wasatch Range. Major access routes are Interstates 15 and 80, U.S. Highways 6, 40, 89, 91, and 189, and State Routes 52, 117, 132, and 210. District Ranger Stations: Heber City, Pleasant Grove, Spanish Fork. Forest Supervisor's Office: 88 West 100 North, Provo, UT 84601, www.fs.fed.us/r4/uinta.

SPECIAL FACILITIES: Winter sports areas; boat ramps; swimming beaches.

SPECIAL ATTRACTIONS: Mt. Nebo National Scenic Byway; Alpine Loop Scenic Backway.

WILDERNESS AREAS: Mt. Timpanogos (10,518 acres); Mt. Nebo (27,010 acres); Lone Peak (30,088 acres, partly in the Wasatch-Cache National Forest).

Immediately east of Salt Lake City are the high mountains of the Wasatch Range. The side of the mountains that faces Salt Lake City is referred to as the Front, and that part of the forest that extends from Salt Lake City's southern suburbs to Nephi is in the Uinta National Forest.

Although there are three wilderness areas in the national forest, excellent highways provide easy access to most of the others features of the forest. Because of its proximity to Salt Lake City and towns and cities along Interstate 15, the Uinta National Forest is heavily visited.

The two dominant mountains in the national forest, Mt. Nebo and Mt. Timpanogos (pl. 53), are the primary features of two wilderness areas, but highways approach so near to them that they are easily observed and visited. The taller of the two is Mt. Nebo, at 11,877 feet. This mountain can be seen from many parts of the surrounding area, but none so close as from the Mt. Nebo National Scenic Byway. The 45-mile byway connects Payson at the northern end with Nephi at the southern end. You may begin your drive from Nephi, starting just a few miles east of town off of State Route 132. Just as you enter the national forest, the pretty Petticoat Cliffs are to the west. The scenic byway follows Salt Creek, then angles right. If you do not take the right fork, you will come to two campgrounds where there are trails to the Mt. Nebo Wilderness.

The scenic byway climbs by switchbacks and suddenly arrives at Devil's Kitchen Geological Area. The vividly colored limestone has eroded into many delicate shapes reminiscent of Bryce Canyon. A short interpretive trail is here. A short distance beyond Devil's Kitchen (fig. 26) is a pull-out on the west side of the highway for a magnificent view of Mt. Nebo.

Figure 26. Devil's Kitchen, Nebo Loop
Scenic Byway.

After several miles in gorgeous scenery, the scenic byway comes to a junction, with the right fork going eventually into Payson Canyon and the left fork going up Santaquin Canyon at the end of the byway. Just as the highway bends around to the north, there is a splendid view south over Nebo Creek. If fishing interests you, Payson Lakes should be your next stop along the byway. Payson Canyon has been carved by Peteetmeet Creek, and the drive through the canyon provides scenic beauty, including waterfalls. The byway then proceeds to Payson. To complete the route, however, take Interstate 15 south to Santaquin and then take the Santaquin Canyon Road. This is another scenic canyon. After emerging from the canyon, there are switchbacks that bring you to an overlook to view folded limestone formations in Santaquin Canyon. After passing tranquil Santaquin Meadows, the highway arrives at the junction with the road to Payson Canyon.

The Alpine Loop is a paved route that completely encircles massive Mt. Timpanogos. This highway also involves two spectacular canyons, American

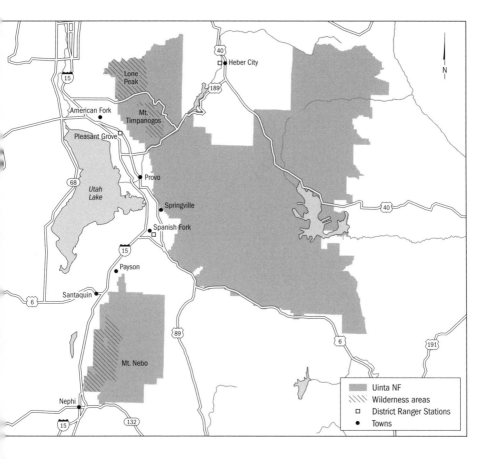

Fork and Provo. Start from the town of American Fork and drive U.S. Highways 89 and 91 to State Route 52 just before the community of Orem. Bridal Veil Falls (pl. 54) will stop you in your tracks. A little farther along are Upper Falls in Davis Canyon. At Wildwood, U.S. Highway 89 continues on, but the Alpine Loop turns left onto State Route 80 and follows the North Fork of the Provo River, zigzagging its way to Aspen Grove. Several trails to the Mt. Timpanogos Wilderness begin here. A very nice one is the Stewart Cascade Trail. This series of cascades drops 75 feet over rocky ledges and is worth the two-mile hike. Another much longer and more difficult trail is the seven-miler up the northeastern flank of Mt. Timpanogos. Emerald Lake is a blue jewel near the edge of the Timp Icefield. Finally, at the top of Mt. Timpanogos, at 11,750 feet, is the Summit House. From the Summit House you have a 360° panoramic view.

The Alpine Loop twists and turns until it gets to the Timpooneke Guard Station where there is the Great Staircase Trail up the northern slope of Mt.

Timpanogos. You will see Scout Falls along the way. Between Aspen Grove and Timpooneke trailheads, there is a side road north off of the Alpine Loop to the Cascade Springs Interpretive Trail. This half-mile paved trail with boardwalks over the wetter areas passes the springs and several cascades. The springs produce over seven million gallons of water daily that flows over ledges of travertine on the way to Provo Deer Creek.

The loop road continues to switch back until it abruptly turns westward and enters the American Fork Canyon. If you elect to turn right on the road to the Silver Lake Summer Home Area, you will come to the two-mile easy trail to Silver Lake in the Lone Peak Wilderness.

It is more than likely that the Squaw Peak Driving Trail is the most thrilling and spectacular drive in the Uinta National Forest. It begins off of U.S. Highway 89 in Provo Canyon and is very crooked as it makes its way south. The road gives access to Rock Creek Overlook, Provo Peak, Horse Mountain, and Little Rock Canyon before ending after 26 miles on the Left Fork of Hobble Creek Road for a 35-mile loop past Red Pine Knoll, Rock Slide Canyon, and then down to U.S. Highway 89 in Spanish Fork Canyon.

U.S. Highway 40 south of Heber passes through Daniels Canyon. If you take the Strawberry River Road into the national forest after Daniels Canyon, you will come to the Trappers Cove Interpretive Trail. This mile-long trail is steep, but there are 30 interpretive stops along the way.

A disconnected part of the Uinta National Forest is southeast of Nephi in the San Pitch Mountains. The rough Chicken Creek Road from Levan to Wales crosses the lower part of the range.

Wasatch-Cache National Forest

SIZE AND LOCATION: Approximately 2 million acres in northeastern and north-central Utah, extending into Wyoming. Major access routes are Interstates 15, 80, and 84, U.S. Highways 40, 89, 91, and 189, and State Routes 36, 39, 65, 66, 101, 152, 167, 190, 196, and 226. District Ranger Stations: Kamas, Logan, Salt Lake City (UT); Evanston, Mountain View (WY). Forest Supervisor's Office: 8236 Federal Building, 125 S. State Street, Salt Lake City, UT 84138, www.fs.fed.us/r4/wcnf.

SPECIAL FACILITIES: Visitor centers; winter sports areas; boat ramps; swimming beaches.

SPECIAL ATTRACTIONS: Big Cottonwood Canyon National Scenic Byway; Lit-

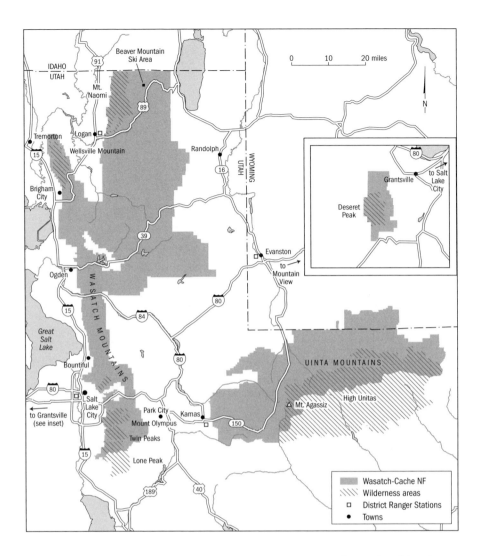

tle Cottonwood Canyon National Scenic Byway; Mirror Lake National Scenic Byway; Ogden River National Scenic Byway; Logan Canyon Scenic Highway.

WILDERNESS AREAS: High Uinta (456,705 acres, partly in the Ashley National Forest); Twin Peaks (11,334 acres); Mount Olympus (15,300 acres); Deseret Peak (25,212 acres); Lone Peak (30,088 acres, partly in the Uinta National Forest); Wellsville Mountain (22,986 acres); Mt. Naomi (44,523 acres).

In 1973, the Cache National Forest east of Ogden and the Wasatch National Forest east of Salt Lake City were combined into a single administrative unit, becoming the Wasatch-Cache National Forest. The Utah part of the Cache National Forest was added to the Wasatch, whereas the Idaho part was absorbed by the Caribou National Forest. To describe the points of interest in the forest better, the forests will be discussed separately.

Wasatch National Forest

The Wasatch National Forest includes much of the high-mountain areas of the Uinta Mountains east from Salt Lake City. The national forest also includes the southern portion of the Wasatch Mountains immediately east of Salt Lake City, as well as two small units in the Stansbury and Sheeprock Mountains west and southwest of Salt Lake City. As there are several top class winter sports areas in the mountains, there is something to do during all seasons of the year.

Every highway leaving Salt Lake City eastward goes through a scenic canyon to get to the Uinta Mountains. The routes through Big Cottonwood Canyon (pl. 55) and through Little Cottonwood Canyon have been designated as National Forest Scenic Byways. Several campgrounds are in each canyon.

Big Cottonwood Canyon National Scenic Byway, which is State Route 152, is in a canyon with sheer rock walls. After 15 miles, it ends in a beautiful high mountain valley. The strongly tilted rocks in the canyon are explained at the Storm Mountain Slide Area pull-out and at another pull-out a half-mile farther on. As you drive along Big Cottonwood Creek, you will see large cottonwood trees and the shrubby chokecherry along the way. Near the Mill B South Fork Picnic Area is the three-mile-long Hidden Falls Trail that enters the Mt. Olympus Wilderness. At first, the trail follows Elbow Creek and passes Hidden Falls and Elbow Canyon Falls. If you continue past the falls, you come to the southwestern flank of Mt. Raymond and the Desolation Trail.

From the Spruces Campground is an easy .3-mile trail to Doughnut Falls where you can see the stream fall through a hole in the rock. The climbing scenic byway is now lined by quaking aspen, Douglas fir, subalpine fir, and Engelmann spruce. At a pull-out it is explained how the valley ahead was formed through glacial action. Two popular ski areas are in the area, as well as many hiking trails, including the Brighton Loop Nature Trail. Some of the plants you will encounter are mountain ash, bearberry honeysuckle, western false hellebore, monkshood, columbine, meadow rue, thimbleberry, which is a kind of raspberry, thinleaf alder, bluebells, quaking aspen, Engelmann spruce, subalpine fir (fig. 27), myrtle pachistema, and black currant.

Little Cottonwood Canyon National Scenic Byway, which is State Route 210, is four miles south of Big Cottonwood Canyon. This 12-mile scenic byway is in a narrow canyon and terminates in the Albion Basin where there is

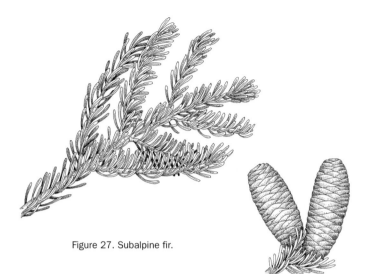

Figure 27. Subalpine fir.

a beautiful high-mountain meadow. The Twin Peaks Wilderness is to the north, and the Lone Peak Wilderness is to the south. Several cascades are visible along the scenic byway. Lisa Falls on the north side is particularly attractive.

At the Tanner Slide Area near the Tanner Flat Campground you can see several avalanche chutes. Two ski areas are ahead, and the one at Snowbird has a tram that goes to the top of Hidden Peak. Because of the high elevation at Albion Basin, the forests surrounding it consist of Engelmann spruce and subalpine fir. One popular trail here is the 0.8-mile trail to Cecret Lake. The Butler Creek Trail crosses the entire Mt. Olympus Wilderness from Little Cottonwood Canyon to Big Cottonwood Canyon.

Five miles north of Big Cottonwood Canyon National Scenic Byway is another beautiful highway up Mill Creek Canyon. The Thayne Canyon Nature Trail from the Box Elder Campground is worth taking. The paved highway up Mill Creek Canyon ends near the Big Water Campground.

North of Mill Creek Canyon, Interstate 80 winds through Parley's Canyon and, a little farther north, State Route 65 has switchbacks as it makes its way up Emigration Canyon. These scenic routes are not in the national forest but go between units of the forest. The next road north of Emigration Canyon passes through Red Butte Canyon. This road ends after about four miles at the foot of Table Mountain. Because the canyon is less visited, it is more pristine than the other canyons, and a large part along the canyon has been designated as the Red Butte Research Natural Area. Red Butte Creek has carved a narrow canyon with abrupt rocky walls. The lower slopes on the south side of the road are heavily shaded, and this moist area supports lush and dense vegetation. Higher on these south slopes is a brush community of Gambel's oak and bigtooth maple. Across the road on the drier, rockier north side is a typical foothill–grassland community of sagebrush, bitterbrush, and blue-bunch wheatgrass. The highest elevations on both the south and north sides support forests of Douglas fir, white fir, subalpine fir, and quaking aspen. Along Red Butte Creek is a riparian community of box elder, bigtooth maple, western birch, chokecherry, red osier dogwood, thinleaf alder, narrowleaf cottonwood, and two or three species of willows. It is necessary to receive permission from the Forest Supervisor's Office to visit this area.

The Weber River and Interstate 80 are in the wide area that separates the Wasatch Mountains from the Uinta Mountains. The Uinta Mountains are unusual in that they are oriented east-west, whereas most mountain ranges in North America are oriented north-south. A scenic drive in the Wasatch National Forest is the Mirror Lake National Scenic Byway, or State Route 150, from Kamas, Utah, to Evanston, Wyoming. Forty-two miles of this byway are in the national forest as it circles its way around the northern edge of the Uinta Mountains. The route provides access to several campgrounds and hiking trails, more than 400 alpine lakes, and countless panoramic vistas.

The Forest Service describes the first few miles of the scenic byway east of Kamas as a drive through a natural arboretum where many of Utah's native trees and shrubs may be seen. Known as the Beaver Creek Natural Arboretum, this is an 8.5-mile-long area along the scenic byway as far as the Soapstone Guard Station. One of the better trails is the Shingle Creek Trail. Four campgrounds are in this stretch so that you can spend as much time as necessary to study the trees and shrubs.

The Slate Gorge Overlook allows you a great view of multicolored sandstone and shale in the gorge. Along the south side of the scenic byway is Upper Provo River Falls, a turbulent but picturesque cascade. The scenic byway continues to climb past alpine meadows, glacial lakes, and huge white boulders. The road twists its way around Trial Lake, Teapot Lake, Lily Lake, and Lost Lake before coming to a couple of overlooks. The first one has a grand view of 11,947-foot Bald Mountain. The second overlook has a scintillating view east into the High Uintas Wilderness, the northern half of which is in the Wasatch National Forest. At this high elevation, the road passes through a forest of lodgepole pine, but a short distance farther on, at Bald Mountain Pass, the 10,775-foot pass is at timberline. Gnarly and very old subalpine firs may be only four feet tall in this bleak habitat. During July, there are a number of pretty alpine wildflowers growing among the rocks. Fortunate observers may see moose, elk, bighorn sheep, and mountain goats. Yellow-bellied marmots are not as common.

Just beyond the pass is picture-perfect Mirror Lake with Hayden Peak often reflected in its blue water. There are several trails to choose from, including a number that enter the High Uinta Wilderness. The trail to Naturalist's Basin is interesting where there are marshy communities of sedges and rushes, alpine wildflower meadows, and several picturesque lakes. Mt. Agassiz looms above the basin on the northwest side. If the trail to Naturalist's Basin is too long for you, try the .7-mile trail to Ruth Lake.

The scenic byway then drops down to a canyon carved by Hayden Fork. A back road from Lily Lake to Christmas Meadows follows Stillwater Creek for four miles and ends at a wonderful wildflower meadow. Bear River Campground marks the end of the scenic byway on national forest land. The long North Slope Road east of the campground is spectacular. It crosses Christmas Tree Creek, then follows the North Fork of Mill Creek for a little while before climbing over Elizabeth Pass. Then after the two Lyman Lakes, one with a campground, the road follows Blacks Fork to the long and narrow Meeks Cabin Reservoir. The Meeks Cabin Historical Site is at the north end of the reservoir.

West of Salt Lake City and south of Interstate 80 are two smaller units of the Wasatch National Forest. Deseret Peak in the heart of the Stansbury Mountains is the center of attraction in the Deseret Peak Wilderness. The peak, at 11,031 feet, may be reached by a four-mile trail from the Upper Narrows Campground at the end of South Willow Canyon Road southwest of Grantville. Outside the wilderness area is a trail up Box Canyon.

A disjunct unit of the national forest is south of Vernon and west of Eureka. The western end of this unit is in the Sheeprock Mountains.

Cache National Forest (Utah Section)

In 1973, the Cache National Forest was split. The northern part in Idaho was made a part of the Caribou National Forest, whereas the larger southern part in Utah was merged with the Wasatch National Forest. The Cache is in the northern end of the Wasatch Range east of Interstate 15 between Ogden and the Idaho border. This is a region of picturesque mountains, cool canyons, sparkling lakes, and colorful rock formations. The Ogden River National Scenic Byway and the Logan Canyon Scenic Highway are good orientation routes for exploring the national forest.

State Route 39 is the Ogden River National Scenic Byway. Ogden Canyon is very narrow, and the scenic byway has several very sharp curves. Sardine Peak is the prominent peak to the south upon entering the national forest. Eventually the scenic byway arrives at Pineview Reservoir, a large body of water with five narrow arms. The shoreline of the reservoir is entirely in the national forest, and there are several campgrounds, resorts, and boat ramps.

Cemetery Point Day-Use Area is at the end of a narrow peninsula that protrudes into the reservoir from the town of Huntsville.

From the Anderson Campground on the south side of the reservoir, the paved State Route 226 climbs by way of several switchbacks to Snow Basin Resort and Ogden Bowl. This is a highly scenic area with Strawberry Peak, DeMoisy Peak, and The Needles surrounding the basin.

After leaving the national forest for about six miles, the scenic byway reenters the Cache National Forest and follows the South Fork of the Ogden River past seven campgrounds in a three-mile stretch. After more private property, the scenic byway comes to Causey Reservoir via a short side road. You may walk all the way around the reservoir on Skullcrack Trail. Some of the rocks are a vivid red. After circling around the western side of Monte Cristo Peak, the scenic byway leaves the national forest at the Monte Cristo Campground.

The Logan Canyon Scenic Highway, U.S. Highway 89, follows Logan Canyon for 39 miles to huge Bear Lake that is outside the national forest. The limestone cliffs along Logan Canyon are vertical in many places and are known as the Chinese Wall. Near the beginning of the scenic byway is a 2.5-mile trail into Spring Hollow, and at the Guinovah-Malibou Campground is a mile-long Riverside Nature Trail along the Logan River. Along the nature trail you will pass the Wind Caves, a series of three natural arches.

Just north of the Preston Valley Campground is a pull-out for a geological interpretive sign. From here take the short Right Fork Road to two unusual rock formations—Tea Pot Rock and Lion's Head.

The Jardine Juniper Trail from the Wood Camp Campground is five miles long and strenuous in some places, but seeing the largest Rocky Mountain juniper in the world is rewarding at the end of the trail. This gnarly tree (pl. 56), more than 3,000 years old, is about 50 feet tall, eight feet in diameter, and with a circumference of 27 feet. If you keep going beyond the juniper, you enter the Mt. Naomi Wilderness.

A rough back road along Temple Fork leads to Old Ephraim's Grave. Old Ephraim was a grizzly bear that was shot in 1923 after poaching sheep and cattle. The Boy Scouts of America have erected a marker where the bear was buried and have written a poem that you can read from the marker.

Rick Springs is a favorite stopping place along the highway. Its peak flow during the summer drains into the Logan River. A nine-mile-long side road to the west goes to Tony Grove Lake and Campground, secluded in the mountains.

The Beaver Mountain Ski Area and lodge is a few miles farther north. The highway now climbs to Bear Lake Summit, which, at 7,800 feet, is the highest elevation along the route. The Old Limber Pine Nature Trail is here and has 16 interpretive stations, the last one bringing you face to face with the

Monster Monarch, the largest limber pine in the world, which has a circumference of nearly 25 feet. This behemoth is thought to be about 2,000 years old. Just beyond the Bear Lake Summit is an overlook for a super view of Bear Lake.

The Wellsville Mountains north of Brigham City are in the Wellsville Mountains Wilderness, an extremely rugged region that the Forest Service calls the steepest grade in the United States. Wellsville Cone and Box Elder Peak are the highest elevation in the wilderness at 9,356 and 9,372 feet, respectively.

NATIONAL FORESTS
IN WASHINGTON

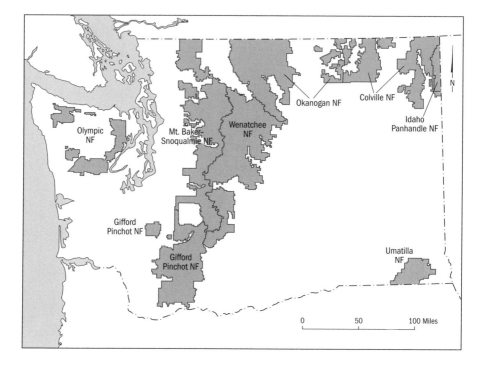

The seven national forests in Washington are in U.S. Forest Service Pacific Northwest Region 6, located at 333 SW 1st Avenue, Portland, OR 97208. The national forests in Washington occupy 7.7 million acres and 21 wildernesses.

Colville National Forest

SIZE AND LOCATION: Approximately 1.1 million acres in northeastern Washington, on either side of the Columbia River and the city of Colville, touching the border of Canada and Idaho. Major access routes are U.S. Highway 395 and State Routes 20, 21, and 31. District Ranger Stations: Metaline Falls, Kettle Falls, Newport, Republic. Forest Supervisor's Office: 765 S. Main Street, Colville, WA 99114, www.fs.fed.us/r6/colville.

SPECIAL FACILITIES: Boat ramps; swimming beaches; winter recreation areas.

SPECIAL ATTRACTIONS: Sherman Pass National Scenic Byway; Kettle Creek National Recreation Trail.

WILDERNESS AREAS: Salmo Priest (50,775 acres, partly in the Kaniksu [ID] National Forest).

People may be attracted to this national forest in what the Forest Service calls "the forgotten corner of Washington" by the many lakes and streams teeming with fish. They may come because of the tranquil, verdant valleys and the mountains that are the foothills of the Rocky Mountains. They may want to enjoy the scenic drives that wind through these mountains. They may just want to enjoy nature and the chance to see bighorn sheep, mule deer, white-tailed deer, elk, black bear (pl. 57), bobcats, cougars, lynx, moose, badgers, wolverines, or even rare grizzly bears and the only herd of caribou below the Canadian border.

The Colville National Forest consists of two distinct ranges, the Selkirk Mountains and the Kettle River Mountains, separated by the Columbia River. Each range is further separated into two units. The Pend Oreille River bisects the Selkirks, whereas Curlew Creek splits the Kettles.

The eastern side of the Selkirk Mountains contains the forest's only wilderness area, the Salmo Priest. Two north-south ridges are the dominant features of the wilderness, connected across the top by the crest of 6,820-foot Salmo Mountain. It is here that lives the only herd of caribou south of the Canadian border. Reduced to only 30 animals in 1974, the population is somewhat larger now because endangered species officials have translocated additional caribou from Canada. Probably not more than two dozen grizzlies also live in the area. The incredible Pewee Falls lies just south of the Canadian border. The falls cascades for 200 feet down a vertical rock face into the Boundary Dam Reservoir.

South of the wilderness area is attractive and large Sullivan Lake surrounded on all sides by dense forests. All water activities may be enjoyed here.

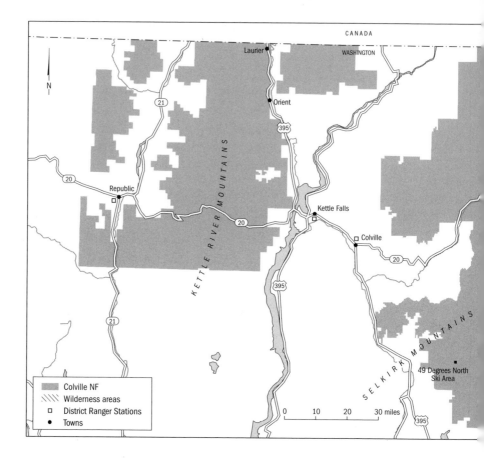

West of the Pend Oreille River, the Selkirk Mountains include picturesque mountains such as Abercrombie Mountain, Huckleberry Mountain, Sherlock Peak, and Granite Peak. Some distance north of Granite Peak is the 2.5-mile Brown's Lake Interpretive Trail. Marble Creek Falls plunges for about 35 feet about 15 miles northeast of Colville. It may be reached by a primitive forest road.

A small unit of the national forest lies south of the Selkirk Mountains and is separated from them by the Little Pend Oreille National Wildlife Refuge. The feature of this small unit is Chewelah Mountain where there is a popular ski area—49 Degrees North.

West of the Columbia River are the Kettle River Mountains. A drive on the Sherman Pass National Scenic Byway provides a fine opportunity to explore this region. The scenic byway, which is State Route 20 from Republic to Kettle Falls, leaves Republic and stays near O'Brien Creek before ascending to Sherman Creek Pass at 5,587 feet. It then levels off somewhat along

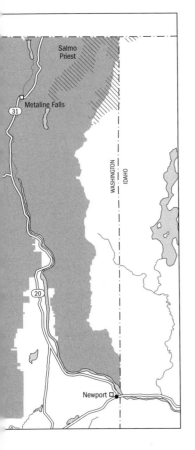

Sherman Creek. The south-facing slopes above O'Brien Creek have a cover of grasses with occasional plants of the shrubby ocean spray and ninebark here and there. As the highway starts to ascend to the pass, Snow Peak looms to the south. At the pass, there is a good view of the surrounding country. An easy half-mile trail to the observation point passes through a serene forest of Douglas fir, subalpine fir, and lodgepole pine. If you are interested in taking a longer hike, Kettle Creek National Recreation Trail crosses the pass. Whether you elect to hike north or south on this trail from the pass, you come to several nice mountain meadows with lots of lupines and paintbrushes (pl. 58). The trail south goes to a meadow that was burned by the White Mountain fire in 1988. Hundreds of fireweeds (pl. 59) are here.

As you descend the east side from the pass on the scenic byway, the highway is routed through a lovely forest of subalpine fir, lodgepole pine, and western larch. The needles of the larch turn a golden yellow in the autumn before dropping to the ground. A forest road that follows Sherman Creek branches off the scenic byway. A short trail leads through a fine forest of grand fir, spruce, and western red cedar. Just before the highway comes to East Portal Interpretive Site, forest roads lead both north and south. The north road goes to Trout Lake where you can hike up Hoodoo Canyon to an observation point overlooking gemlike Emerald Lake. The south road climbs up Bangs Mountain where there are several views of the Columbia River.

Toward the Canadian border, another scenic drive exits U.S. Highway 395 halfway between Laurier and Orient and takes a circuitous route around Marble Mountain before joining the West Deer Creek Road that winds up in Curlew.

West of Curlew Creek is a small range with Bodie Mountain at its center. A scenic road from Pollard on the edge of Curlew Reservoir parallels Trout Creek before heading for Bodie Mountain.

Gifford Pinchot National Forest

SIZE AND LOCATION: Approximately 1.3 million acres on the western slopes of the Cascade Mountains from the southern border of Mt. Rainier National Park to the Columbia River. Major access routes are U.S. Highway 12 and State Routes 14, 141, 503, 504, and 706. District Ranger Stations: Randle, Trout Lake. Forest Supervisor's Office: 10600 NE 51st Circle, Vancouver, WA 98682, www.fs.fed.us/r6/gpnf.

SPECIAL FACILITIES: Visitor centers; winter sports areas.

SPECIAL ATTRACTIONS: Mount St. Helens National Volcanic Monument; Spirit Lake Memorial Highway.

WILDERNESS AREAS: Trapper Creek (5,970 acres); Tatoosh (15,750 acres); Mount Adams (46,626 acres); Indian Heaven (20,960 acres); Glacier View (3,123 acres); Goat Rocks (108,279 acres, partly in the Wenatchee National Forest); William O. Douglas (168,157 acres, partly in the Wenatchee National Forest).

Most visitors to the Gifford Pinchot National Forest head for Mt. St. Helens to see the destruction and recovery from the explosive eruption of Mt. St. Helens at 8:32 A.M. on Sunday, May 18, 1980. At that moment, the north face of the 9,677-foot mountain collapsed during an earthquake that measured 5.1 on the Richter scale. An avalanche of massive rock and ice debris forcibly struck Spirit Lake, crossed a ridge nearly 1,400 feet high, and roared tumultuously down the North Fork of the Toutle River for 14 miles. Pressurized gases within the volcano exploded through the avalanche and produced a rock-laden wind so powerful that all the trees within 230 square miles were either knocked down or left standing upright but dead. In the wake of the explosion, gray ash fell over much of eastern Washington and adjacent areas, rocks and mudslides cascaded down on all sides of the mountain, and pumice poured out of the volcano. In nine hours, the eruption was over, but the nature of the landscape had been changed forever. Volcanic debris filled Spirit Lake, and the north slope of Mt. St. Helens had the appearance of the surface of the moon. Trees with diameters as much as six feet lay like jackstraws across the terrain. Mud slides buried many areas with several feet of mud. The summit of Mt. St. Helens had been reduced to 8,400 feet.

In 1982 the Mt. St. Helens National Volcanic Monument was designated, and 110,000 acres surrounding the mountain were set aside for the public to see the devastation and watch while the forces of nature began to repair the area. Eruptions continued until 1986, building a lava dome inside the crater

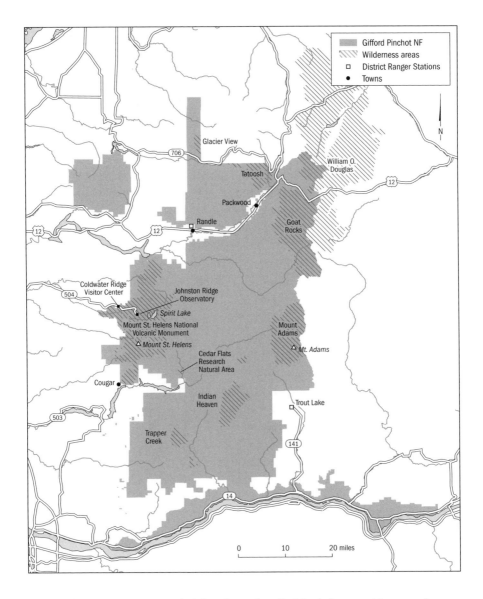

Gifford Pinchot NF
Wilderness areas
□ District Ranger Stations
● Towns

Glacier View

706

William O.
Douglas

12

Tatoosh

Packwood

Randle

12

12

Goat
Rocks

Coldwater Ridge
Visitor Center

504

Johnston Ridge
Observatory

Spirit Lake

Mount St. Helens National
Volcanic Monument

△ Mount St. Helens

Mount
Adams

△ Mt. Adams

Cedar Flats
Research
Natural Area

Cougar

Indian
Heaven

Trout Lake

503

Trapper
Creek

141

14

N

0   10   20 miles

that has now grown to a height of 920 feet. I visited the area 10 years after that explosive Sunday morning to see where any plants might have begun to recolonize the devastation zone.

Many trails have been laid out to permit exploration of the area around the volcano, including some that go to the perimeter of the crater. Anyone wishing to hike above 4,800 feet must obtain a permit from the Forest Service Supervisor's Office that administers the national monument.

State Route 504 from Castle Rock has been named the Spirit Lake Memorial Highway. Thirty miles east of Castle Rock is the Hoffstadt Bluffs Viewpoint. Although this viewpoint is 14 miles from the volcano, it is the point where the avalanche of rocks and ice that roared down the North Fork of the Toutle River came to a halt. As you look around in the adjacent forest, you will see dead trees standing, a stark reminder of that spring day in 1980. Nine miles farther east along State Route 504 is Elk Rock and the entrance to the Mt. St. Helens National Volcanic Monument. The mountain you see looming up due east of you is Mt. Adams, some 40 miles away. In another four miles is the Coldwater Ridge Visitor Center where you may learn about all aspects of this mighty eruption and the recovery of the vegetation going on at this very moment. The quarter-mile loop Winds of Change Interpretive Trail at the visitor center explains what you are seeing, and there is a stunning view of the Toutle River Valley. Nearby Coldwater Lake was formed as a result of the eruption. A quarter-mile trail along the shoreline has a boardwalk out into the lake. This interpretive trail describes the recent origin of the lake. A little beyond the visitor center is the Johnston Ridge Observatory where you can look into the mouth of the crater and see the lava dome.

Forest Road 25 south from the community of Randle goes into the northeastern side of the national monument. Forest Road 99 goes to the old miner's car that was buried in ash near Meta Lake. From the miner's car is a quarter-mile paved trail to the lake. Probably the most interesting trail here is the seven-mile-long Truman Trail that goes into the devastation area of the Spirit Lake Basin. Longer trails are for the more adventurous hiker.

A few of the trails north of Mt. St. Helens but outside the blast zone are worth considering. A short distance south of Cispus Creek is a half-mile trail along Quartz Creek that meanders through a forest of giant trees. The largest of these are Douglas firs that have diameters of about 10 feet. The contrast here is remarkable because just a mile farther south is the blast area where the forests are completely gone. The Woods Creek Information Station is about two miles south of Randle. A 1.5-mile trail leads through a mixed hardwood forest. Two trails lead from the Iron Creek Campground: one is a quarter-mile trail through an old-growth forest, the other is a 1.5-mile path beneath large Douglas firs and western red cedars growing along the river.

The south side of Mt. St. Helens features lava flows from ancient eruptions and mudflows from the 1980 explosion. Although the Lava Canyon Trail at the end of a forest road is long and difficult when it reaches the canyon, the first 1.5 miles are fairly easy and cross the Muddy River where you can see the extensive mudflow that consumed the river. While you are in that area, drive to the Cedar Flats Research Natural Area where a dense old-growth forest of western red cedars and Douglas firs are draped with mosses and lichens. Only some fallen trees are reminders of the volcanic violence that grazed this area. A one-mile-long trail is lined with pendant fern fronds.

You may drive to the southern tip of the Mt. St. Helens National Volcanic Monument to find one of the most unusual features in the national forest. Ape Cave is the largest lava tube in the country. This old tube is an incredible 2.5 miles long. Two layers are within the cave. The 0.75-mile trail in the upper layer goes across a level mudflow. You see a huge lava ball wedged in the ceiling. The trail in the lower cave is 1.5 miles long, and you have to pick your way over rock piles and eventually scale an eight-foot lava wall. A Forest Service partner, Northwest Interpretive Association, rents lanterns to those who wish to explore Ape Cave. By the way, there are no apes in the cave. The name was given to the cave by the Mt. St. Helens Ape Club, a group of rugged outdoor adventurers.

One mile from the entrance to Ape Cave is the Trail of Two Forests that crosses a 2,000-year-old lava flow on a boardwalk.

From Carson on State Route 14, the Wind River Highway enters the southern end of the Gifford Pinchot National Forest. All points of interest in the southern and southeastern parts of the forest may be accessed from this highway. Six miles north of Carson is the Panther Creek Road. Take it to Panther Creek Falls where there are two 175-foot falls side by side, one dropping from Panther Creek and the other from Big Creek.

After passing the Carson National Fish Hatchery, the wild area to the west is the Trapper Creek Wilderness. Here is a good example of subalpine forests with subalpine meadows interspersed. The forests of silver fir, noble fir, and western hemlock, with scattered Douglas firs and mountain hemlocks mixed in, are impressive. Particularly nice are the open, treeless areas, some covered by the low, tangled growth of mountain heather and Cascade huckleberry, others by a good mixture of wildflowers. Wildflowers include broad-leaved lupine, Cascade aster, scarlet paintbrush, magenta paintbrush, spreading phlox, bracted lousewort, western pasqueflower, fan-leaved cinquefoil, and, in wetter seepage areas, green false hellebore and Sitka valerian.

A short spur road to the east stops at the Lower Falls Creek Trailhead where there is a pleasant trail along scenic Falls Creek. In 1.25 miles you come to the finest waterfall in the national forest. Falls Creek Falls is a triple falls whose total drop is 150 feet.

Five miles east of Falls Creek Falls is the Big Lava Bed that is nine miles long and four miles wide. The volcanic crater that caused the lava flow is 500 feet deep and may be seen near the northern end of the lava flow.

Indian Heaven Wilderness is a short distance north and northwest of the Big Lava Bed. This wilderness is on a plateau with a good growth of silver fir, noble fir, and subalpine fir. The Forest Service counts at least 150 small lakes in the wilderness. Of historical interest is the Indian Racetrack, reached by a long, uphill, eroded trail to near the southern boundary of the wilderness.

You may still make out the beaten-down tracks in a 2,000-foot-long field where Indians reputedly held horse races.

About five miles west of the Mt. Adams Ranger Station on the eastern edge of the Gifford Pinchot National Forest are the Ice Caves that are so cold that they are able to maintain their icy formations during the summer. Wooden steps lead into the caves, but the entire area may be slippery. Ice from the caves once was sold in The Dalles, Oregon, 60 miles away. Approximately 14 miles northwest of the ranger station on the Trout Creek Road and Forest Road 88 is a short trail to Langfield Falls where Mosquito Creek drops for 100 feet. You can hear the thunder of the falls long before you can see them.

Mount Adams, whose 12,276-foot summit is the second highest in the state, is the dominant feature of the Mt. Adams Wilderness. Ten glaciers are on the flanks of Mt. Adams. Experienced climbers flock to this area. Twenty-one miles of the Pacific Crest National Scenic Trail cross the wilderness.

Two wilderness areas are adjacent to Mt. Rainier National Park at the north end of the national forest. Glacier View Wilderness is a tiny wilderness where, from the 5,450-foot summit of Glacier View Peak, you may get an unobstructed view of the glaciers on the western side of Mt. Rainier. A heavy growth of forest is in this wilderness as well as several sky-blue alpine lakes. Tatoosh Wilderness consists of two parallel ridges, the Tatoosh Range and Backbone Ridge. These exceptionally rocky ridges spawn numerous streams that ultimately flow in the Cowlitz River and Butter Creek. Part of the Goat Rocks Wilderness is in and managed by the Gifford Pinchot National Forest. Just outside the western border of the wilderness is scenic Packwood Lake where water-based activities may be enjoyed.

Mount St. Helens

More than 20 years ago, on May 18, 1980, a mighty explosion blew out the north side of Mt. St. Helens, one of the highest peaks in the Northwest's volcanic Cascade Range. On the mountain's northern flanks, trees on every exposed slope within a radius of 6.5 miles were pulverized or blown away, and the denuded earth was covered with more than six feet of ash and rock. For as much as 10 miles beyond this blast zone, trees fell like matchsticks, stripped of their branches. Needles on standing pine and fir trees up to 1,000 feet beyond this blow-down area were singed orange by the heat.

In the blast zone, a landslide and avalanche of volcanic debris dumped soil, fragmented plants, rocks, and trees into Spirit Lake, raising the water level 120 feet and the water temperature about 74 degrees F. The upstream channel of the North Fork of the Toutle River was obliterated, and a giant mudflow, fed by melting glaciers and snowpacks, washed downstream. Smaller mudflows invaded streams to the east, south, and west. The Washington Department of Game estimates that 11,000,000 fish, 27,000 grouse, 11,000 hares, 6,000 black-tailed deer, 5,200 elk, 1,400 coyotes, 300 bobcats, 200 black bears, and 15 mountain lions perished (and despite forewarning, 57 people also lost their lives). Subsequent explosions on May 25, June 12, and July 22 spread thick layers of pumice.

Now designated a National Volcanic Monument, Mt. St. Helens and its surroundings are managed by Washington's Gifford Pinchot National Forest. A visitor center near Castle Rock provides an orientation for travelers, complete with a model of the volcano and short film about the 1980 eruptions. Part of the area north of the crater is closed to the public, but visitors may view the devastation from Windy Ridge, five miles from the crater and accessible by car. From there a trail south to Lahar skirts the closed area. Climbers can make their way to the summit from the south side of the volcano, which is constantly monitored for signs of hazardous activity.

I visited Mt. St. Helens in 1990. Recovery of the vegetation and animals on Mt. St. Helens had varied during the decade following the eruption according to whether the areas affected were originally above timberline, forested, or clear-cut; whether they were covered by snowpack; and whether they were buried under layers of ash, pumice, rocky debris, or mudflows. The biological communities will never be the same as before the eruption, but the plants and animals are gradually building toward reasonably stable ecosystems.

In the May 1982 issue of *Natural History,* biologist James A. MacMahon reported on the plight of animals in the area. The ones that survived the initial blast were generally those that live in underground burrows, such as pocket gophers. However, in the first few months after the eruption, observers noted occasional muskrats, minks, deer mice, tailed frogs, Pacific giant salamanders, newts, trout, ants, spiders, beetles, millipedes, and centipedes. Most animals that migrated into the devastated area were faced with the problem of finding something to eat. Only wood-boring insects readily found sustenance in the downed trees. Today the animal populations have begun to return, and according to wildlife biologist Vaughan Marable, they include grouse, hares, black-tailed deer, elk, coyotes, bobcats, black bears, and mountain lions.

Biologist Robert C. Wissmar and his colleagues studied the effects of the eruption on Spirit Lake and smaller area lakes. Immediately after the May 18 eruption, they found a 33-fold increase in suspended material in the water, which turned it light brown to black. The reduced light destroyed or impaired the lake algae, whereas bacteria increased by more than a million times. The bacteria lived off the sugars, amino acids, and other compounds from the vegetation and soils that had entered the water following the eruption. Some of these bacteria were potential pathogens. Four months after the eruption, the water lacked oxygen, but there were high concentrations of methane and carbon dioxide. The rotten-egg odor of hydrogen sulfide was easy to detect.

As the water has gradually become less cloudy, chlorophyll-bearing algae have returned, altering the mineral concentration in the water. Only after considerable changes in the chemistry of the water took place was regrowth of vegetation in the lakes possible.

The recovery of plant life on Mt. St. Helens has depended on the local changes wrought by the eruption, the presence or absence of snowpack at the time, and the degree of subsequent erosion. In general, regrowth is least evident where deposits of rock, ash, pumice, and mud were heavy, with little subsequent erosion.

In the blast zone, plants that were protected by an insulating cover of snow apparently did best. Outside the blast zone, however, botanists Joseph A. Antos and Donald B. Zobel found that thick layers of ash on top of the snow seemed to inhibit immediate growth of plants.

Within the area where trees were blown down or scorched, plant material that was transported and redeposited by the eruption did not regenerate. Buds on the underground portions of plants often survived, however. By 1982, 70 percent of the plants growing in these areas were regenerated from such buds, including thimbleberry, salmonberry, and elderberry. The greatest number of species appeared on moderately steep slopes, where erosion created gullies in the ash layer and exposed the original nutrient-rich soils.

In contrast, where mudflows or debris flows left thick deposits, few plants arose from underground buds for several years after the eruptions. Broken branches of willows and black cottonwoods, which are able to form roots and sprouts, did regenerate quickly on these flows, especially if they got caught along the way on a log snag or a clump of soil.

According to botanist A. B. Adams and his colleagues, revegetation was fairly vigorous at the end points of mudflows, where clumps of roots, soil, and forest fragments came to rest. The regeneration came from tubers, as in the western starflower, corms in the trillium-leaved sorrel, rhizomes in bear grass, and taproots in pearly everlasting. However, most plants that eventually appeared on debris flows were recolonized from seeds that had blown in from some distance away.

Extensive deposits of pumice immediately north of the crater, known as the Pumice Plains, have been studied by ecologists Roger del Moral and David M. Wood. The pumice is deep but contains some large mudflow channels and small gullies. Most of the plants that have colonized this forbidding area are herbaceous perennials with windblown seeds, such as pearly everlasting and fireweed. The vegetation is concentrated on moist sites with irregular topography, where seeds can become lodged. However, one common plant found here is lupine, whose seeds disperse poorly; possibly, some lupine seeds in the area managed to survive the eruptions.

Before the events of 1980, Mt. St. Helens was 9,677 feet high, about 1,500 feet higher than today. Scien-

Figure 28. Black cottonwood.

tists had observed that the mountain was botanically different from other, nearby high volcanic mountains, such as Mt. Adams to the east (also in Gifford Pinchot National Forest), Mt. Rainier to the north, and Mt. Hood to the south. This was due, at least in part, to the relative youth of Mt. St. Helens's surface soil, as the volcano had erupted as recently as 1857.

In the subalpine regions below timberline, other high mountains in the northwestern United States usually have a parklike forest of mountain hemlock, subalpine fir, Alaska cedar, and whitebark pine, interspersed with meadows of colorful wildflowers. On Mt. St. Helens, this type of community was sparse, and the tree composition consisted primarily of lodgepole pine, noble fir, Pacific silver fir, subalpine fir, and mountain hemlock. In addition, the deciduous black cottonwood, a plant normally found at low elevations, was quite common. The timberline, above which trees are stunted or do not grow at all, was irregular, not showing the sharp transition usually apparent on other high mountains. Among the stunted trees above the timberline on Mt. St. Helens was black cottonwood (fig. 28), as well as Douglas fir, lodgepole pine, and subalpine fir.

Botanist Arthur Kruckeberg has noted that above timberline before the 1980 eruption, ridges extending to the north, covered by rock volcanic debris from earlier eruptions, supported a meager assemblage of herbaceous wildflowers. The only flowering plant able to grow in the unstable pumice on either side of these ridges was the pyrola-leaved buckwheat, whose elongated root system anchors the plant in rocky terrain. Kruckeberg also noted that on Mt. St. Helens there were no endemic plants—species or varieties found on this mountain and nowhere else—whereas most of the higher mountains nearby usually had one or more endemics. This could be traced to the past eruptions, as plants had not been established in Mt. St. Helens's young soils long enough for new species to arise.

Because these differences were still apparent more than a century after the previous recorded eruptions of Mt. St. Helens, scientists studying the mountain should still be observing the effects of the great 1980 eruption for decades.

Mount Baker–Snoqualmie National Forest

SIZE AND LOCATION: Approximately 1.7 million acres along the western slopes of the Cascade Mountains from the Canadian border to Mt. Rainier National Park. Major access routes are Interstate 90, U.S. Highway 2, and State Routes 20, 92, 410, 530, and 542. District Ranger Stations: Darrington, North Bend, Sedro-Woolley, Skykomish. Forest Supervisor's Office: 21905 64th Avenue West, Mountlake Terrace, WA 98043, www.fs.fed.us/r6/mbs.

SPECIAL FACILITIES: Winter sports areas; boat ramps; swimming beaches; public information centers.

SPECIAL ATTRACTIONS: Mount Baker National Scenic Byway; Mountain Loop National Scenic Byway.

WILDERNESS AREAS: Boulder River (49,132 acres); Mt. Baker (119,398 acres); Noisy-Diobsud (14,443 acres); Alpine Lakes (362,670 acres, partly in the Wenatchee National Forest); Glacier Peak (527,308 acres, partly in the Wenatchee National Forest); Henry M. Jackson (100,337 acres, partly in the Wenatchee National Forest); Norse Peak (952,180 acres, partly in the Wenatchee National Forest); Clearwater (14,514,acres).

From the Canadian border to the northern edge of Mt. Rainier National Park, the Mt. Baker–Snoqualmie National Forest occupies the entire western slopes of the Cascade Mountains. This national forest was originally two separate forests, but they were combined into one in 1974. A small portion of the Mt. Baker–Snoqualmie is administered by the Okanogan and Wenatchee National Forests.

This is an area of dramatic landscapes, with numerous mountains above 7,000 feet, some with glaciers, alpine meadows, glacial lakes, and old-growth forests. Two volcanoes are here—Mt. Baker and Glacier Peak. When you see steam and smell sulfur being emitted from Sherman Crater on Mt. Baker, you know this volcano is not totally dormant. Because of the thick volcanic soils that have filtered into the valleys, Douglas firs and western hemlocks become huge as part of the old-growth forests. Higher in the mountains are stands of silver fir and mountain hemlock that, because of their shape, are able to shed snow and avoid the brunt of the wind better than other species. Alpine meadows are covered by shrubs, such as huckleberry and heather, that grow close to the ground.

Mount Baker, at 10,781 feet, is the dominant feature of the Mt. Baker

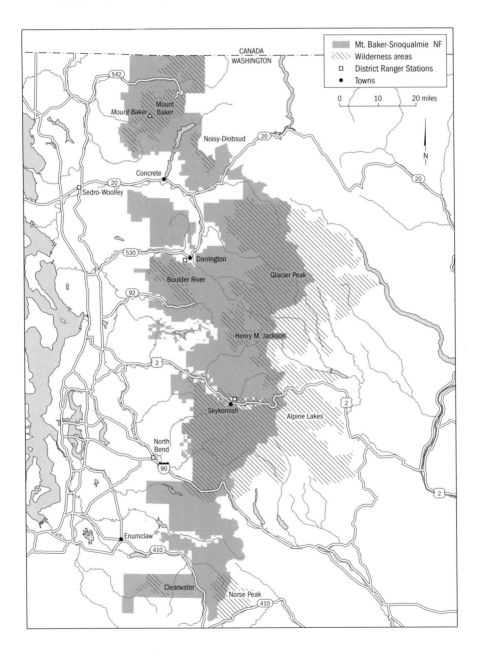

Ranger District. Lying within the Mt. Baker Wilderness, glaciers on this mountain flow into the Nooksook and Baker rivers.

Glacier Peak is the other volcanic mountain in the national forest. At a height of 10,568 feet, and 50 miles south of Mt. Baker, this volcano most recently showed some signs of life 200 years ago. All the other mountains in the national forest have rocks dating back 600 million years.

You may want to start your exploration of the Mt. Baker–Snoqualmie National Forest in the Mt. Baker area and work your way south from there. State Route 542, the Mt. Baker National Scenic Byway from Bellingham, some 60 miles from Mt. Baker, will take you to Artist Point, a spectacular place to see Mt. Baker and Mt. Shuksan, eight miles to the southwest and not accessible by road. Artist Point is in the heart of the Heather Meadows Recreation Area, and there are several places where wildflowers form dense masses during July and August. Several self-guiding nature trails are here as well as a visitor center. The Picture Lake Trail and the Fire and Ice Trail offer pleasant half-mile hikes from trailheads in the area.

You can also pick up information about the area at the Glacier Public Service Center in the village of Glacier where State Route 542 first enters the national forest. If you take the Glacier Creek Road from here to its end, you will come to one of the best viewpoints of Mt. Baker, which is only five miles in front of you. In between you and the mountaintop are the massive Coleman and Roosevelt glaciers.

Seven miles east of Glacier on the scenic byway is a viewpoint of thunderous Nooksack Falls. About one mile beyond the falls, the scenic byway follows the southern boundary of North Fork Nooksack Research Natural Area for 3.5 miles. Located on a steep slope above the North Fork of the Nooksack River, this is an excellent example of an old-growth forest dominated by Douglas fir, western hemlock, western red cedar, and silver fir. At higher elevations are mountain hemlock, whereas down along the river are red alder and black cottonwood. On the densely shaded forest floor are Pacific yew, vine maple, Cascade Oregon grape, holly fern, twin-flower, evergreen violet, pipsissewa, and snowline pyrola.

Baker Lake and national forest lands all around it may be reached via the Baker Lake Highway (Forest Road 11) off of State Route 20 about four miles east of Hamilton. You may launch your boat from several places at Baker Lake, and there are picturesque secluded inlets such as Horseshoe Cove and Maple Grove. At the end of Baker Lake Road, you find the trailhead to the 14-mile Baker Lake Trail.

From the south end of Baker Lake is a road that follows Anderson Creek before winding up to Lilypad Lake and finally to a parking area below Anderson Buttes. Trails here lead to the Anderson and Watson Lakes as well as toward Mt. Watson in the Noisy-Diobsud Wilderness.

Darrington is a small community on State Route 530 on a flat between the Sauk River and the North Fork Stillaguamish River. Three miles south on the Mountain Loop Highway and then three miles on Clear Creek Road is scenic Asbestos Creek Falls. The Boulder River Trail is 8.6 miles along a road west of Darrington off of Route 530, then 3.7 miles to the trailhead. A one-mile hike brings you to a falls that roars 80 feet over a cliff into Boulder River. The falls are within the Boulder River Wilderness.

Another small town in the forest is Verlot, situated on the banks of the South Fork Stillaguamish River on the Mountain Loop National Scenic Byway. Before embarking on the Mountain Loop Road, you may want to visit a couple of pristine natural areas near Verlot. A 2.7-mile trail leads from the South Fork Stillaguamish River to Lake Twenty-two. Although the 2,460-foot elevation of the lake is relatively low, the area is reminiscent of a much higher subalpine lake. An ice field is at one edge of the lake and remains until mid-summer or sometimes all year long.

Four miles northeast of Lake Twenty-two is the Long Creek Research Natural Area situated on a ridge that rises nearly 2,400 feet above the South Fork Stillaguamish River. This is a virgin forest of western hemlock with large western red cedars, silver firs, and Douglas firs mixed in.

For several miles the Mountain Loop Road follows the route of the Everett and Monte Cristo Railroad that was built to transport ore from the mountains along the river. As you drive along, you will see signs of the past: old foundations from long-ago towns and burned-down hotels, mining remnants, and abandoned railroad grades and trestles. From the Gold Basin Campground you may walk around the site of a little town that was here and that boasted a population of 72 in 1910. You will be able to see the mill pond, a log chute, the boiler house, and remnants of the old mill.

The Big Four Picnic Area is on the site of the old Big Four Resort Hotel where only a fireplace and foundations remain. Today, you may take a one-mile walk on a wheelchair-accessible trail from here to the Ice Caves during the summer season. After crossing a marshy area on boardwalks, the trail wanders through a woods and then crosses the South Fork of the Stilliguamish River on a footbridge. After next walking beneath a dense forest of western hemlock, silver fir, and western red cedar, you come face to face with a big ice field on the north side of Big Four Mountain. In the ice field at the base of the 4,000-foot rock wall are the Ice Caves. Because ice from the ceilings of the caves may fall at any time and avalanches are always a threat, it is hazardous to enter the caves.

At Barlow Pass on the Mountain Loop Road is a trail to the old railroad grade. A gated road from the pass ends at Monte Cristo in 4.4 miles, an important mining town that had a population of 200 in 1902. At the time of the census of 1920, there was only a single inhabitant.

The Mountain Loop Road continues beyond Barlow Pass and follows the South Fork of the Sauk River until that river eventually joins the North Fork. To the east of the junction is pretty North Fork Falls. The falls are a quarter-mile from the road.

Some travelers may begin their fun in the forest by taking U.S. Highway 2 over Stevens Pass on the border with the Wenatchee National Forest. A popular ski area is at Stevens Pass. The highway follows the route of the old "railroad through the Cascades," as it was described by John F. Stevens who designed this engineering marvel. The railroad went through narrow, steep canyons and required nine switchbacks and a grade of 4 percent to cross Stevens Pass. Eventually two tunnels, one 7.8 miles long, were bored through the mountain near the pass, and portions of the old switchback railroad grade were converted into an automobile route in 1925. From the west side of the pass you can see portions of the original railroad grade. The Iron Goat Trail takes visitors on an interpretive walk past old snowsheds and tunnels on the grade.

After descending from the pass, U.S. Highway 2 follows the Tye River, coming to Deception Falls. From the Deception Falls Picnic Area is a short trail from which to observe the 60-foot cascade.

Nine miles west of the tiny community of Skykomish is Eagle Falls. Park near Milepost 39 and take a short hike to see the 70-foot falls. In another 2.5 miles, take the Mt. Index Road that crosses the South Fork of the Skykomish River. A 1.5-mile trail leads to remarkable Bridal Veil Falls, which descends 200 feet in a series of four falls. If you are ready for another hike, there is a nice nature trail at the Troublesome Creek Campground 12 miles up valley from Index along the North Fork Skykomish Road.

Interstate 90 winds through the southern part of the Mt. Baker–Snoqualmie National Forest. Beyond the north end of large Keechelus Lake at Snoqualmie Pass are four popular ski areas. In 1868, a wagon road was constructed over Snoqualmie Pass that served as the major route to the Puget Sound area. One mile of this old wagon road has been preserved, beginning at the Denny Creek Campground two miles off of Interstate 90. Points of interest include the remains of red cedar puncheon that formed bridges over boggy terrain, decaying stumps of trees that had to be felled for the right-of-way, and unearthed iron tracks of an ore-hauling narrow gauge railroad. At the end of the trail is the 70-foot drop of fabulous Franklin Falls. At the Snoqualmie Pass summit area is Gold Creek Pond, which features a paved one-mile loop trail, a picnic area, and great views of the Alpine Lake Wilderness.

About 23 miles east of the old Enumclaw Ranger Station, State Route 410 enters the Mt. Baker–Snoqualmie National Forest and for eight miles goes through magnificent old-growth forests until the highway enters Mt. Rainier National Park at Crystal Mountain Boulevard. The first mile of this highway through an old-growth forest is the Mather Memorial Parkway, named for Stephen T. Mather, the first director of the National Park Service. A little farther south, at the lower end of The Dalles Campground, a nature trail winds through the John Muir Grove, a forest with some of the most outstanding trees around, including a Douglas fir estimated to be more than 700 years old and more than nine feet in diameter. About eight miles south of The Dalles Campground is a forest highway that climbs up Silver Creek to Crystal Mountain. Crystal springs bubble out of the ground at the Silver Creek Campground, and Crystal Mountain is a ski destination. Take a lift to the top of Crystal Mountain for one of the best views of Mt. Rainier.

Lake Twenty-two Natural Area

Although Lake Twenty-two is only at an elevation of 2,460 feet, it is a subalpine lake surrounded by subalpine forests and meadows. Located near the top of a rugged mountain in the northern Cascades of Washington in the Mt. Baker–Snoqualmie National Forest, Lake Twenty-two is so subalpine that there is a permanent snowfield situated near the water's edge at one point around the lake. Lake Twenty-two is at the heart of the 790-acre Lake Twenty-two Natural Area.

The trailhead to Lake Twenty-two is about two miles east of the old Verlot Ranger Station along a Forest Highway. The trail begins from the South Fork Stillaguamish River and makes a steady climb to Lake Twenty-two where the trail ends. En route, the trail increases in elevation a total of 1,360 feet. The mountain slopes that surround the lake rise for another 1,140 feet.

Lake Twenty-two is the result of glaciation where water has filled a cirque to a maximum depth of 53 feet. The rocks that outcrop here and there along the trail are mostly sedimentary. At the higher elevations around the lake, most of the outcroppings are granular.

The area surrounding Lake Twenty-two is cool, with summer highs averaging only about 79 degrees. Average annual precipitation is just over 80 inches, including 47.5 inches of annual snowfall. Some rain usually falls throughout the summer, so that the area around Lake Twenty-two is usually kept fairly moist. The steep, north slope around the lake is even cooler and wetter. Lake Twenty-two drains into Lake Twenty-two Creek, and the water eventually drops into the South Fork Skillaguamish River. The trail to Lake Twenty-two sometimes comes near the creek.

Lake Twenty-two Creek itself is beautiful as the clear water flows over boulders of all sizes. The vegetation along the creek is lush, with huge plants of lady fern mixed in with mountain boykinia and pig-a-back plant. Blue currant and salmonberry are also common and conspicuous. Of course, the presence of devil's-club sometimes makes exploring and walking along the creek difficult.

Although the natural area appears heavily forested, nearly one-fourth of it consists of open meadows, rocky cliffs, and talus slopes. The most impressive communities are the old-growth forests of western red cedar and western hemlock at the lower elevations. Western red cedars here are as much as 12 feet in diameter, with the average size ranging from five to eight feet. Western hemlocks are smaller, but some are more than four feet in diameter. Because most of the young growth in the forest consists of western hemlock, this species appears to be the one that will eventually dominate these forests. Mixed in with the western red cedar and western hemlock in the canopy are occasional specimens of Pacific silver fir and Sitka spruce. These old-growth forests are so lush with vegetation that it is difficult to walk under the trees except where the trail is located. This lush vegetation consists of shrubby plants such as Alaska blueberry, oval-leaved blueberry, rusty menziesia, all of the heath family, and five-leaved bramble. Bunchberry (pl. 60) is the common wildflower, and deer fern the dominant fern. Great masses of mosses, including a species of *Sphagnum,* grow in profusion.

The higher mountain slopes around Lake Twenty-two have a different forest community with the dominant trees being Pacific silver fir, mountain hemlock, and Alaska cedar. Some of these stands are also old growth, with the largest and oldest trees more than three feet in diameter. The shrub layer is similar to that below the western red cedar and western hemlock forest, but with the addition of salmonberry and copper-bush. The nonwoody dominants in this community are deer fern, rosy twisted stalk, and beadruby.

At several places along the trail to the lake, the forests are interrupted by shrubby thickets where vine maple, Sitka alder, elderberry, and swamp gooseberry are common. Below these shrubs is a rich and diverse flora of lady fern, rock brake fern, bracken fern, goatsbeard, montia, bedstraws, and pig-a-back plant. A significant stand of bigleaf maple occurs at one place along the trail.

In the moist areas that surround the lake, a variety of wetland species occurs, including several kinds of sedges, as well as bistort, marsh marigold, green false hellebore, Sitka valerian, and alpine lady fern. Here and there among several kinds of blueberries are mountain ashes. Surprisingly, two species that normally occur above timberline are found on the shore of the lake: mountain heather and partridgefoot.

Zoologists report nearly 50 species of mammals may frequent the natural area, including snowshoe hares (pl. 61), pikas (pl. 62), mountain beavers, mountain lions, bobcats, black bears, elk, and mountain goats.

Okanogan National Forest

SIZE AND LOCATION: 1,706,000 acres in north-central Washington, from the Canadian border to Lake Chelan and from North Cascades National Park to 30 miles east of Oroville. Major access roads are U.S. Highway 97 and State Routes 20, 21, 30, and 153. District Ranger Stations: Tonasket, Leavenworth, Winthrop. Okanogan Valley Office: 215 Melody Lane, Wenatchee, WA 98801, www.fs.fed.us/r6/okanogan.

SPECIAL FACILITIES: Winter sports areas; boat ramps; visitor center; swimming areas.

SPECIAL ATTRACTIONS: North Cascades National Scenic Byway; Ross Lake National Recreation Area.

WILDERNESS AREAS: Lake Chelan–Sawtooth (151,435 acres, partly in the Wenatchee National Forest); Pasayten.

The Okanogan National Forest has two different appearances. The western side has the jagged peaks, deep valleys, cathedral forests, and alpine tundra of the Cascade Mountains. The eastern side is rolling hills, parklike ponderosa pine forests, and tranquil meadows.

The huge Pasayten Wilderness is the epitome of the Cascades. Cathedral Peak near the Canadian border, Nanny Goat Mountain, Frosty Pass, and the Three Pinnacles conjure up images of this range. The Pacific Crest National Scenic Trail goes over several passes on its route south from the Canadian border. Favorite trailheads to the wilderness area are from the Iron Gate Campground where there is a five-mile-long trail to broad and beautiful Horseshoe Basin with tundra-tipped peaks towering above. The other trailhead is from Harts Pass Campground where it is two miles to Buffalo Pass, three miles to Windy Pass, and four miles to Foggy Pass. However, the campground is not accessible by passenger cars.

The North Cascades National Scenic Byway goes through the Cascade Mountains just below the Pasayten Wilderness. From Winthrop to Mazama,

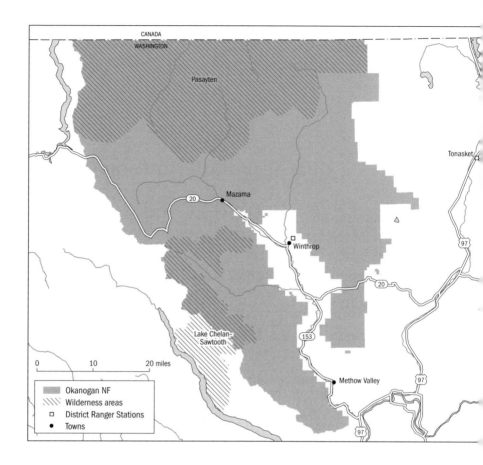

the highway stretches along the Methow River valley. After Methow, the mountains close in on the highway. One mile beyond Mazama, the scenic byway takes a due westerly course. After about three miles, look for a forest road to the south. Take this road to the Cedar Creek Trailhead for a 45-minute walk over fairly flat terrain to the spectacular Cedar Creek Falls, a series of 30-foot cascades.

Back on the scenic byway, and after passing the two narrow peaks of The Needles to the north and Silver Star Mountain to the south, you come to Washington Pass Overlook. At 5,480 feet, you can look down into Early Winters Creek and Snagtooth Ridge beyond, and you can look across to Copper Basin, which is surrounded by the impressive Kangaroo Ridge. Immediately across the highway from Washington Pass is Liberty Bell Mountain. Before you get too comfortable in your vehicle, it is time to get out at the Whistler Mountain Viewpoint. The flanks of Whistler Mountain have large patches of alpine meadows. A little farther on is a neat mile-long wheelchair-accessible

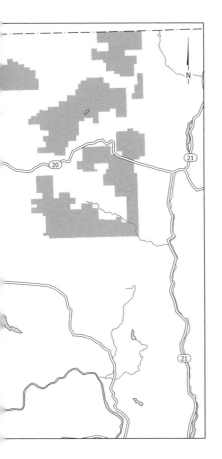

nature trail at Rainey Pass that culminates in a viewing deck at the edge of Rainey Lake. This is subalpine country with glaciated peaks and several waterfalls visible from the trail. In the spring you may be treated with the blossoms of wood trillium, false Solomon's-seal, wild lily-of-the-valley, and vanilla leaf, the last with its strangely shaped leaves. By early summer, the hairy Indian paintbrush and lavender beardstongue are in their full glory. The descent of the highway from Rainey Pass goes down to Granite Creek and eventually to Ross Lake National Recreation Area.

Back at Mazama, the Harts Pass Scenic Road heads northwest from its junction with the North Cascades National Scenic Byway. This extremely narrow and sometimes scary road follows the route of the 1890 gold prospectors. You see the remains of gold-mining operations when you get to Harts Pass, 23 miles away. Impressive 2,000-foot-tall Goat Ridge stands guard above the Methow River for four miles just north of Mazama. At the north end of Goat Ridge, walk up Gates Creek for a quarter mile, then try to climb a treacherous talus slope to get a view of a 100-foot waterfall rushing down the side of Goat Ridge.

Toward the end of the road, at Harts Pass Campground, is Slate Peak. The views of the Cascade Range from the top of the peak are unbeatable. Beyond the campground, the road becomes even more difficult, but it will bring you close to the old gold-mining communities of Barron and Chancellor.

There is plenty to see and do along the Chewuch Road north of Winthrop. Three miles beyond Cub Pass is a side road to Sweetgrass Butte. From the top of the butte is a 360° view of Silver Star and Gardner Mountains, both far to the southwest. If you stay on the Chewuch Road all the way to the end of the road just before reaching the Thirty Mile Campground, you will have the opportunity to see two more impressive waterfalls. The lower part of Falls Creek Falls and Chewuch Falls both drop 30 feet. The latter is reached by a three-mile trail from Thirty Mile Campground.

If you still have not seen enough waterfalls, take the Gold Creek Road south of Twisp to the Foggy Dew Campground and follow Foggy Dew Road for four miles. From the end of this road is a 2.5-mile trail to Foggy Dew Falls, which rushes for 100 feet down a very narrow chute.

From the community of Conconully toward the eastern side of the Okanogan National Forest, the Salmon Creek Road heads north to Salmon Meadows Campground. After looking at the wildflowers in the nearby meadow, you may wish to walk a quarter mile to Salmon Falls, which drop in four different falls over a distance of a quarter mile. Should you want to spend the rest of the day driving a very twisting back road that is steep in places, continue past Salmon Meadows over Lone Frank Pass to Tiffany Springs, Tiffany Meadows, and tiny Roger Lake before eventually returning to Conconully.

East of the Okanogan River are a few small units of the Okanogan National Forest. Most of the points of interest are in the vicinity of Bonaparte and Lost Lakes. The road along Bonaparte Creek off of State Route 30 goes to Bonaparte Lake, Beaver Lake, and Lost Lake, all with good boating facilities. Between Beaver Lake and Lost Lake is the Big Tree Botanical Area where a quiet nature trail goes beneath some of the tallest conifers in this part of Washington.

Olympic National Forest

SIZE AND LOCATION: 632,324 acres in northwestern Washington, north and west of Olympia. U.S. Highways 101 and 119 encircle the forest. District Ranger Stations: Forks, Quilcene. Forest Supervisor's Office: 1835 Black Lake Boulevard, SW, Olympia, WA 98512, www.fs.fed.us/r6/olympic.

SPECIAL FACILITIES: Boat ramps.

SPECIAL ATTRACTIONS: Quinault Rain Forest.

WILDERNESS AREAS: The Brothers (16,682 acres); Wonder Mountain (2,349 acres); Colonel Bob (11,961 acres); Buckhorn (44,258 acres); Mount Skokomish (13,015 acres).

Except for parts of Hawaii and Alaska, the northwestern corner of the State of Washington receives more rainfall, an average of 12 feet annually, than any other place in the United States. Because of this high amount of precipitation, this region of Washington is occupied by some of the most lush temperate rainforests you will ever see. This corner of the state is occupied by the Olympic Peninsula, a 6,500 square mile area surrounded on three sides by

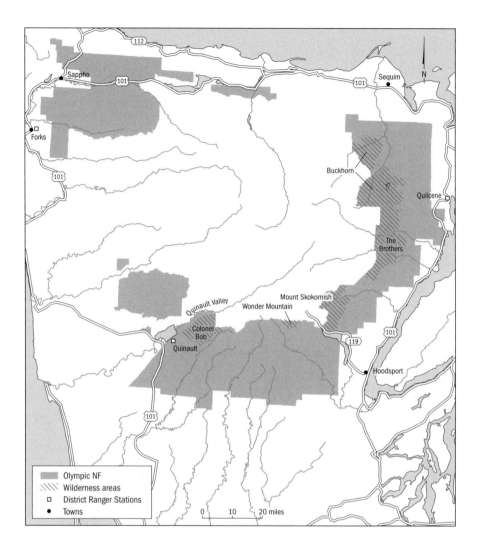

Olympic NF
Wilderness areas
□ District Ranger Stations
• Towns

0   10   20 miles

saltwater: the Pacific Ocean to the west, Strait of Juan de Fuca to the north, and the inland waters of Puget Sound to the east. The interior of the peninsula has rugged mountainous terrain. Approximately 900,000 acres near the center of the peninsula are in the Olympic National Park, which is, in turn, surrounded on three sides by the 632,324 acres of the Olympic National Forest.

U.S. Highway 101 forms a complete loop around the Olympic Peninsula, and all points of interest in the national forest may be accessed off this route.

The signature features of the Olympic National Forest are the temperate rainforests with their huge moss- and lichen-draped trees. The best of these forests are in the Hoh, Queets, and Quinault Valleys, although the last is the only one in the national forest. In addition to the rainforests are rugged mountains, some high enough to harbor alpine vegetation. The countless streams that drain the mountains are a delight to fishermen. Some of the streams are associated with picturesque waterfalls.

The most convenient access to the national forest from mainland Washington is from the capital city of Olympia. Twenty-one miles north of Olympia, off of U.S. Highway 101, take your first side trip on Skokomish Valley Road, a 40-mile jaunt that follows the impressive South Fork of the Skokomish River. A 2.5-mile-long gravel forest road to the north will bring you to the impressive Steel Bridge standing 470 feet above rocky Vance Creek Canyon. It was built by the Simpson Timber Company in 1929 as part of their logging railway. Returning to Skokomish Valley Road, proceed to the Brown Creek Campground and the pleasant mile-long Beaver Pond Nature Trail. Where the road dead-ends, the small Wonder Mountain Wilderness is a short distance northeast. Although the summit of 4,848-foot Wonder Mountain supports a dense forest of western hemlock, Douglas fir, and silver fir, there are no marked hiking trails, and you are literally on your own.

Back on U.S. Highway 101, continue to Hoodsport and take another road that follows the eastern shore of Lake Cushman before entering the national forest. At the Big Creek Campground, there is an easy one-mile self-guiding nature trail that forms a loop. If you want to stretch your legs a little more, there is also a four-mile loop that meanders through the Big Creek basin. From the campground take the secondary forest road east, climbing switchbacks and finally coming to Elk Lake. A marvelous trail here goes through majestic old-growth forest. Armed with a good forest service map, follow a forest road from Elk Lake to the Hamma Hamma River. If you continue along the river west to the Lena Creek Campground, you will find Lena Lake Trail, a strenuous trail that leads north to picturesque Lena Lake and then on to the twin peaks of The Brothers in The Brothers Wilderness. A second equally difficult trail, Mt. Skokomish Trail, leads to Mt. Skokomish in a wilderness that bears its name. Both of these wilderness areas have rocky-topped mountains that are ideal for advanced rock climbers and for observing mountain goats and marmots. The lower elevations in these wildernesses have magnificent old-growth forests.

The next stop is at the Interrorem Guard Station off another side road from 101. The guard station is an old forest service building that is still used. This is a quiet area along the Duckabush River Road. A self-guiding nature trail passes cedar stumps that are as much as nine feet in diameter. Beyond the guard station is a rough road along Murhut Creek. A trail less than one mile long goes to Murhut Falls.

Four-and-one-half miles south of Quilcene is a gravel road to the top of

Mt. Walker. To get to the summit, take the right fork at Walker Pass. If you are there in June or July, you will be treated to a wonderful display of wild rhododendrons (fig. 29) in bloom. From the 2,800-foot summit of the mountain you will be able to see the Olympic and Cascade Mountains, Puget Sound, the Hood Canal, Mt. Baker, and Mt. Rainier. If you choose the left fork at Walker Pass, you will soon come to the Fallsview Campground where a very short trail at the south end of the campground leads to a viewpoint overlooking the falls, which drop 80 to 120 feet into the Big Quilcene River. A fence protects visitors from falling into the canyon floor.

Figure 29.
Rhododendron.

If you visit the Quilcene Ranger Station, you can pick up brochures and cassette tapes to take with you on two scenic drives. One climbs over Bon Jon Pass and goes to Mt. Zion where there is a hiking trail. The other passes over Bear Mountain. Both tours have a stop at Lords Lake.

U.S. Highway 101 continues to the community of Sequim and then runs west and out of the forest, following the northern boundary of Olympic National Park. After passing the Fairholm Campground at the western end of Lake Crescent in the national park, the highway reenters the national forest. Immediately take the Soleduck Road south for 12 miles to the Soleduck Hot Springs Resort where the road ends. Hike the Soleduck Falls Trail from here. In one mile you come to a footbridge crossing the river, and soon there is a perfect view of scintillating Soleduck Falls whose split falls plunge 60 feet.

At the Klahowya Campground is the Pioneer's Path Nature Trail, a quarter-mile interpretive loop through a forest of tall western hemlock and tangles of vine maple. The forest floor beneath the hemlocks is carpeted with soft mosses. Graceful ferns and the shamrock-leaved wood sorrel protrude from the carpet.

At the village of Sappho is Beaver Creek Road, a secondary road north for two miles to Beaver Creek. Just beyond the bridge is a short trail to the base of Beaver Falls where three separate cataracts a total of 80 feet wide plummet down a rocky cliff.

U.S. Highway 101 turns south after Sappho and makes its way out of the forest through the community of Forks and crosses the Hoh River. For a while the highway is at the edge of the Pacific Ocean before turning eastward. After finally getting to Quinault Lake, you are back in the Olympic National Forest for the best thrill yet, because here is the Quinault Rain Forest. You can observe the lush vegetation at its finest on a half-mile nature trail through the Big Tree Grove. For more of the same there is a four-mile Quinault Loop Trail from the Falls Creek Campground.

You can drive along the Quinault River all the way into Olympic National Park. Just before entering the park, the road circles around the northern end of Colonel Bob Wilderness. With average rainfall 120 inches annually, there are outstanding old-growth forests here. Northeast of the 4,492-foot Colonel Bob Peak is Fletcher Canyon, surrounded by sheer cliffs. The canyon is only one mile by trail from the forest highway. U.S. Highway 101 follows outside the southern boundary of the national forest back to Olympia.

Olympic Rain Forest

Many years ago I was surprised to learn that an area in the Olympic Peninsula of northwestern Washington State was referred to as a rain forest, a label that for me conjured up a picture of a steaming jungle. Because the peninsula is at least 2,000 miles north of the tropics, I knew this image was inappropriate, but I could not imagine what a temperate rain forest would be like or how it had developed in that particular place. To satisfy my curiosity, I planned a trip there.

My first introduction to the Olympic rain forest came as I hiked along a nature trail a short distance from the Quinault Ranger Station in Olympic National Forest. Quinault is one of three valleys harboring prime temperate rain forest. A foglike mist hung heavily in the air, enshrouding tall trees draped with countless epiphytes, or air plants. I began to tick off the similarities between what I was seeing and the typical rain forest of the tropics: many trees soared to heights of more than 200 feet; the tallest trees in the canopy were evergreens; there was a middle layer primarily of deciduous trees; abundant epiphytes clothed the trees; and the air was moist. (A check with the nearest weather station revealed that rainfall in the valleys is 120 to

180 inches each year, the highest amount anywhere in the contiguous United States.)

Despite many similarities, however, there were differences between this temperate rain forest and a tropical one. The tall canopy of a tropical rain forest is created by many species of broad-leaved, usually slender-trunked evergreens, whereas Olympic is dominated by only four species, all needle-bearing conifers with exceptionally thick trunks. An abundance of woody vines, or lianas, hang from the tropical giants, whereas vines are practically absent in Olympic. The epiphytes on the branches of the tropical trees are mostly flowering plants, primarily orchids and bromeliads, whereas mosses, lichens, liverworts, and ferns grow on the trunks and branches of the temperate trees. The lower layer of woody plants in a tropical rain forest consists of tree ferns and bamboos; in Olympic, the shrub layer is sparse and consists mainly of vine maple, huckleberry, and devil's-club. And whereas the tropics are hot, Olympic is cool and mild.

Quinault Valley and the more northerly Queets and Hoh Valleys lie on the southwestern flank of the Olympic Mountains, a part of the Coast Ranges, which parallels the Pacific Ocean shore from British Columbia to California. The broad, U-shaped valleys were carved out thousands of years ago by glaciers that edged slowly down the mountains. Today, the Quinault, Queets, and Hoh rivers, which flow through these valleys, are fed by remnants of the Ice Age, the glaciers that persist on the highest ridges.

Mount Olympus (7,965 feet), which caps the range, is only 32 miles from the Pacific Ocean. As moisture-laden clouds blow inland from the ocean, they meet the mountain barrier and are swept upward, where the chill air causes them to condense and release their water. The resultant 10 to 15 feet of rainfall each year is reflected in the phenomenal sizes of the four conifers that dominate the Olympic rain forest—Sitka spruce, western hemlock, Douglas fir, and western red cedar. The Sitka spruces account for about one-third of the trees. They are much larger here than anywhere else in their range, sometimes reaching heights of 250 feet, with trunk diameters of 13 feet. Western hemlocks are the next most abundant; they stretch more than 200 feet and their trunks may exceed 8 feet in diameter. Douglas firs grow nearly 300 feet tall; one mammoth has a diameter of 15 feet. Western red cedars may be only half as tall, but they often have greater girths. One measures nearly 22 feet in diameter.

Over time, some of these large trees crash to the forest floor and set off a chain of events destined to last several hundred years. The fallen trees are quickly invaded by borers searching for food and a place to lay their eggs. The holes made by these insects provide a niche for fungal spores to lodge. Soon hundreds of fungi penetrate the fallen log, causing decay. Before long mosses and lichens find a home on the downed tree, and the seeds of herbs, shrubs, and trees sprout there. The tree seedlings seem to benefit from the presence on their roots of fungi that convert nutrients to usable forms.

Because they help promote the growth of a new generation of trees, the fallen conifers are known as nurse logs. The tree seedlings on a nurse log develop slowly at first, as their roots are not long enough to encircle the fallen trunk and reach the soil, but once the roots touch ground, the seedlings grow rapidly.

Owing to competition, only a small number of the hundreds of seedlings that sprout on a nurse log reach maturity. After scores of years, with the old nurse log nearing total decay, perhaps as many as 15 trees will survive, arranged as a colonnade. When a nurse log has totally decomposed, each living tree in the colonnade seems to be standing on stilts, where its roots grew around the old log on their way to the soil. Not until many years later are the openings beneath the stilts filled in. Although thousands of tree seedlings also germinate on the forest floor, most of them are unable to compete with the dense growth of mosses and herbs that carpet the ground, so most of the forest's tree regeneration depends on nurse logs.

Tall trees also topple to the ground in tropical rain forests, and the gaps they leave in the canopy admit sunlight essential to the growth of certain seedlings. Often these seedlings develop from seeds that lodge in the bark of fallen trees. Although some tropical trees decay rapidly in the intense heat and humidity, many of them have extremely hard wood that impedes decay. These then become tropical nurse logs, much like the fallen trees in the temperate Olympic rain forest.

Wenatchee National Forest

SIZE AND LOCATION: Approximately 1.2 million acres in west-central Washington, from Chelan and Wenatchee to the crest of the North Cascade Mountains. Major access roads: Interstate 90, U.S. Highways 2, 12, 97, and 97A, and State Routes 207, 209, and 410. District Ranger Stations: Chelan, Cle Elum, Entiat, Naches. Forest Supervisor's Office: 215 Melody Lane, Wenatchee, WA 98801, www.fs.fed.us/r6/wenatchee.

SPECIAL FACILITIES: Boat ramps; swimming beaches; winter sports areas.

SPECIAL ATTRACTIONS: Chiwawa Valley Scenic Drive; White River National Scenic Byway.

WILDERNESS AREAS: Lake Chelan–Sawtooth (151,435 acres, partly in the Okanogan National Forest); Henry M. Jackson (100,337 acres, partly in the Mt. Baker–Snoqualmie National Forest); Glacier Peak (572,338 acres, partly in the Mt. Baker–Snoqualmie National Forest); Alpine Lakes (362,670 acres, partly in the Mt. Baker–Snoqualmie National Forest); Norse Peak (52,180 acres, partly in the Mt. Baker–Snoqualmie National Forest); William O. Douglas (168,157 acres, partly in the Gifford Pinchot National Forest); Goat Rocks (108,279 acres, partly in the Gifford Pinchot National Forest).

It would be very difficult to find anywhere in the United States with more electrifying scenery than in the North Cascade Range, where cathedral forests fill deep valleys, where waterfalls tumble and mountain streams rumble, where snow-capped and glacier-bearing high peaks are punctuated by deep blue lakes and glacial tarns. The land from the crest of these mountains east to the Columbia River is in the Wenatchee National Forest, although it is sometimes difficult to know when you have left the Wenatchee and entered the Okanogan, Mt. Baker–Snoqualmie, and Gifford Pinchot National Forests, which are contiguous on most sides.

From Wenatchee, scenic U.S. Highway 2 begins a steady climb up the east face of the North Cascades. At the very popular ski area at Stevens Pass, the highway enters the Mt. Baker–Snoqualmie National Forest.

Begin your adventure in the Wenatchee by taking one of the Lake Chelan Boat Company's vessels along the 55 miles of Lake Chelan to Stehekin, which is located at the southern edge of the Lake Chelan National Recreation Area administered by the National Park Service. Stehekin is the most remote community in the lower 48 states not reached by a road. The boat makes stops on national forest land at Prince Creek, Lucerne, and Moores Point. If you disembark at Prince Creek or Moores Point, there are nice campgrounds where the Prince Creek and Fish Creek enter Lake Chelan. From the campgrounds you may hike a long way along the creeks toward Sawtooth Ridge, which forms the border between the Wenatchee and Okanogan national forests. The most popular disembarkation point along the way is at Lucerne on the west side of Lake Chelan. Not only are there campgrounds here, but it is close to the jumping-off place to explore Domke Lake at the foot of Domke Mountain. From the south end of the lake you can follow a creek to where it plunges 50 feet into Lake Chelan. You can also see Domke Falls from the boat on its return trip to Chelan. Heading due west for eight miles is a roadbed that parallels Railroad Creek to the remains of Holden Mine, the state's largest copper, gold, and zinc mine until it closed in 1957. From Holden there is a trail into Glacier Peak Wilderness. In seven miles this trail passes Hart Lake and comes to pretty Crown Point Falls before it begins its

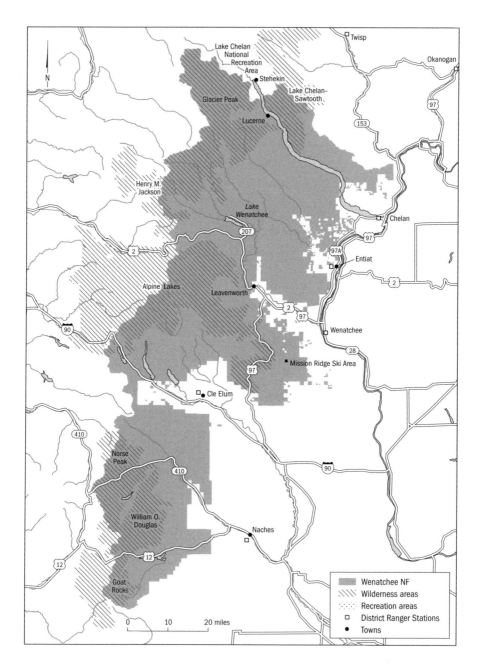

ascent to Suiattle Pass. At the pass you can pick up the Pacific Crest National Scenic Trail. North on this trail goes through the Wenatchee National Forest to North Cascades National Park; west on the trail goes through the Mt. Baker–Snoqualmie National Forest on its way eventually to Glacier Peak.

If you have plenty of time when you get to Stehekin, go to the far northwest corner of Lake Chelan National Recreation Area to the National Park Service's High Bridge Campground. It is only a mile from the campground into Agnes Gorge in the Wenatchee. The most rugged and experienced hiker can hike from Cascade Pass in the national park to Dome Glacier and Dome Peak in the Wenatchee, passing through the extremely rugged Ptarmigan Traverse.

From Chelan, U.S. Highway 97 follows the course of the Columbia River circling far to the south to Sunnyslope before turning northeast to stay near the Wenatchee River. Some 18 miles south of Chelan is the community of Entiat. Just one mile beyond is the tiny village of Brief. A short walk brings you to Preston Falls that bounces thunderously on its way over a protruding rocky mountainside for 100 feet.

In another nine miles along the Entiat River Road is Silver Falls. The half-mile Silver Falls National Recreation Trail comes to the base of the cliff where you can watch in awe the 140-foot drop of the falls.

Back on U.S. Highway 2 in Leavenworth, there is more adventure. State Route 209, the Chumstick Road, follows Chumstick Creek through dense stands of ponderosa pines. In two miles, the Entiat Ridge Road goes on a breathtaking side trip that ultimately comes to Sugarloaf Peak. You have a 360° unobstructed view from here.

Just outside Leavenworth, U.S. Highway 2 makes its way through deep and dark Tumwater Canyon with steep rock walls closing in on either side of the highway. After 16 miles, take State Route 207 to the eastern end of Lake Wenatchee. Two options exist for scenic driving. The first route is the Chiwawa Valley Scenic Drive. As you travel it, look for evidence of gold prospecting and homesteads of years gone by. If you take this 24-mile road to its terminus at Trinity, you will pass 15 campgrounds and 22 hiking trails. After crossing the Chiwawa River, there is a mountain meadow known as Chikamin Flats where huckleberry-picking draws many people during the season. Four miles past the flats is a clearing where trees were knocked down during a debris flow triggered by a torrential downpour in 1990. In a quarter mile is the wooden shed and flagpole, all that remains of the old 1913 Rock Creek Guard Station. At the end of the road is Trinity, where there are a few buildings remaining from a large mining town of the 1920s. Trinity is the trailhead for the Buck Creek Pass Trail in the Glacier Peak Wilderness. It is nine miles from the end of the road to the pass.

The White River National Scenic Byway is the other driving route from Lake Wenatchee that should not be missed. From the northwest end of the lake, the scenic byway never strays far from this wild river. A stone and concrete cross marks the grave of pioneer Eunice Harris. A very rough side road goes to Dirtyface Peak where you can see in all directions including down, where the White River is 900 feet below. Next is the historic Napeequa Crossing where Wenatchee Indians forded the White River. Huckleberry-picking is great in season. Near the end of the 24-mile drive is White River Falls and an adjacent campground. Hemmed in by narrow rock walls, the river falls 132 feet, ending with a tumultuous white splash.

If you wish to explore the Wenatchee National Forest south of U.S. Highway 2, backtrack to Leavenworth. Drive south past the Leavenworth Fish Hatchery and take the Icicle Creek Road if you plan to do some hiking and camping in the Alpine Lakes Wilderness. Several options exist. To reach the incomparable Enchanted Lakes, you need to be in top physical condition and have ample time. Branch off the Icicle River Road at the Bridge Creek Campground. At the end of this three-mile road is the trailhead to the Enchanted Lakes. Allow three days just to get in and out of the area, but when you do make it, you are treated to such outstanding features as pointed Prusik Peak with rock-strewn Gnome Tarn at its base. Several scenic trails lead from an Alpine Wilderness trailhead at the end of Icicle River Road.

Between Leavenworth and Cashmere, U.S. Highway 97 branches south from U.S. Highway 2, for a while staying near Peshastin Creek. Just after passing the Ruby Creek side road, you will come to an old miner's arrastra where gold was prospected between 1870 and 1890. U.S. Highway 97 twists and turns and climbs to Swauk Pass. A campground once was at the pass that is adjacent to wildflower-laden Tronsen Meadow, but the presence of the spotted owl has closed this area to visitation.

If you are in the mood for a more exciting route over the crest of the Wenatchee Mountains, leave U.S. Highway 97 just south of the miner's arrastra and take the remarkably crooked Scotty Creek Road over Blewett Pass. From the Mineral Spring Campground a little farther south, there is a forest road to Red Top Mountain that is a must for avid rock hunters. Colorful rocks and minerals of many types may be sought all over the mountain. After U.S. Highway 97 leaves the national forest land, it is only a short distance to Interstate 90 and Cle Elum.

Back on U.S. Highway 2, there are roads to the Mission Ridge Ski Area of the national forest. The more lengthy and circuitous route follows Mission Creek from Cashmere to its junction with the Squilchuk River Road. The more direct route is to take the Squilchuk River Road from Wenatchee.

The extreme southwestern corner of the Wenatchee National Forest is adjacent to the eastern side of Mt. Rainier National Park and the eastern edge of Gifford Pinchot National Forest. The region is actually in the old Snoqualmie National Forest but is administered by the Wenatchee. State Route 410 from Yakima and U.S. Highway 12 from Naches are the major access roads. After entering the national forest about 30 miles west of Naches is the

Tieton Ranger Station in the heart of a popular recreation area that features Rimrock Lake. The lake is ideal for all kinds of water activities. Several campgrounds are along the Tieton River, as well as interesting scenery such as Goose Egg Mountain, Soup Creek, and Kloechman Rock. Just off the highway at the northeast end of Rimrock Lake is a series of vertical basalt columns known as the Wildcat Post Pile. It is worth seeing.

A different highway from the ranger station circles around the eastern side of Goose Egg Mountain and Rimrock Lake and follows the South Fork of the Tieton River until it ends at the Klickton Divide. From an observation point south of the Grey Creek Campground is a view of Blue Slide, a colorful prehistoric rock and earth slide.

Forty miles northwest of Yakima on State Route 410 is the Rattlesnake Creek Road that twists its way to small McDaniel Lake. From the north side of the lake is a steep and narrow trail to the top of Meeks Table, an isolated flat-topped butte that stands 500 feet above the rolling terrain that surrounds it. Vertical cliffs 300 feet high are on all sides of the butte, with talus slopes near the base. Nice stands of Douglas fir are at the east and west ends of Meeks Table. Associated with the Douglas fir are western larch and grand fir. Most of the forest floor beneath these trees is covered by pinegrass and elk sedge, with occasional yellow-blooming heart-leaved arnica present. Along the southern edge of the butte, the forest composition changes with ponderosa pine the dominant tree, although there are scattered Douglas firs. Pinegrass, elk sedge, and broad-leaved lupine are in the understory. About half of the top of the butte, however, is treeless where there are numerous rocky outcrops and a minimum of soil. More than half of these treeless openings are without any vegetation. Where soil is a little more plentiful, Sandberg's bluegrass and Columbia needlegrass are prevalent. Mixed in with these grasses are Douglas' buckwheat, sedum, spreading phlox, and Kellogg's smartweed, along with the shrubby stiff sage.

Continuing north on State Route 410 there is a pull-out for the Boulder Creek Picnic Area. From the picnic grounds is a half-mile trail to Devil Creek Falls and Boulder Cave. The trail goes to the base of the 30-foot falls that can be viewed from a shell-like amphitheater. Downstream is Boulder Cave, a 400-foot-long cavern in basalt, carved by Devil Creek. Bring a flashlight if you wish to explore the cave.

From the Cross Creek Campground on U.S. Highway 12 is a long dead-end road that primarily follows Sand Creek through dense forests. Near the end of the road is the Raven Roost Viewpoint where there are unobstructed views of Mt. Rainier and the saw-toothed summits of Fifes Peaks.

ART CREDITS

We wish to express our gratitude to the Forest Service staff who put us in touch with photographers, helped to locate good images, or donated their images. Peter Keller, the wilderness program manager of the Pacific Southwest Region, directed us to Roxane Scales, public affairs officer, and Anne Bradley, regional botanist. They in turn spread the word about this project among their counterparts in other regional offices. Elton (Sonny) Cudabac of the Southern Region helped us to locate images in their Digital Image Library. Thanks also to David Palazzolo of the Intermountain Region and Tom Iraci of the Pacific Northwest Region.

Plates

Erwin and Peggy Bauer/Wildstock, 10

Brother Alfred Brousseau, Saint Mary's College, 21

California Academy of Sciences, 28, 45

Gerald and Buff Corsi, California Academy of Sciences, 15

Joseph Dougherty, 47

William Follette, 18, 22, 23, 27, 41, 46

John Game, 5, 26

Joyce Gross, 42

Tanya Harvey, 3, 38, 40, 43, 58, 60, 62

Kirk Keogh, www.first2lastlight.com, 30, 32, 34, 35

Russ Kerr, 24

Shue-Huei Liao, 13

Robert Mohlenbrock, 4, 6, 11, 12, 49, 52, 59

Lynn and Donna Rogers, bear.org, bearstudy.org, 61

Scott T. Smith, ScottSmithPhoto.com, 8, 9, 19, 20, 29, 31, 33, 36, 37, 48, 53, 54, 55, 56

USDA Forest Service, Pacific Southwest Region, 16

USDA Forest Service, Pacific Southwest Region, Ken DeCamp, 25, 39

USDA Forest Service, Pacific Southwest Region, Roger and Donna Aitkenhead, 44

USDA Forest Service, Intermountain Region, Teresa Prendusi, 7, 14, 17, 50, 51

USDA Forest Service, Pacific Northwest Region, Thomas Iraci, 1, 2

USDA Forest Service, Southern Region, Bill Lea, 57

Figures

Jennie Haas, 14

Jeanne R. Janish, 20

Andrea J. Pickart, Title page, 2, 4, 5, 9, 10, 11, 12, 13, 15, 16, 17, 21, 22, 23, 24, 27, 28, 29

Robert Mohlenbrock, 6, 7

P. S. Sobol, courtesy of the Uinta National Forest, 26

University of California Press, 8

USDA Forest Service, Pacific Northwest Region, Ted Demetriades, 18, 19

USDA Forest Service, Pacific Northwest Region, 3

USDA Forest Service, Pacific Northwest Region, Thomas Iraci, 1

C. Webber, California Academy of Sciences, 25

INDEX OF PLANT NAMES

fritillary, purple *(Fritillaria atropurpurea)*, 122

fuzzy rattail *(Ivesia argyrocoma)*, 129

galleta *(Hilaria jamesii)*, 295

gentian
    dwarf *(Gentiana affinis)*, 34
    dwarf *(Gentiana parryi)*, 34
    mountain bog *(Gentiana calycosa)*, 182
    Sierra *(Gentianopsis holopetala)*, 122
    spreading *(Gentiana affinis)*, 290

geranium, wild *(Geranium caespitosum)*, 25, 71

gilia
    scarlet *(Gilia aggregata)*, 263, 265
    stiff *(Gilia rigidula)*, 36

ginger, wild *(Asarum caudatum)*, 123, 176

goatsbeard *(Aruncus sylvester)*, 328

goldeneye, showy *(Viguiera multiflora)*, 288

goldenglow *(Rudbeckia laciniata)*, 27, 74

goldenrod
    California *(Solidago californica)*, 156
    dwarf *(Solidago ciliosa)*, 34
    low *(Petradoria pumila)*, 278

goldthread, cutleaf *(Coptis laciniata)*, 227

gooseberry *(Ribes lacustre)*, 25, 33
    swamp *(Ribes laxiflorum)*, 182, 328

grama, blue *(Bouteloua gracilis)*, 295

grape
    Cascade Oregon *(Berberis nervosa)*, 227, 239, 324
    Oregon *(Berberis aquifolium)*, 169
    Piper's *(Berberis piperiana)*, 241

grass
    Alaska bent *(Agrostis alaskana)*, 7
    alpine bent *(Agrostis humilis)*, 278
    Arizona wheat *(Agropyron arizonicum)*, 45
    beach *(Ammophila arenaria)*, 246
    beavertail *(Calochortus coeruleus)*, 122
    bluebunch wheat *(Agropyron spicatum)*, 184, 196, 225, 306
    bluejoint *(Calamagrostis canadensis)*, 6, 7, 8, 25, 155
    Columbia needle *(Stipa columbiana)*, 343
    dwarf alkali *(Puccinellia phryganodes)*, 7
    fall panic *(Panicum dichotomiflorum)*, 27
    golden-eyed *(Sisyrinchium californicum)*, 246
    hair *(Deschampsia caespitosa)*, 25, 277
    Idaho bent *(Agrostis idahoensis)*, 25
    Indian rice *(Oryzopsis hymenoides)*, 295
    June *(Koeleria micrantha)*, 25, 287
    Lettermann's needle *(Stipa lettermannii)*, 288
    mutton *(Poa fendleriana)*, 295
    needle *(Stipa comata)*, 268, 295
    needle *(Stipa occidentalis)*, 268, 295

onion *(Melica bulbosa)*, 278

orchard *(Dactylis glomerata)*, 74

Patterson's blue *(Poa pattersonii)*, 278

pine *(Calamagrostis rubescens)*, 163, 166, 184, 343

purple reed *(Calamagrostis purpurascens)*, 277

salt *(Distichlis spicata)*, 98, 203

Sandberg's blue *(Poa sandbergii)*, 163, 218, 225, 343

seashore blue *(Poa macrantha)*, 246, 247

yellow-eyed *(Sisyrinchium longipes)*, 27

grass-of-Parnassus *(Parnassia californica)*, 122

grass-of-Parnassus *(Parnassia kotzebuei)*, 18, 19

greasewood *(Sarcobatus vermiculatus)*, 98

groundsel
    arrowleaf *(Senecio triangularis)*, 263, 264
    Bolander's *(Senecio bolanderi)*, 241
    butterweed *(Senecio serra)*, 163
    Layne's *(Senecio laynei)*, 88
    San Francisco Peak *(Senecio franciscana)*, 28, 34
    tundra *(Senecio atropurpureus)*, 19
    Uinta *(Senecio uintahensis)*, 279

groundsmoke
    dwarf *(Gayophytum humile)*, 264, 265
    spreading *(Gayophytum diffusum)*, 263, 265

hackberry
    desert *(Celtis pallida)*, 69, 70
    netleaf *(Celtis reticulata)*, 48, 58

harebell, American *(Campanula rotundifolia)*, 25, 45

hawthorn *(Crataegus douglasii)*, 169

hazelnut
    beaked *(Corylus cornuta)*, 230, 239
    California *(Corylus californica)*, 151

heather, mountain *(Phyllodoce aleutica)*, 317, 329

heather, mountain *(Phyllodoce breweri)*, 317, 329

heather, mountain *(Phyllodoce coerulea)*, 317, 329

heather, mountain *(Phyllodoce empetriformis)*, 317, 329

hedgehog, Fendler's *(Echinocereus fendleri)*, 62

hellebore, false green *(Veratrum viride)*, 182, 263, 304, 317, 328

hemlock
    mountain *(Tsuga mertensiana)*, 203, 234, 263, 266, 317
    western *(Tsuga heterophylla)*, 225, 244, 258, 317, 334

hibiscus *(Hibiscus coulteri)*, 47

hibiscus *(Hibiscus denudatus)*, 47
honeysuckle
    bearberry *(Lonicera involucrata)*, 25, 304
    desert *(Lonicera interrupta)*, 46, 48
    pink *(Lonicera hispidula)*, 156
hopsage, spiny *(Grayia spinosa)*, 98
hop hornbeam
    eastern *(Ostrya virginiana)*, 36, 37
    Knowlton's *(Ostrya knowltonii)*, 31,
      36, 37
horsetail
    common *(Equisetum arvense)*, 7
    giant *(Equisetum telmateia)*, 182
    meadow *(Equisetum pratense)*, 6
    swamp *(Equisetum fluviatile)*, 7
huckleberry *(Vaccinium parvifolium)*, 11,
    323, 337
    black *(Vaccinium membranaceum)*, 238
    blue *(Vaccinium membranaceum)*, 187
    Cascade *(Vaccinium deliciosum)*, 317
    dwarf *(Vaccinium caespitosum)*, 187
    evergreen *(Vaccinium ovatum)*, 247

Indian pipe *(Monotropa uniflora)*, 173, 227
inkberry *(Rhamnus crocea)*, 87
inside-out flower *(Vancouveria planipetala)*,
    144, 239
iris
    Missouri *(Iris missouriensis)*, 98
    Rocky Mountain *(Iris missouriensis)*, 25
    wild *(Iris innominata)*, 122, 241
ivesia, hidden *(Ivesia cryptocaulis)*, 207
ivy
    poison *(Toxicodendron radicans)*, 74
    redwood *(Vancouveria planipetala)*, 144

Jacob's-ladder *(Polemonium foliosissimum)*,
    34
Jacob's-ladder *(Polemonium viscosum)*, 279
jasmine, rock *(Androsace septentrionalis)*,
    278
jatropha *(Jatropha cardiophylla)*, 70
jewelflower, milkwort *(Streptanthus
    polygaloides)*, 87
jojoba *(Simmondsia chinensis)*, 60, 62, 68, 75
Joshua tree *(Yucca brevifolia)*, 127, 203
juniper
    alligator *(Juniperus deppiana)*, 24, 43, 71,
      99, 295
    California *(Juniperus californica)*, 130
    creeping *(Juniperus communis)*, 33
    mountain *(Juniperus scopulorum)*, 263
    one-seeded *(Juniperus monosperma)*, 53,
      71, 295
    Rocky Mountain *(Juniperus scopulorum)*,
      53, 281, 290, 308
    Utah *(Juniperus osteosperma)*, 24, 53, 163,
      206, 283

western *(Juniperus occidentalis)*, 33, 93,
    117, 124, 215

kalmia, bog *(Kalmia polifolia)*, 16
kalmiopsis *(Kalmiopsis leachiana)*, 234, 235
kinnikinnick *(Arctostaphylos uva-ursi)*, 169,
    175
kitten-tails *(Synthyris ranunculina)*, 206

Labrador tea *(Ledum groenlandicum)*, 9, 16,
    122, 182
lady's-slipper
    clustered *(Cypripedium fasciculatum)*,
      170, 172
    yellow *(Cypripedium calceolus)*, 19, 175
larch, western *(Larix occidentalis)*, 170, 225,
    249, 313, 343
larkspur
    dwarf blue *(Delphinium nuttallianum)*,
      122
    Menzies' *(Delphiniuim menziesii)*, 263
    red *(Delphinium nudicaule)*, 123
    tall *(Delphinium glaucum)*, 18
    tall blue *(Delphinium exaltatum)*, 27, 278
laurel
    alpine *(Kalmia microphylla)*, 262
    California *(Umbellularia californica)*, 239,
      241
    Sierra *(Leucothoe davisii)*, 123
ledum, western *(Ledum glandulosum)*, 187
lewisia, three-leaf *(Lewisia triphylla)*, 264
lily
    bead *(Clintonia uniflora)*, 176, 263
    Bolander's *(Lilium bolanderi)*, 144
    cobra *(Darlingtonia californica)*, 121, 144,
      237
    corn *(Veratrum caifornicum)*, 25, 156
    giant fawn *(Erythronium oreganum)*, 123,
      241
    pale fawn *(Erythronium grandiflorum)*,
      264, 279
    San Luis Mariposa *(Calochortus
      obispoensis)*, 111, 265
    Sego *(Calochortus gunnisonii)*, 175
    Sierra *(Lilium kelleyanum)*, 123
    Tolmie's Mariposa *(Calochortus tolmei)*,
      241
    Vollmer's *(Lilium pardalinum vollneri)*,
      144
    Washington *(Lilium washingtonianum)*,
      123, 156
    water *(Nymphaea odorata)*, 85
lily-of-the-valley
    false *(Maianthemum dilatatum)*, 244
    wild (*Maianthemum dilatatum)*, 331
linanthus, Baldwin Lake *(Linanthus killipii)*,
    128
locoweed, Jones' *(Oxytropis jonesii)*, 282

ocean spray *(Holodiscus discolor)*, 169, 173, 226, 313
ocotillo *(Fouquieria splendens)*, 32, 43, 62, 127
onion
  coast flat-stemmed *(Allium platycaule)*, 144
  Sanford's wild *(Allium sanfordii)*, 88
orchid
  Alaska rein *(Platanthera unalascensis)*, 241
  bog *(Platanthera saccata)*, 25, 27, 182, 298
  calypso *(Calypso bulbosa)*, 26
  coralroot *(Cypripedium mertensiana)*, 26, 74
  coralroot *(Cypripedium striata)*, 227
  fairy-slipper *(Calypso bulbosa)*, 55
  ladies'-tresses *(Spiranthes romanzoffiana)*, 41
  phantom *(Cephalanthera austiniae)*, 172, 173
  western rattlesnake plantain *(Goodyera oblongifolia)*, 26, 122, 154, 227, 240
owl-clover, mountain *(Orthocarpus imbricatus)*, 263, 265
oxalis, Oregon *(Oxalis oregana)*, 239, 240
oxytheca, Cushenbury *(Oxytheca parishii goodmaniana)*, 129, 130

pachistema *(Pachystema myrsinites)*, 175, 304
paintbrush
  ashy-gray *(Castilleja cinerea)*, 129
  frosted *(Castilleja pruinosa)*, 144, 241
  hairy Indian *(Castilleja hispida)*, 331
  harsh *(Castilleja hispida)*, 263, 265
  Indian *(Castilleja confusa)*, 27, 73, 175, 184, 200
  magenta *(Castilleja oreopola)*, 317
  marsh *(Castilleja miniata)*, 98
  scarlet *(Castilleja miniata)*, 277, 317
palm, California fan *(Washingtonia filifera)*, 127
palo verde *(Cercidium microphyllum)*, 60, 73
  blue *(Cercidium floridum)*, 48, 69
  foothill *(Cercidium floridum)*, 62, 69
parsley
  few-flowered desert *(Lomatium martindalei)*, 263
  hemlock *(Conioselinum scopulorum)*, 25
  spring *(Cymopterus beckii)*, 282
parsnip
  cow *(Heracleum maximum)*, 7
  sheep *(Lomatium macrocarpum)*, 123
partridge-foot *(Leutkia pectinata)*, 264, 329
pasque-flower, western *(Anemone occidentale)*, 317

peach, desert *(Prunus andersonii)*, 98
penstemon
  ash *(Penstemon cinicola)*, 117, 281
  broad-leaved *(Penstemon ovatus)*, 170
  Cardwell's *(Penstemon cardwellii)*, 265
  cliff *(Penstemon rupicola)*, 263, 265
  Parry's *(Penstemon parryi)*, 60
  small-flowered *(Penstemon procerus)*, 263, 265
phacelia
  serpentine *(Phacelia corymbosa)*, 144
  virgate *(Phacelia heterophylla)*, 263
phlox
  cushion *(Phlox hoodii)*, 278
  daggerleaf *(Phlox rigida)*, 282
  long-leaved *(Phlox longifolia)*, 47
  spreading *(Phlox diffusa)*, 317, 343
pig-a-back plant *(Tolmeia menziesii)*, 328
pine
  Apache *(Pinus latifolia)*, 43
  bristlecone *(Pinus longaeva)*, 33, 90, 200, 273
  Chihuahua *(Pinus chihuahuana)*, 43, 44
  coast *(Pinus contorta)*, 247
  Coulter *(Pinus coulteri)*, 83, 111, 125
  digger *(Pinus sabiniana)*, 86, 114, 134, 135
  foxtail *(Pinus balfouriana)*, 101, 114, 133
  Jeffrey *(Pinus jeffreyi)*, 83, 91, 117, 125, 203, 235
  knobcone *(Pinus attenuata)*, 125, 144, 235
  limber *(Pinus flexilis)*, 56, 80, 165, 201, 290, 309
  lodgepole *(Pinus contorta)*, 9, 99, 117, 127, 159, 164
  Mexican pinyon *(Pinus cembroides)*, 43, 295
  pinon *(Pinus edulis)*, 202, 273, 283, 292
  ponderosa *(Pinus ponderosa)*, 24, 56, 60, 61
  prince's *(Chimaphila umbellata)*, 227, 263, 265
  single-leaf pinon *(Pinus monophylla)*, 288, 295
  southwestern white *(Pinus strobiformis)*, 43, 44, 55
  sugar *(Pinus lambertiana)*, 117, 137, 215, 230, 233
  western white *(Pinus monticola)*, 106, 170, 203, 235, 265
  whitebark *(Pinus albicaulis)*, 101, 176, 191, 203, 215, 321
pipsissewa *(Moneses uniflora)*, 26, 324
  little *(Chimachilla menziesii)*, 227
plantain *(Plantago major)*, 88
  water *(Alisma plantago-aquatica)*, 123

polemonium
leafy *(Polemonium foliosissimum)*, 288
showy *(Polemonium pulcherrimum)*, 265
poppy
Matilija *(Romneya coulteri)*, 111
prickly *(Argemone mexicana)*, 24
potentilla
five-finger *(Potentilla arguta)*, 6, 7
powderpuff bush *(Mimosa grahamii)*, 45
prickly pear *(Opuntia phaeacantha)*, 27, 37,
43, 47, 73, 287
cliff *(Opuntia erinacea)*, 47
clock-faced *(Opuntia chlorotica)*, 47
Engelmann's *(Opuntia engelmannii)*, 45,
47, 62
purple *(Opuntia santa-rita)*, 47
Santa Rita *(Opuntia santa-rita)*, 40
primrose
cave *(Primula specuicola)*, 298
Parry's *(Primula parryi)*, 34, 278
purslane, marsh *(Ludwigia palustris)*, 123
pussy-toes, 207
arching *(Antennaria arcuata)*, 278
rosy-flowered *(Antennaria rosea)*, 45
small-leaved *(Antennaria aprica)*, 45
pyrola
green *(Pyrola virens)*, 45
snowline *(Pyrola minor)*, 324
white-veined *(Pyrola picta)*, 227, 263

quinine bush *(Cowania mexicana)*, 48

rabbitbrush *(Chrysothamnus nauseosus)*, 98,
117, 285
redwood
California *(Sequoia sempervirens)*, 93,
171, 238
coast *(Sequoia sempervirens)*, 112, 141,
144, 238, 239
rhododendron
Kamschatka *(Rhododendron camtscha-
ticum)*, 4
Pacific *(Rhododendron macrophyllum)*,
238
western *(Rhododendron macrophyllum)*,
247
rhynchosia, trailing *(Rhynchosia edulis)*, 49,
50
rose
Arizona *(Rosa arizonica)*, 71
baldhip *(Rosa gymnocarpa)*, 226, 265
wood *(Rosa gymnocarpa)*, 99, 288
rose-bay *(Rhododendron macrophyllum)*,
226, 239, 240
California *(Rhododendron
macrophyllum)*, 144
Lapland *(Rhododendron lapponicum)*, 18

rosemary, bog *(Andromeda polifolia)*, 9, 16
rosewood, Arizona *(Vaquelinia californica)*,
44, 75
rye
beach *(Elymus mollis)*, 7, 8
blue *(Elymus glaucus)*, 163
wild *(Elymus salinus)*, 290

sage, white *(Artemisia ludoviciana)*, 125,
220, 343
sagebrush
big *(Artemisia tridentata)*, 9, 8, 117, 163,
218, 292
black *(Artemisia nova)*, 117, 163, 290, 292
coastal *(Artemisia pycnocephala)*, 84
desert *(Artemisia nova)*, 218, 222
low *(Artemisia arbuscula)*, 218, 287, 288
shrubby *(Artemisia tripartita)*, 215, 295
silver *(Artemisia cana)*, 117
stiff *(Artemisia rigida)*, 185
saguaro *(Carnegiea gigantea)*, 47, 60, 62, 69,
70, 73, 75
salal *(Gaultheria shallon)*, 11, 18, 226, 238,
239, 240, 247
salmonberry *(Rubus spectabilis)*, 6, 7, 9, 264,
320, 328
sandwort
Bear Valley *(Arenaria ursina)*, 129
Howell's *(Minuartia howellii)*, 144
rock *(Arenaria saxosa)*, 34
rosy-flowered *(Arenaria rubella)*, 207
saxifrage
bog *(Saxifraga oregana)*, 122
dotted *(Saxifraga bronchialis)*, 19
matted *(Saxifraga bronchialis)*, 265
prickly *(Saxifraga tricuspidata)*, 16
tufted *(Saxifraga caespitosa)*, 19, 265
western *(Saxifraga occidentalis)*, 264
scurfpea, wedgeleaf *(Psoralea tenuiflora)*, 48
sedge
elk *(Carex geyeri)*, 166, 343
Henderson's *(Carex hendersonii)*, 172
large-headed *(Carex macrocephala)*, 247
Lyngby *(Carex lyngbyei)*, 7
Ross's *(Carex rossii)*, 219
Sitka *(Carex sitchensis)*, 6, 7
slough *(Carex obnupta)*, 246
threeway *(Dulichium arundinaceum)*, 155
sedum *(Sedum divergens)*, 343
dwarf *(Parvisedum congdonii)*, 88
sequoia *(Sequoiadendron giganteum)*, 131,
137, 141, 151
serviceberry, alder-leaved *(Amelanchier
alnifolia)*, 99, 169, 182, 230, 288
shadbush, Utah *(Amelanchier utahensis)*, 50
shepherd's-purse *(Capsella bursa-pastoris)*,
74

white *(Trillium ovatum)*, 239
wood *(Trillium chloropetalum)*, 331
turkey-peas *(Orogenia fusiformis)*, 264
twin-flower *(Linnaea borealis)*, 227, 240, 324
twisted stalk *(Streptopus amplexifolius)*, 6, 123, 154
rosy *(Streptopus roseus)*, 239, 328

valerian, Sitka *(Valeriana sitchensis)*, 263, 264, 317, 328
vanilla leaf *(Achlys triphylla)*, 227, 239, 263, 265, 331
vetch
American *(Vicia americana)*, 263, 265
Cushenbury milk *(Astragalus albens)*, 129, 130, 207
milk *(Astragalus alpinus)*, 19, 207
sweet *(Vicia americana)*, 288
vetchling *(Lathyrus palustris)*, 7
violet
evergreen *(Viola sempervirens)*, 324
marsh *(Viola nephrophylla)*, 41
northern bog *(Viola nephrophylla)*, 123
northern white *(Viola canadensis)*, 74
redwoods *(Viola sempervirens)*, 172, 227, 239, 240, 263, 265
stream *(Viola glabella)*, 263
yellow *(Viola glabella)*, 88, 128, 207

wait-a-minute bush *(Mimosa binucifera)*, 48
wallflower *(Erysimum concinnum)*, 241
rough *(Erysimum asperum)*, 263, 265
walnut, Arizona *(Juglans major)*, 27, 36, 43, 58, 71
watercress *(Nasturtium officinale)*, 123
water hemlock, Douglas' *(Cicuta douglasii)*, 122
waterleaf *(Hydrophyllum occidentalis)*, 288
Fendler's *(Hydrophyllum fendleri)*, 264
Pacific *(Hydrophyllum tenuipes)*, 264
water shield *(Brasenia schreberi)*, 123

waterwort *(Elatine brachysperma)*, 25
whipplevine *(Whipplea modesta)*, 240
whisker brush *(Linanthus ciliatus)*, 88
whitlowgrass, breaks *(Draba subalpina)*, 282
whortleberry, grouse *(Vaccinium scoparium)*, 175, 187, 222
willow
willow herb *(Epilobium hornemannii)*, 98, 155
dwarf *(Epilobium latifolium)*, 19
windflower
three-leaved *(Anemone oregana)*, 263
white *(Anemone parviflora)*, 278
wintergreen
alpine *(Gaultheria humifusa)*, 182
one-sided *(Pyrola secunda)*, 263, 265
pink-flowering *(Pyrola asarifolia)*, 18
white-veined *(Pyrola picta)*, 122
woodbine *(Parthenocissus incerta)*, 74
wood sorrel
red *(Oxalis oregana)*238
shamrock-leaved *(Oxalis oregana)*, 336
woolly-weed *(Hieracium scouleri)*, 222

yampah, broad-leaved *(Perideridia gardneri)*, 123
yarrow *(Achillea millefolium)*, 7
northern *(Achillea sibirica)*, 16
yellow show *(Amoreuxia palmatifida)*, 47
yew
California *(Taxus brevifolia)*, 151
Pacific *(Taxus brevifolia)*, 17, 169, 230, 249, 324
western *(Taxus brevifolia)*, 226
yucca
banana *(Yucca baccata)*, 27
Mojave *(Yucca schidigera)*, 130

zinnia
desert *(Zinnia grandiflora)*, 69
springleaf *(Zinnia pumila)*, 48

GENERAL INDEX